Springer Proceedings in Mathematics & Statistics

Volume 363

Springer Proceedings in Mathematics & Statistics

This book series features volumes composed of selected contributions from workshops and conferences in all areas of current research in mathematics and statistics, including operation research and optimization. In addition to an overall evaluation of the interest, scientific quality, and timeliness of each proposal at the hands of the publisher, individual contributions are all refereed to the high quality standards of leading journals in the field. Thus, this series provides the research community with well-edited, authoritative reports on developments in the most exciting areas of mathematical and statistical research today.

More information about this series at http://www.springer.com/series/10533

Jan Awrejcewicz
Editor

Perspectives in Dynamical Systems II: Mathematical and Numerical Approaches

DSTA, Łódź, Poland December 2–5, 2019

Springer

Editor
Jan Awrejcewicz
Department of Automation, Biomechanics
and Mechatronics
Lodz University of Technology
Lodz, Poland

ISSN 2194-1009 ISSN 2194-1017 (electronic)
Springer Proceedings in Mathematics & Statistics
ISBN 978-3-030-77309-0 ISBN 978-3-030-77310-6 (eBook)
https://doi.org/10.1007/978-3-030-77310-6

Mathematics Subject Classification: 28DXX, 34Cxx, 37-XX, 46LXX, 65-XX, 70-XX, 74-XX, 76-XX

This Springer imprint is published by the registered company Springer Nature Switzerland AG
The registered company address is: Gewerbestrasse 11, 6330 Cham, Switzerland

Preface

Fifteenth International Conference "Dynamical Systems—Theory and Applications" (DSTA 2019) took place in Lodz, Poland from 2nd to 5th December 2019. It was the 15th edition in the series of conferences organized every 2 years in Lodz by the Department of Automation, Biomechanics and Mechatronics of the Lodz University of Technology.

Its Scientific Committee composed of 64 scientists has been hand-picked over the years by originator and main organizer of the whole DSTA Conference Series—Prof. Jan Awrejcewicz.

For this edition, they had to review over 360 submitted topics to choose 200 that were to be presented during the DSTA 2019 by participants representing 40 countries from all over the world.

It resulted in the program of conference that covered both theoretical and experimental approaches to widely understood dynamical systems, including topics devoted to bifurcations and chaos, control in dynamical systems, asymptotic methods in nonlinear dynamics, stability of dynamical systems, lumped mass and continuous systems vibrations, original numerical methods of vibration analysis, nonsmooth systems, dynamics in life sciences and bioengineering, as well as to the engineering systems and differential equations.

All papers included in the following book were submitted and presented during DSTA 2019. They contribute partially to the diverse approaches and topics covered by a wide scope of dynamical systems.

What follows is a brief description of the book content.

Comparison study between the Proportional-Integral Derivative and Linear-Quadratic-Regulator control schemes while employing the fully functional and nonlinear simulation model of two-wheel self-balancing human transporter is presented in chapter "Nonlinear Modelling and Control of Self-Balancing Human Transporter."

Chapter "Nonlinear Tourist Flows in Barcelona" is focused on the mathematical model of nonlinear differential equations, which allows to study the dynamic interaction between the main factors that affect the tourist flows of Barcelona using nonlinear ordinary differential equations to represent the interactions between residents, tourists, and investors and political/economic decisions and geopolitical factors as external forces.

Chapter "Convergence of Dual Infinity Series" presents the solution of the system of the partial differential motion equations describing the movement of plate element by Fourier's series for function expressed by-product of three or four functions of the particular variables.

Application of the full spectrum analysis for the existence of the new backward whirl phenomena that immediately appear after the passage through the critical forward whirl rotational speed in accelerated intact and cracked rotor-disk systems is illustrated in chapter "Full Spectrum Analysis for Studying the Backward Whirl in Accelerated Rotor Systems."

In chapter "Switched Reluctance Motor Dynamic Eccentricity Modelling," the dynamic eccentricity occurring in the brushless electric motor built of iron when the center of the rotor is not at the center of rotation and minimum air gap revolves with the rotor has been simulated using the finite element method of the FEMM software.

Using modified Tikhonov regularization for calculating the transmissibility function matrices of a complex, flexible structure like a commercial wine refrigerator, Hörtnagel et al. in chapter "Harmonic Transfer Path Analysis of a Wine Refrigerator" developed novel and multi-stage robust algorithm improving the stability of the estimation process by making the selection of the regularization parameter more robust.

In chapter "Risk Related Prediction for Recurrent Stroke and Post-stroke Epilepsy Using Fractional Fourier Transform Analysis of EEG Signals," application of the Fractional Fourier Transform to analyze component of the EEG signals, for detection of pre-stroke events in the EEG signal in cases of recurrent stroke and post-stroke epilepsy, is demonstrated.

Glushkov et al. in chapter "Chaos, Bifurcations and Strange Attractors in Environmental Radioactivity Dynamics of Some Geosystems" proposed application of fractal sets, chaos, and dynamical systems theories to analyze, predict, and compute a temporal chaotic dynamics of arbitrary chaotic radioactive geosystems (ecosystems) and to provide accurate numerical modelling and analysis of temporal dynamics of the atmospheric pollutants.

In chapter "Dynamics of Chains as a Tool to Study Thermomechanical Properties of Proteins," application of the methodology based on the chain dynamics to study thermodynamic protein properties exhibited by the non-Markovian processes is presented.

Based on the Lagrange equations of the second kind mathematical model of the crane supported flexibly, numerical calculations of the influence of the jib's flexibility and load's mass on the maximum stresses due to the deformations of the actuators at a given crane's working moment are performed in chapter "Evaluation of the Crane's Actuators Strength Based on the Results Obtained from Dynamics Model."

Chapter "Nonlinear Dynamics of Atomic and Molecular Systems in an Electromagnetic Field: Deterministic Chaos and Strange Attractors" presents a novel computational approach to studying deterministic chaos and strange attractors exhibited by nonlinear processes in atomic and molecular systems embedded into electromagnetic field.

Results of modelling, analysis, and forecasting of the dynamics for relativistic backward-wave tube with accounting for relativistic effects, dissipation factor, and an effect of presence of the space charge are presented in chapter "Deterministic Chaos, Bifurcations and Strange Attractors in Nonlinear Dynamics of Relativistic Backward-Wave Tube."

In chapter "Detection of Chaotic Behavior in Dynamical Systems Using a Method of Deformable Active Contours," problem of automatic detection of the chaotic behavior of the dynamical system is numerically studied with a special Hamiltonian structure and using Poincaré sections composed of point clouds in chaotic cases with the help of active contour method.

Dynamic version of the principle of virtual displacements, the modified couple-stress theory, and the third-order theory of laminated composite plates and shells was applied by Barulina et al. in chapter "Dynamics of Sensing Element of Micro- and Nano-electromechanical Sensors as Anisotropic Size-Dependent Plate" to obtain the differential equations of motion and natural boundary conditions for nanoelectromechanical sensor.

In chapter "Dynamic Analysis and Damage of Composite Layered Plates Reinforced by Unidirectional Fibers Subjected Low Velocity Impact," solid-based model and shell-based model are used for dynamic analysis and investigation of low velocity impact response of composite layered plates reinforced by unidirectional fibers.

Anish and Shankar in chapter "Identification of Nonlinear Joint Interface Parameters Using Instantaneous Power Flow Balance Approach" proposed novel approach of using combined acceleration matching and instantaneous power flow balance as objective functions in time domain for identification of the nonlinear joint interface parameters.

Numerical procedure for the sensitivity analysis of hybrid systems based on the evaluation of adjoint equations which are consistent with discrete-time equations resulting from the numerical integration of the governing equations by an implicit Runge–Kutta method is presented in chapter "Numerical Procedure for the Sensitivity Analysis of Hybrid Systems."

Mozyrska et al. in chapter "Asymptotic Stability of Fractional Variable-Order Discrete-Time Equations with Terms of Convolution Operators" formulated new asymptotic stability conditions while analyzing the stability of the linear systems with the Caputo fractional and variable-order difference operators of convolution type.

Chapter "Dynamics of Circular Plates Under Temperature and Mechanical Loadings" is devoted to the study of nonlinear vibrations of the mathematical model of a heated plate subjected to harmonic loading based on the application of the geometrically nonlinear Reissner-Mindlin plate theory.

Theoretical and numerical investigation of a fractional-order version of the Rulkov neuronal model, involving Caputo fractional variable-order differences of convolution type are presented in chapter "A Rulkov Neuronal Model with Caputo Fractional Variable-Order Differences of Convolution Type."

In chapter "Electrostatically Actuated Initially Curved Micro Beams: Analytical and Finite Element Modelling," equilibria forms branching for various initial curvature and geometry parameters of initially curved microbeam loaded by a nonlinear, configuration dependent and subjected to electrostatic forces is investigated by a model order reduction technique and numerical continuation methods.

Dohnal et al. in chapter "Numerical and Analytical Investigation of Chatter Suppression by Parametric Excitation" present concept of increasing process stability during milling the time-periodic modulation of the tool support. For this purpose, the numerical results of stability charts are discussed in terms of spindle speed and cut depth and show classic chatter lobes that are modified by the parametric excitation.

In chapter "Nonlinear Study of a Pneumatic Artificial Muscle (PAM) Under Superharmonic Resonance Condition Using Method of Multiple Scales," nonlinear behavior, dynamic stability, and bifurcation exhibited by a one degree-of-freedom system consisting of nonlinear pneumatic artificial muscle with additional external spring are modelled and analyzed.

Two-mode long-wave low-frequency approximation of the full dispersion relations incorporating both the fundamental mode and the first harmonic for anti-plane shear deformation of a high-contrast three-layered laminate of an asymmetric structure are presented in chapter "Two-Mode Long-Wave Low-Frequency Approximations for Anti-Plane Shear Deformation of a High-Contrast Asymmetric Laminate."

Saeed and Al-Shudeifat in chapter "A Study on the Coefficient of Restitution Effect on Single-Sided Vibro-Impact Nonlinear Energy Sink" offer studies on the effect of a ratio of restitutive to deformative impulses during impact on the capability to irreversibly transfer-induced impulsive energy out of the fundamental, highly energetic mode for the single-sided vibro-impact nonlinear energy sink.

DSTA Conferences are aimed to provide a common platform for exchange of new ideas and results of recent research in the field of scientific and technological advances in modern dynamical systems. Over the last 25 years, both approaches and understanding of sciences significantly evolved to include new ideas and trends, but the traditional views are still present and provide the basic understanding. Therefore, both as Head of Organizing and Scientific Committees of DSTA 2019 and as the Editor of volume of Springer Proceedings, I hope that this book will provide the readers with both answers to their problems and ideas for their novel approaches to study nonlinear dynamical systems.

I greatly appreciate the help of the Springer Editors, Elizabeth Leow and Dahlia Fisch, as well as Springer Project Coordinator—Murugesan Tamilsevan in publishing this volume of the Springer Proceedings in Mathematics and Statistics. I would also like to express my gratitude to Scientific Committee of DSTA 2019 and all reviewers for their help and professional support during the book preparation.

Lodz, Poland
December 2020

Jan Awrejcewicz

Contents

Non-linear Modelling and Control of Self-Balancing Human Transporter

Saransh Jain, Sarthak Jain, and Mohit Makkar

Abstract Various modelling and control strategies have been developing in quest for efficiently managing non-linear systems, which is majorly done by incorporating maximum possible aspects of behaviour of a system into mathematical equations and then implementing the control schemes to track the desired trajectories. Though some of these developed control schemes are still struggling to produce satisfying results when it comes to controlling non-linear systems, Proportional-Integral-Derivative (PID) and Linear-Quadratic- Regulator (LQR) are the two very efficient control schemes known for their stability properties and optimal control when applied to non-linear systems. Self Balancing Human Transporter (SBHT) is one such non-linear system which is widely used and needs to be effectively controlled to maintain uniform speed and dynamic stability. It is very crucial to work on both, the dynamics and efficient control of two wheel SBHT. This article will show the design and analysis of more advanced and recently developed algorithms of the above mentioned control schemes being applied on the new, more precise, fully functional and non-linear simulation model of two wheel SBHT. Comparison study between the two has also been done on various parameters.

Keywords PID control · LQR control · SBHT · Non-linear systems · Non-linear control

1 Introduction

Two wheel SBHT is a highly unstable system consists of a platform between the two wheels for a human being to stand on it and a lever connected to the platform and wheels to control velocity of SBHT. The movement of the lever guides the direction and speed of SBHT. The lever or SBHT for that matter needs to be controlled to

S. Jain · S. Jain · M. Makkar (✉)
The LNM Institute of Information Technology, Jaipur, India
e-mail: mohit.makkar@lnmiit.ac.in

J. Awrejcewicz (ed.), *Perspectives in Dynamical Systems II: Mathematical and Numerical Approaches*, Springer Proceedings in Mathematics & Statistics 363,
https://doi.org/10.1007/978-3-030-77310-6_1

maintain its balance and the balance of the human being. The dynamics of the system revolves around wheel velocity, lever velocity, wheel position and lever position.

The dynamics of the two wheel SBHT is very closely related to the model of inverted pendulum. Prasad et al. [10], Garzón et al. [4] and [9] discussed in detail the modeling of an inverted pendulum. The assumptions regarding the system dynamics are taken from [9]. Muhammad et al. [9] assumed that the wheels of the vehicle always remain on the road and the tires retain the grip, sticking to the road. The driving lever is balanced at an upright position, the friction of the tires included is by due analysis and is an approximate value. Further, model of self balancing robot in [2, 6, 12] and [14] was studied followed by the study of model of mobile robot with alternative use of human transporter in [1, 3, 5] and [8].

PID control is one of the most used control methods. It works on efficient tuning of proportional, integral and derivative constants depending on the required desired performance. By tuning the PID gain system can be stablize to the desired states. Various researchers used PID in controlling two wheel SBHT [7, 11]. Arvidsson and Karlsson [1] achieved a stable configuration using PID controller. PID control scheme used by Jamil et al. [6] and Villacres et al. [13] gave good results.

LQR control on the other hand has very unique properties; e.g. it provides the optimal control law, the best gain matrices for the particular initial and the final states of the system. It also provides the design flexibility in terms of Q and R matrices. Switching the values of R matrix can check the control input of the system and by tweaking the Q matrices one can penalize the change of the states if they are not at the reference states. Many researchers such as [10, 14] and [1] implemented LQR controller on the system similar to two wheel SBHT.

In this paper, a new non-linear simulation model of two wheel SBHT is taken which is to be controlled by specially designed control algorithms of PID and LQR for the new model. The model in this paper is closest to the actual scenario with very few acceptable assumptions. The control schemes developed are also accurate and exact in their behaviour. All the simulation results were obtained on MATLAB / Simulink. This article also compares the performance of PID and LQR controllers with reference to control of two wheel SBHT.

Section 2 will discuss the mathematical model of TW-SBHT followed by the simulation results of its open kinematic model in section III. Section IV and V will show the design and development of control schemes of PID and LQR controller respectively which will also be validated by respective simulation results. In section VI, the comparison of the two, PID and LQR is done. Section VII will conclude and talk about future prospects.

2 Mathematical Model

In this section, we introduce the kinematics and the dynamical behaviour of the self-balancing human transporter. This mathematical model is further developed on

Table 1 Parameters of the system

Parameter	Symbol	Units
Gravitational constant	g = 9.81	m/s^2
Wheel weight	m = 4.6	Kg
Radius of wheel	$R_w = 0.24$	metre
Wheel inertia	$J_w = mR_w^2$	kgm^2
Transporter body weight	Mv = 30.05	Kg
Mass of driver	Mr = 80	kg
Total mass	M = Mr+Mv	Kg
Body height	H = −0.03	Metre
Length of driver	Lr = 1.8	Metre
Distance between centre of mass and wheel axle	L = 0.7155	Metre
Inertia of driver	J_r = MrLr2/3	kgm^2
Body pitch Inertia	$J_b = 87.89$	kgm^2
Inertia of DC motor	$J_m = 0.0075$	kgm^2
Resistance of motor	$R_m = 0.14$	Ω
Back EMF constant of DC motor	Kb = 0.72	Vs/rad
Torque constant of DC motor	Kt = 0.833	Nm/A
Gear Ratio of DC motor	n = 14	–
Friction coefficient axle and bearings	$F_m = 0.3$	–
Friction between tyres and ground	$f_w = 0.5$	–
Input voltage	V = 24	Volt

Table 2 Coordinates of the system

Parameter	Symbol
x_l, y_l, z_l	Position coordinates of left wheel
x_r, y_r, z_r	Position coordinates of right wheel
x_b, y_b, z_b	Position coordinates of centre of mass

the various already existing models [1–5, 13] and [8]. The equation of motion of SBHT is derived using the Lagrangian mechanics. The model derived in this article includes the dynamics of pitch angle, the mass of the rider, height of the driver and frictional forces involved. It also considers the chassis inertia as the part of body inertia.

$$\psi = \mathrm{Pitchangleofthedrivinglever}$$

$$\theta = \mathrm{Averageangleofthecorrespondingwheel}$$

The expression of the total energy is derived from the Lagrangian function. The total energy can be written as the sum of translation kinetic energy (T_1), rotational kinetic energy (T_2) and potential energy (U) (all the values of symbols are mentioned in Table 1 and the coordinates in the Table 2).

$$T_1 = \frac{1}{2}m(\dot{x}_l^2 + \dot{y}_l^2 + \dot{z}_l^2) + \frac{1}{2}m(\dot{x}_r^2 + \dot{y}_r^2 + \dot{z}_r^2) + \frac{1}{2}M(\dot{x}_b^2 + \dot{y}_b^2 + \dot{z}_b^2) \tag{1}$$

$$T_2 = \frac{1}{2}J_W\dot{\theta}_l^2 + \frac{1}{2}J_W\dot{\theta}_r^2 + \frac{1}{2}J_W\dot{\psi}^2 + n^2 J_m(\dot{\theta}_l - \dot{\psi})^2 + \frac{1}{2}n^2 J_m(\dot{\theta}_r - \dot{\psi})^2 \tag{2}$$

$$U = mgz_l + mgz_r + Mgz_b \tag{3}$$

Coordinates can be expressed as:

$$(x_m, y_m, z_m) = (\int R_w\dot{\theta}dt, 0, R_w)$$

$$(x_l, y_l, z_l) = (x_m, \frac{W}{2}, z_m)$$

$$(x_r, y_r, z_r) = (x_m, -\frac{W}{2}, z_m)$$

$$(x_b, y_b, z_b) = (x_m + Lsin(\psi), y_m, z_m + Lcos(\psi))$$

Substituting the Lagrangian L in Eqs. (4) and (5):

$$\frac{d}{dt}(\frac{\partial L}{\partial \dot{\theta}}) - (\frac{\partial L}{\partial \theta}) = F_\theta \tag{4}$$

$$\frac{d}{dt}(\frac{\partial L}{\partial \dot{\psi}}) - (\frac{\partial L}{\partial \psi}) = F_\psi \tag{5}$$

By solving equations (4) and (5), the final two equations of two wheel SBHT comes out to be:

$$((2m+M)R_w^2+2J_w+2n^2J_m)\ddot{\theta}+(MLR_wcos(\psi)-2n^2J_m)\ddot{\psi}-MLR_w\dot{\psi}^2sin(\psi) = F_\theta \tag{6}$$

$$(MLR_wcos(\psi) - 2n^2J_m)\ddot{\theta} + (ML^2 + J_b + 2n^2J_m)\ddot{\psi} - MgLsin(\psi) = F_\psi \tag{7}$$

The generalized forces F_θ and F_ψ are calculated by considering the DC motor torque, viscous friction between the body and motor axle and the friction between the wheels and ground.

By applying Kirchoff's Law on the DC motor we get the final equations of F_θ and F_ψ as:

$$F_\theta = a(V) - 2(b + F_w)\dot{\theta} + 2b\dot{\psi} \tag{8}$$

$$F_\psi = -a(V) + 2b\dot{\theta} - 2b\dot{\psi} \tag{9}$$

$$a = \frac{nK_t}{R_m}$$

$$b = \frac{nK_t K_b}{R_m} + F_m$$

Now, substituting (8) and (9) into (6) and (7) and then solving for the angular acceleration of wheel $\ddot{\theta}$ and the angular acceleration of the driving lever $\ddot{\psi}$, we get:

$$\begin{aligned}\ddot{\theta} = (&J_b Va + 2J_b b\dot{\psi} - 2J_b b\dot{\theta} - 2J_b F_w \dot{\theta} + L^2 MVa + 2L^2 Mb\dot{\psi} - 2L^2 Mb\dot{\theta} - 2L^2 MF_w \dot{\theta} \\ &-4J_m F_w n^2 \dot{\theta} + L^3 M^2 R_w \dot{\psi}^2 sin(\psi) - L^2 M^2 R_w g cos(\psi) sin(\psi) \\ &+J_b LMR_w \dot{\psi}^2 sin(\psi) + 2J_m LMgn^2 sin(\psi) + LMR_w Va cos(\psi) \\ &+2LMR_w b\dot{\psi} cos(\psi) - 2LMR_w b\dot{\theta} cos(\psi) \\ &+2J_m LMR_w n^2 \dot{\psi}^2 sin(\psi))/(2J_b J_w + L^2 M^2 R_w^2 + 2J_w L^2 M + J_b MR_w^2 \\ &+2J_b J_m n^2 + 4J_m J_w n^2 + 2J_b R_w^2 m + 2J_m L^2 Mn^2 + 2J_m MR_w^2 n^2 \\ &+2L^2 MR_w^2 m + 4J_m R_w^2 mn^2 - L^2 M^2 R_w^2 cos(\psi)^2 + 4J_m LMR_w n^2 cos(\psi))\end{aligned} \tag{10}$$

and

$$\begin{aligned}\ddot{\psi} = -(&2J_w Va + 4J_w b\dot{\psi} - 4J_w b\dot{\theta} + MR_w^2 Va + 2MR_w^2 b\dot{\psi} - 2MR_w^2 b\dot{\theta} + 2R_w^2 Vam \\ &+4R_w^2 bm\dot{\psi} + 4J_m f_w n^2 \dot{\theta} - 4R_w^2 bm\dot{\theta} - LM^2 R_w^2 g sin(\psi) \\ &-2J_w LMg sin(\psi) - 2J_m LMgn^2 sin(\psi) - 2LMR_w^2 gm sin(\psi) \\ &+LMR_w Va cos(\psi) + L^2 M^2 R_w^2 \dot{\psi}^2 cos(\psi) sin(\psi) \\ &+2LMR_w b\dot{\psi} cos(\psi) - 2LMR_w b\dot{\theta} cos(\psi) - 2LMR_w f_w \dot{\theta} cos(\psi) \\ &-2J_m LMR_w n^2 \dot{\psi}^2 sin(\psi))/(2J_b J_w + L^2 M^2 R_w^2 + 2J_w L^2 M + J_b MR_w^2 \\ &+2J_b J_m n^2 + 4J_m J_w n^2 + 2J_b R_w^2 m + 2J_m L^2 Mn^2 + 2J_m MR_w^2 n^2 \\ &+2L^2 MR_w^2 m + 4J_m R_w^2 mn^2 - L^2 M^2 R_w^2 cos(\psi)^2 + 4J_m LMR_w n^2 cos(\psi)).\end{aligned} \tag{11}$$

3 Kinematics of Two Wheel SBHT

θ, $\dot{\theta}$, ψ and $\dot{\psi}$ are the four states of the system. There responses are plotted in Figs. 1, 2, 3 and 4.

The top most position of the driving lever is considered to be zero radian. Figure 3 clearly explains the instability of the lever. Initially the lever completes two circles and then comes to rest due to the friction while in the case of wheel, speed become

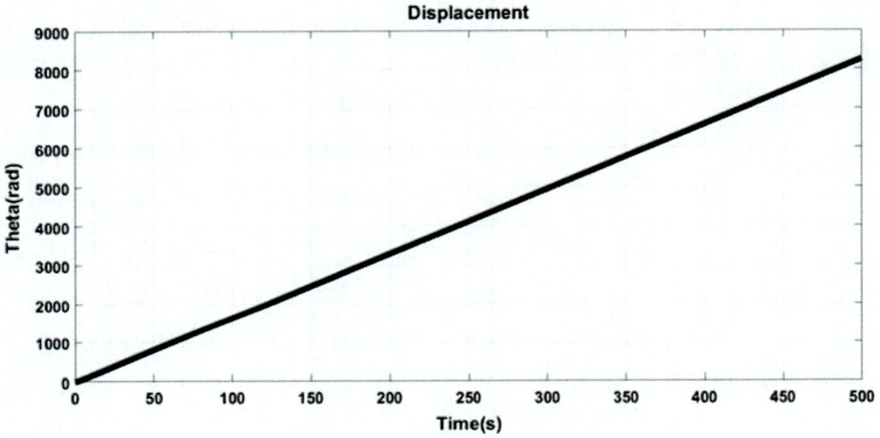

Fig. 1 Displacement

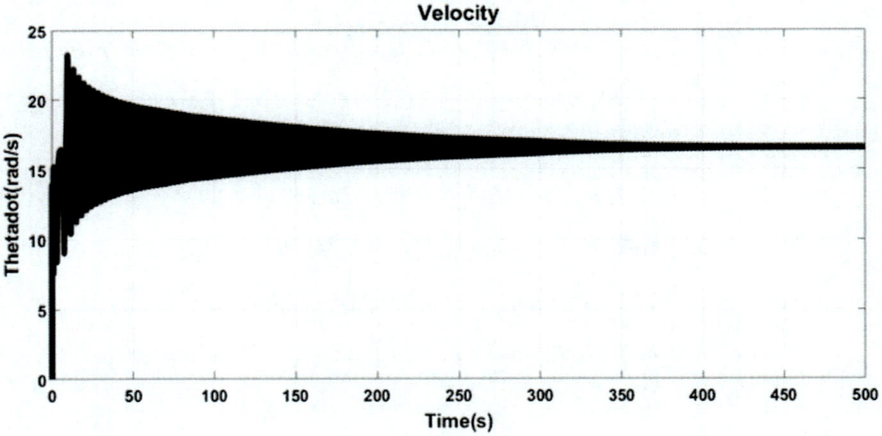

Fig. 2 Velocity

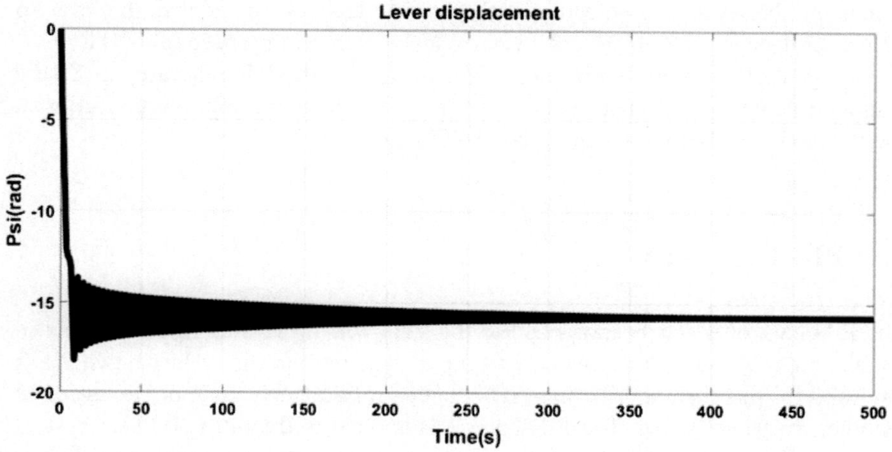

Fig. 3 Lever displacement

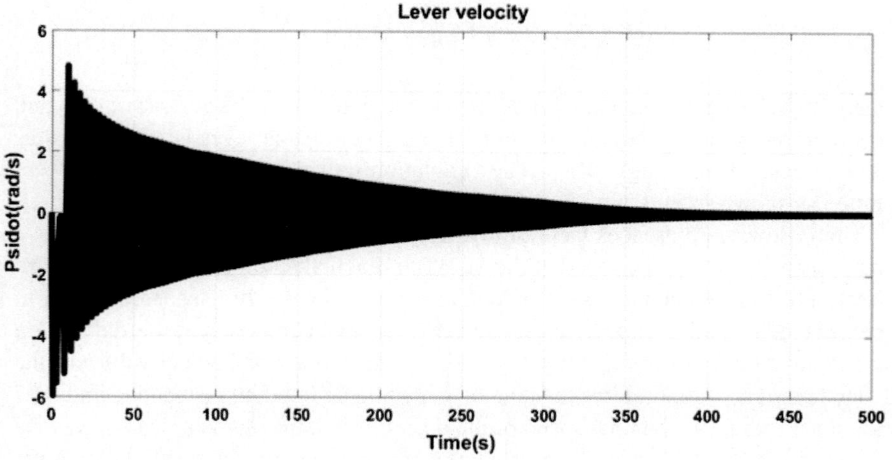

Fig. 4 Lever velocity

constant after some time because we consider the DC motor connected to power supply of 24 V model while simulating. The to and fro motion of the lever makes the movement of the wheels oscillatory as shown in the graph of Fig. 1

3.1 Controller Design

As shown in Fig. 3, the lever is unstable at operating point (zero radian). SBHT is to be controlled using Non-linear closed loop controller for speed control and balancing the lever. The objective of the control design is to stabilize the lever so that rider does not fall off from the vehicle and to control the speed of SBHT.

The goal of the controller is to maintain a constant speed. In this case, the SBHT can be brought to a constant speed when lever maintains the constant deflection at particular angle after giving an initial disturbance.

4 PID Controller

the PID controller has been applied to the state which plays an important role in stabilizing the system and works according to the control design. Hence, two PID's are used for this system. The first PID is to control the pitch angle of the lever and another for wheel speed. The PID control is applied to the full non-linear system and Simulink model on MATLAB is designed for the purpose. The control law of the PID controller is given by:

$$u(t) = K_p e(t) + K_i \int e(t)\,dt + K_d \frac{de}{dt} \tag{12}$$

where, K_p is the proportional gain, K_i is integral gain, K_d is the derivative gain and $e(t)$ represents the tracking error. Initial pitch angle of the lever is taken to be as 18° and desired final angle of 6°. The simulation was performed, and results were plotted as shown in Figs. 5, 6, 7 and 8

The following inferences were drawn from the simulation outcomes: There is a large positive notch in the speed of the SBHT, initial deflection of 18° in the forward direction is the reason behind the sudden increase in the resulting speed. In order to achieve the final value of pitch angle of 6° the Tw-SBHT moves in forward direction to minimize the error in the angle between the desired and the actual values of the deflections. The initial reference value of $\dot{\theta}$ is set to be 0 radian per second and 0.27 radian per second as the final speed of the wheel. When the pitch angle achieves the desired value 6°, at the same time value of $\dot{\theta}$ becomes constant that is 0.9 radian per second. The value of $\dot{\theta}$ in Fig. 6 does not approach to zero because deflection of 6° in is still present in forward direction hence it is moving in forward direction with constant speed. With the above simulation results we can claim that PID control technique works effectively as a non-linear controller.

5 LQR Controller

Optimization is necessary to make the system fast and user friendly. LQR is the best possible solution for a given constraint. The control system works in a sequence in which the control signal passes from a controller to a process (plant) and then diverges into two sections out of which one gives the result and one is feedback

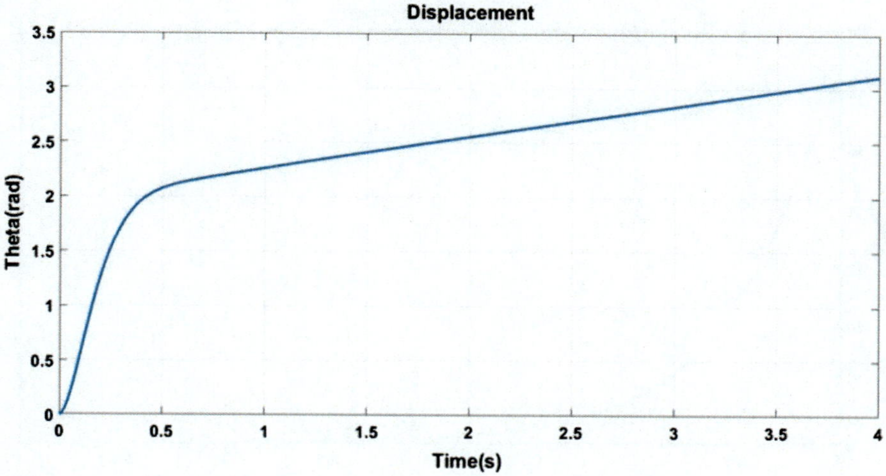

Fig. 5 PID: SBHT displacement

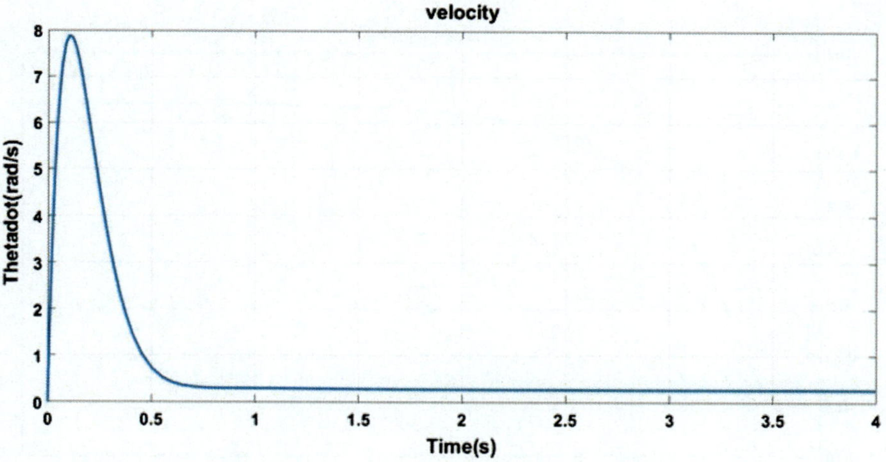

Fig. 6 PID: SBHT velocity

which overcomes some physical constraints to give us the maximum performance which is determined by performance index (PI) or cost function. The problem statement is to locate and optimize the given system at a value or follow some state variable and simultaneously get the maximum value of performance index by various tests.

The LQR controller is optimal version of the pole placement method. The LQR controller provides the best control law which moves the system to the desired eigenvalue [12]. In order to get the system matrices A and B, linearization is applied on the full nonlinear system at the operating point, the value of all the four states (θ

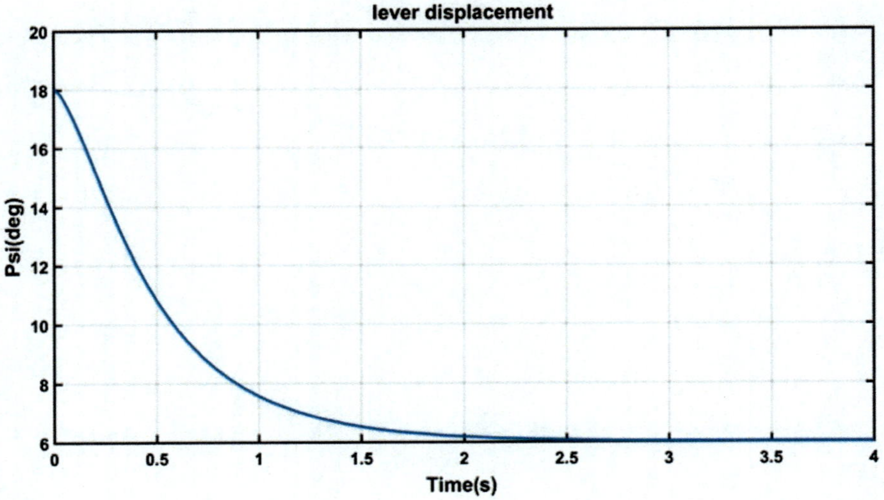

Fig. 7 PID:lever displacement

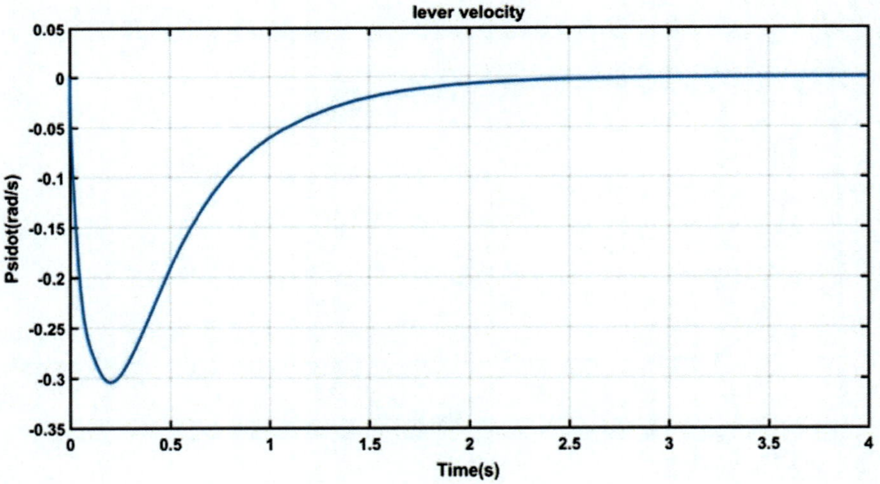

Fig. 8 PID:lever velocity

$\dot{\theta}$ ψ and $\dot{\psi}$) is taken as $x_0 = [0, 0, 0, 0]^T$. The controllability matrix (A, B) is full rank. Hence, the system is controllable. The linear state space is obtained as:

$$\dot{x} = Ax - Bu \tag{13}$$

where

$$x = [\theta, \dot{\theta}, \psi, \dot{\psi}]^T. \tag{14}$$

The full state feedback controller, $U = -Kx$ where K is the gain matrix. K is computed by minimizing the cost function. The cost function indicates how bad the system is if states are not at the reference state.

$$J = \int (x^T Q x + u^T R u) \tag{15}$$

In Eq. (15), Q and R are the positive semi-definite and positive symmetric matrices, respectively. The gain matrix of control law are Eq. (16):

$$\dot{x} = (A - BK)x. \tag{16}$$

The gain matrix is calculated by solving the Eq. (17):

$$K = R^{-1} B^T P. \tag{17}$$

In Eq. (17), P is a positive semi-definite matrix which is calculated by solving Algebraic Riccati Equation (ARE) Eq. (18)

$$A^T P + PA - PBR^{-1} B^T P + Q = 0. \tag{18}$$

System matrix A and B after linearization of the non-linear model are computed as:

$$A = \begin{bmatrix} 0 & 1 & 0 & 0 \\ 0 & -15.4409 & -8.6018 & 15.4901 \\ 0 & 0 & 0 & 1 \\ 0 & 2.4999 & 6.0160 & -2.4998 \end{bmatrix}$$

$$B = \begin{bmatrix} 0 \\ 10.6735 \\ 0 \\ -1.7225 \end{bmatrix}$$

and the weight matrices are computed by hit and trial method, which comes out to be:

$$Q = \begin{bmatrix} 6 & 0 & 0 & 0 \\ 0 & 1e8 & 0 & 0 \\ 0 & 0 & 1e6 & 0 \\ 0 & 0 & 0 & 32 \end{bmatrix} \quad and$$

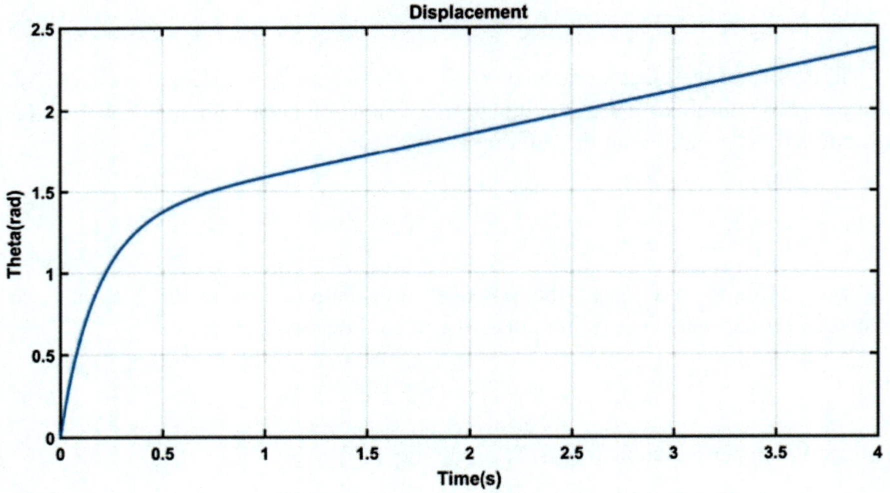

Fig. 9 LQR:SBHT displacement

$$R = \begin{bmatrix} 0.001 \end{bmatrix}$$

After computing A, B, Q and R matrices, eigenvalue (gain value) K was calculated with the help of MATLAB. The desired state of system was taken similar to PID controller, which made the comparison study easier. The desired value of states (θ, $\dot{\theta}$, ψ and $\dot{\psi}$) are taken as $x_d = [4\ .27\ \pi/30\ 0]$respectively. Initial condition is taken as $x_0 = [0\ 0\ \pi/18\ 0]$. The control input U is fed to the full nonlinear system thereafter simulation is performed, and graphs were plotted.

Response of all the four states (θ, $\dot{\theta}$, ψ and $\dot{\psi}$) were plotted as shown in Figs. 9, 10, 11, and 12. ψ (Fig. 11) was initially taken as 18° and after few seconds it smoothly attains the desired value (6°). In Fig. 10, the initial overshoot of velocity is due to the positive deflection of pitch angle in forward direction. According to control objective it should move in forward direction and attain the desired angle which is clearly reflected from Fig. 11. SBHT moves with constant velocity in forward direction because pitch angle has the constant deflection of 6° in positive direction. These simulation results justify that LQR fulfills all the control objectives and works according to the control strategy.

6 Comparison of PID Controller and LQR Controller

After the simulation of PID and LQR controller, the system state responses namely SBHT wheel velocity ($\dot{\theta}$) and lever displacement (ψ) were observed and compared. The performance of these controllers was compared in terms of the settling time

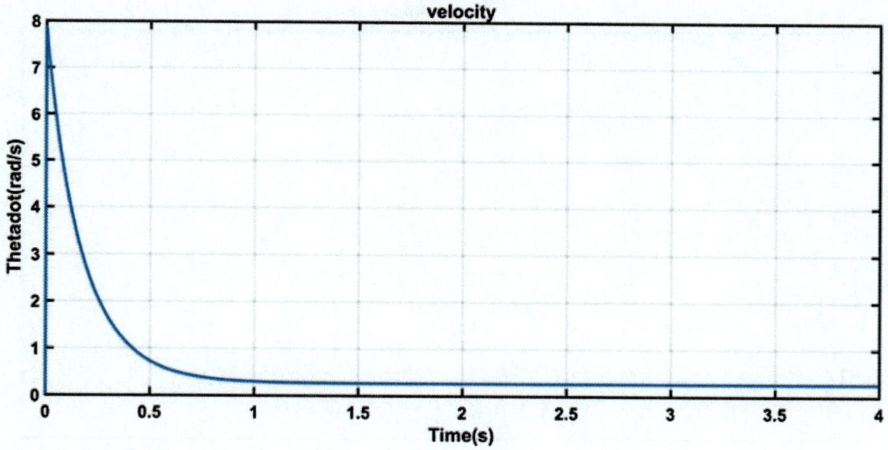

Fig. 10 LQR:SBHT velocity

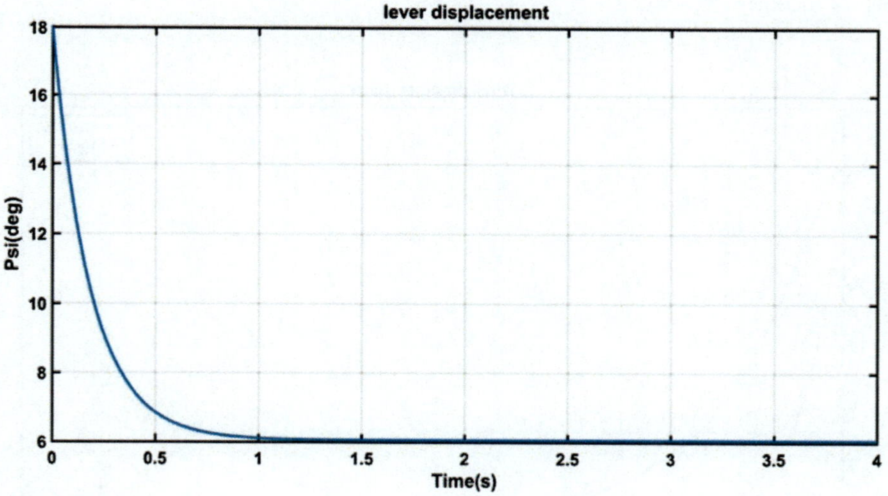

Fig. 11 LQR:lever displacement

of the response, percentage of overshoot and the steady state error of both the controller. The data was taken from the graphs and mention in the table for better assessment. Firstly, the state ψ, the pitch angle of the lever is compared and graphs were plotted. As seen from the simulation results Fig. 13, the LQR controller achieves the final state of control objective little bit smoother as compared to the PID controller. The percentage overshoot in both the controller is almost zero but steady state error in LQR controller is slightly greater than the PID controller. The settling time of the LQR controller is lesser as compared to the PID controller. The data in the Table 3 shows that LQR controller respond faster than PID controller. But in PID

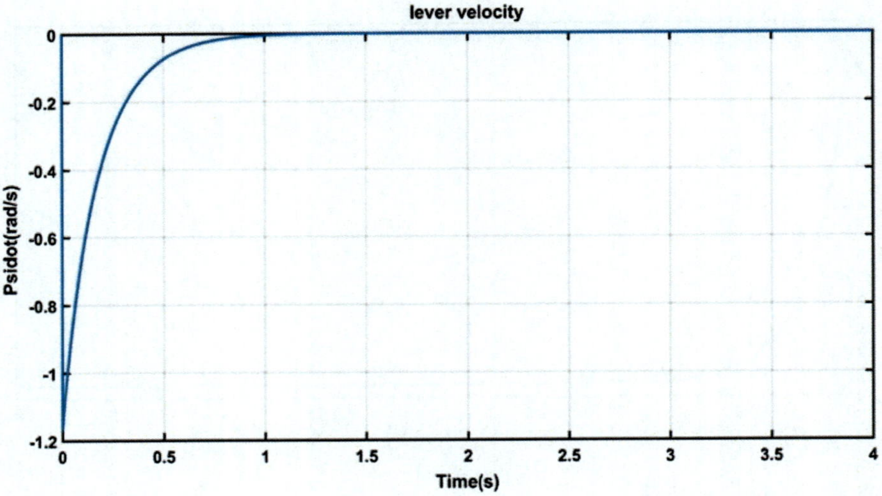

Fig. 12 LQR:lever velocity

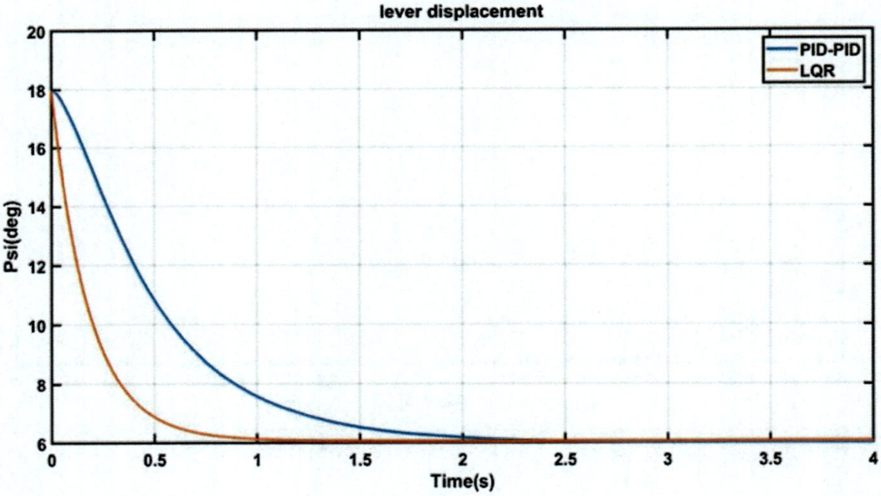

Fig. 13 PID VS LQR: ψ

Table 3 PID VS LQR: $\dot{\theta}$

Parameter	LQR controller	PID controller
Settling time	0.2440	0.2627
Steady state error	2.703%	0%
Rise time	−4.47	0.614

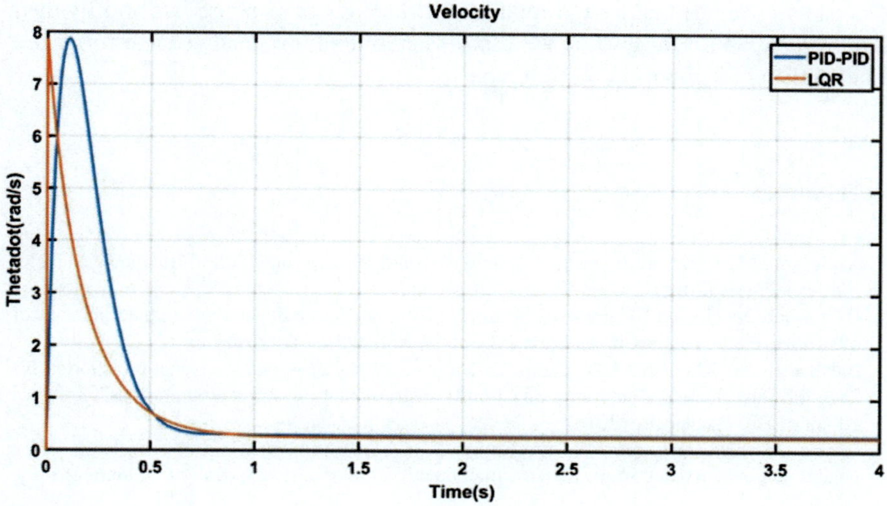

Fig. 14 PID VS LQR: $\dot{\theta}$

Table 4 PID VS LQR: ψ

Parameter	LQR controller	PID controller
Settling time	5.4229	5.4015
Steady state error	0%	−0.299%
Rise time	−3.8527	−0.4891

controller there is no steady state error while in LQR controller steady state error of 2.703% is present. The comparison graph of $\dot{\theta}$ was plotted in Fig. 14, the PID controller take very long time to reach the final reference value of $\dot{\theta}$. However, in LQR controller, the settling time is 0.2440 sec (Table 4), which is less as compared to the PID controller. Although PID controller respond slowly but its steady state error is lesser than the LQR controller. Their is a trade off in notch and settling time if settling time is reduced, the notch will reduce as well. The notch is set similar to LQR controller, which make the comparison easier.

7 Conclusion

In this paper performance of the conventional control design like PID controller is compared with the modern nonlinear control technique like LQR controller. The control design is evaluated on the basis of settling time etc. As observed from the graphs and the data, it shows that there is the trade off in the performance of LQR and PID controller. The LQR response time is better than PID controller on the other hand in terms of accuracy and steady state error PID controller outperforms then the LQR controller. However, the accuracy of LQR is better than the PID controller.

This paper conclude that for the application where faster response is a priority then LQR is preferable. However, in the cases where minimum steady state error is key concern than PID controller is preferable.

References

1. Arvidsson, M., Karlsson, J: Design, Construction and Verification of a Self Balancing Vehicle. Master's Thesis (2012)
2. Babazadeh, R., Gogani Khiabani, A., Azmi, H.: Nonlinear modeling and optimal output control of two wheeled balancing transporter. J. Comput. Rob. **8**(2), 1–7 (2015)
3. Babazadeh, R., Khiabani, A.G., Azmi, H: Optimal control of segway personal transporter. In: 2016 4th International Conference on Control, Instrumentation, and Automation (ICCIA), pp. 18–22. IEEE, Piscataway (2016)
4. Garzón, M.A.H., Hernández, W.O.C., Reyes, A.P.G., Quintero, O.E.C.: Two-wheeled inverted pendulum robot nxt lego mindstorms: mathematical modelling and real robot comparisons. 1Revista Politécnica **36**(1), 12 (2015)
5. Gocmen, A.: Design of two wheeled electric vehicle PhD Diss., Atilim University (2011)
6. Jamil, O., Jamil, M., Ayaz, Y., Ahmad, K.: Modeling, control of a two-wheeled self-balancing robot. In: International Conference on Robotics and Emerging Allied Technologies in Engineering (iCREATE) (2014). https://doi.org/10.1109/iCREATE.2014.6828364
7. Jose, A., Augustine, C., Malola, S.M., Chacko, K: Performance study of PID controller and LQR technique for inverted pendulum. World J. Eng. Technol. **3**(02), 76 (2015)
8. Kwak, S., Choi, B.-J: Design of fuzzy logic control system for segway type mobile robots. Int. J. Fuzzy Logic Intell. Syst. **15**(2), 126–131 (2015)
9. Muhammad, M., Buyamin, S., Ahmad, M., Nawawi, S.: Dynamic modeling and analysis of a two-wheeled inverted pendulum robot. In: 2011 Third International Conference on Computational Intelligence, Modelling and Simulation (CIMSiM), pp. 159–164. IEEE, Piscataway (2011)
10. Prasad, L.B., Tyagi, B., Gupta, H.O.: Optimal control of nonlinear inverted pendulum system using PID controller and LQR: performance analysis without and with disturbance input. International Journal of Automation and Computing **11**, 661–670 (2014)
11. Ren, T.-J., Chen, T.-C., Chen, C.-J: Motion control for a two-wheeled vehicle using a self-tuning pid controller. Control Eng. Practice **16**(3), 365–375 (2008)
12. Sun, L., Gan, J: Researching of two-wheeled self-balancing robot base on LQR combined with PID. In: 2010 2nd International Workshop on Intelligent Systems and Applications (ISA), pp. 1–5. IEEE, Piscataway (2010)
13. Villacres, J., Viscaíno, M., Herrera, M., Camacho, O": Real-time implementation of different controllers for a two-wheeled inverted pendulum. Int. J. Circ. Syst. Sig. Process. **10**, 281–290 (2016)
14. Yamamoto, Y: Nxtway-gs model-based design-control of self-balancing two-wheeled robot built with lego mindstorms nxt. Cybernet Systems Co., Ltd (2008)

Nonlinear Tourist Flows in Barcelona

Enric Trullols, Immaculada Massana, Joana d'Arc Prat, Josefina Antonijuan, and Gerard Olivar

Abstract Tourism is not only a source of wealth, but also a positive way of knowing and mixing local and foreign cultures. However, overexploitation of natural resources and inadequate behaviours (among other factors) can lead to a conflict between tourists and locals that makes the tourist economy not sustainable. The modelling of tourist flows is a good tool to face and overcome this problem, looking for the balance between natural and socioeconomic resources, the interests and rights of tourists and locals. This paper proposes a mathematical model of nonlinear differential equations, which allows to study the dynamic interaction between the main factors that affect the tourist flows of Barcelona. We have used non-linear Ordinary Differential Equations to represent the interactions between residents, tourists and investors. Political/economic decisions and geopolitical factors have been added as external forces.

Specific aspects of Barcelona have been taken into account, such as the impossibility of extending the city, the lack of regulation (and excessive regulation), the affluence of investors and the increase in prices (which pressures residents outside from the city).

Our results show that sustainability, that is, the positive values of the population of tourists, locals and investors, in the long run is possible under the appropriate economic and social decisions.

Keywords Tourism · Bifurcations · Nonlinear dynamics

E. Trullols (✉) · I. Massana · J. d'Arc Prat · J. Antonijuan
Department of Applied Mathematics, Universitat Politècnica de Catalunya/SARTI research group, Barcelona, Spain
e-mail: enric.trullols@upc.edu; immaculada.massana@upc.edu; joana.darc.prat@upc.edu; josefina.antonijuan@upc.edu

G. Olivar
Department of Natural Sciences and Technology, University of Aysen, Coyhaique, Chile
e-mail: gerard.olivar@uaysen.cl

J. Awrejcewicz (ed.), *Perspectives in Dynamical Systems II: Mathematical and Numerical Approaches*, Springer Proceedings in Mathematics & Statistics 363,
https://doi.org/10.1007/978-3-030-77310-6_2

1 Introduction

Barcelona, the capital of Catalonia, is one of the most popular European destinations for international tourism. The number of overnight stays in Barcelona reached 9.5 million in 2017. The international expense of visitors to Barcelona has increased progressively in recent years, in line with the growing volume of tourists arriving in the city. Passenger traffic to Barcelona airport has increased more than doubled since 2000 to reach 47.3 million by 2017 (50 million by 2018). The city is also affected by cruise tourism, with more than 700 cruises arriving at the Port of Barcelona every year. Approximately 2.7 million cruise passengers embarked, landed or moved to the port in 2017 [1–3].

Some residents are not satisfied with the strong increase in tourism and describe Barcelona as a saturated city where prices have skyrocketed and locals have trouble finding a flat at an affordable price. The first confrontations between tourists and locals have arrived, and the city council takes drastic measures such as the freezing of tourist licenses [4].

Dynamic interactions between populations and natural resources has been studied by several authors, starting with the predator-prey seminal work of Lotka and Volterra [5, 6]. Brander and Taylor [7] presented a simple predator–prey model of renewable resource use which simulates the history of the Easter Island Civilisation, describing the presence of feast–famine cycles. They showed that the overexploitation of natural resources caused a sharp reduction in the human population. Several other authors have developed this model taking into account additional aspects such as institutions, property rights and technical progress [8].

2 The Model

Regarding tourist flows and the impact on the environment Rinaldi and Casagrandi [9] proposed a model referring to a non-specific site that includes tourism (T), environment (E) and capital (C). Tourists (T) and capital (C) impact negatively on environmental quality (E), while environmental quality and infrastructure are attractive for tourists. The positive flow of T to C represents the investment of part of the benefits associated with tourism in new facilities for visitors.

Our work presents a model adapted to Barcelona, which includes only two actors, tourists (T) and residents (R). In some way the model can be seen as a simplification of the model of Rinaldi and Casagrandi [9] in which the residents (R) play the role of environment (E). We have assumed that the capital (C) is proportional to the tourist (T) as follows $C = aT - bT$ where a is the average spend per tourist and b is related to taxes and local investments. As capital (C) is proportional to tourists (T), it seems appropriate to proceed only with two variables (T and R).

Let's consider the number of tourists T (t) and residents R (t). Both are functions of time. More specifically, T and R are the number of daily overnight stays.

According to the literature [7] we have assumed that natural growth can be characterized by simple logistic dynamics. The associated equation modelizes a first step of rapid growth proportional to the population (P) and a second step of deceleration proportional to the square of the population in which the population tends to stabilize reaching its maximum capacity.

$$\dot{P} = \alpha P \left(1 - \frac{P}{k}\right) \tag{1}$$

Barcelona is one of the most densely populated cities in Europe, limited by the mountains and the sea and has no possibility to expand. Even more, Barcelona has achieved its maximum capacity (around two million beds including tourists and locals). This is a notorious difference with respect to other cities and this fact must be taken into account in our model.

With regard to the tourist equation (2), the first term is associated with the novelty and the attraction of the city by itself and driven by the shared positive experiences of tourists. The second term, of slowdown, is related to negative experiences of the tourists associated with the massification, price increase, loss of originality and crimes among others [10]. A new term has been added to the logistics equation to model the direct action of the residents against the tourists (or in favor).

$$\dot{T} = \alpha_1 T \left(1 - \frac{T}{k_1}\right) - \alpha_2 R \tag{2}$$

where,

- α_1 is the natural growth factor (the area become known by word-a-mouth recommendations of the tourists).
- α_2 is a factor related to the actions and policies in favour or against the tourist.
- k_1 is the asymptotic maximum (carrying capacity).

In the absence of tourists, the population of residents R can also be modeled with a logistics equation. The model (3) assumes an initial growth of the population associated with new opportunities and free spaces and a process of slowdown until it reaches its maximum capacity. In presence of a parallel population of tourists, who share the same spaces in the city, the logistic equation must be modified. The first modification deals with the maximum capacity of the city, as explained above, Barcelona has no possibility to increase its capacity and the increase of tourism population implies a decrease of the residents population. A second modification deals with the fact that a small population of residents does not support the tourist pressure and tends to disappear (Alee effect). Investors mobbing against locals could be included in these Allee effect.

$$\dot{R} = \alpha_3 R \left(1 - \frac{R}{(k - k_1)}\right) \left(\frac{R}{k2} - 1\right) - \alpha_4 RT \tag{3}$$

where,

- α_3 is the natural growth factor (intrinsic regeneration rate).
- α_4 is a factor related to the actions and policies in favour or against the tourist.
- k_1 is the asymptotic maximum (carrying capacity).
- $k - k_1$ is the asymptotic maximum, being k the capacity of the city.
- k_2 is the survival number (values under this tends to extinguish the population).

Consequently, the system that relates tourists to residents, is given by,

$$\begin{cases} \dot{T} = \alpha_1\, T\, \left(1 - \frac{T}{k_1}\right) - \alpha_2\, R \\ \dot{R} = \alpha_3\, R\, \left(1 - \frac{R}{(k-k_1)}\right) \left(\frac{R}{k2} - 1\right) - \alpha_4\, R\, T. \end{cases} \tag{4}$$

3 Static Solutions and Their Stability

The proposed Eqs. (4) are time-independent when $\dot{T}$ and $\dot{R}$ are simultaneously equal to zero.

$$\begin{cases} 0 = \alpha_1\, T\, \left(1 - \frac{T}{k_1}\right) - \alpha_2\, R \\ 0 = \alpha_3\, R\, \left(1 - \frac{R}{(k-k_1)}\right) \left(\frac{R}{k2} - 1\right) - \alpha_4\, R\, T. \end{cases} \tag{5}$$

Discussion about the stability of the solution is done according the values of the eigenvalues of Jacobian matrix,

$$J = \begin{pmatrix} \alpha_1 - 2\, T\, \frac{\alpha_1}{k_1} & -\alpha_2 \\ -\alpha_4\, R & \alpha_3 \left(-1 + \frac{2\, R}{k_2} + \frac{2\, R}{k-k_1} - \frac{3\, R^2}{k-k_1}\right) - \alpha_4\, T \end{pmatrix} \tag{6}$$

- For $T = 0$, $R = 0$,

$$J = \begin{pmatrix} \alpha_1 & -\alpha_2 \\ 0 & -\alpha_3 \end{pmatrix} \tag{7}$$

  This solution (zero tourists and zero residents) is unstable because the eigenvalue α_1 (the natural growth ratio of tourists) is a real positive number.
- For $T = k_1$, $R = 0$,

$$J = \begin{pmatrix} -\alpha_1 & -\alpha_2 \\ 0 & -\alpha_3 - \alpha_4\, k_1 \end{pmatrix} \tag{8}$$

In this case, stability depends on the sign of the parameters $-\alpha_3 - \alpha_4 k_1$. Describes a scenario without residents, being the city only for tourists, which we call "theme park".

- For ($R \neq 0$ and $\dot{R} = 0$), we obtain,

$$T = \frac{\alpha 3}{\alpha_4}\left(\frac{k - k_1 + k - 2}{k_2\,(k - k_1)} R - 1 + \frac{R^2}{k_2\,(k - k_1)}.\right) \tag{9}$$

and a 4° polynomial, that,

- May not have any real solution. Then, there would be only two stationary solutions, the null and so-called "theme park".
- May have two real solutions and two complexes. There would be 4 stationary solutions.
- May have four real solutions. In total there would be 6 stationary solutions.

4 Results and Discussion

There is some freedom in the choice of parameters and the results strongly depend on these values. The authors would like to emphasize the methodology and the discussion of the results, rather than the specific results, which have been understood as one of the possible scenarios. In our model, natural growth factors ($\alpha_1 = 0.08$ and $\alpha_3 = 0.025$) are taken from the literature and previous works by the authors [11]. K-parameters are reasonable values for Barcelona, related to their capacity to hold residents and tourists $k = 2e6$, $k_1 = 2e5$, $k_2 = k_1$.

A bifurcation analysis have been done varying some meaningful parameters like α_2 and α_4 (starting values of $\alpha_2 = 0.00005$, $\alpha_4 = 0.0000002$ have been selected in order to study the range around the bifurcation points).

There are at least six real time independ solutions of the dynamical system that can be found solving the Eqs. (4).

1. $T = 0$, $R = 0$.
2. $T = k_1$, $R = 0$.
3. $T = 125.22$, $R = 200{,}225.43$.
4. $T = 1130.11$, $R = 1{,}797{,}963.20$.
5. $T = 199{,}210.99$, $R = 1{,}257{,}434.16$.
6. $T = 199{,}533.68$, $R = 744{,}377.21$.

Figure 1 shows the time independent solutions for tourist and residents for a range of α_4 values. The stationary solution T = 200,000, R = 0 ("theme park") is also represented. Only the "theme park" solution stands to the right of the limit point (LP = 2.2e-7). Between the branch point (BP = -1.2e-6) and limit point (LP = 2.2e-7) there are two stable stationary solutions.

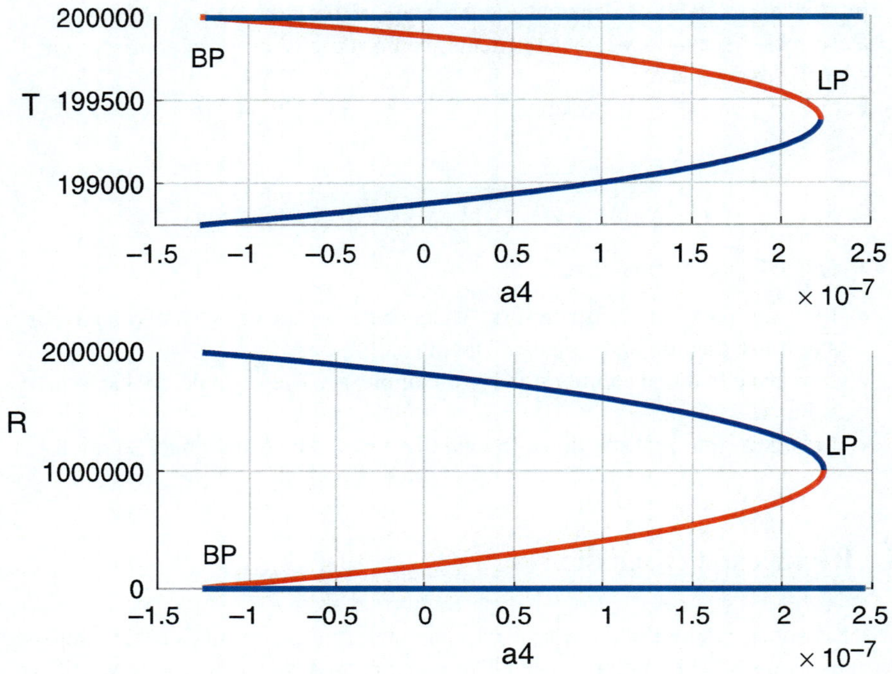

Fig. 1 Time independent solutions for tourists (T) and residents (R) for a range of the parameter α_4. BP stands for Branch Point and LP for Limit Point. Red and blue are for unstable and stable solutions respectively

LP bifurcation is associated to the disappearance of sustainable equilibrium, leading to the extinction of a population. It usually appears when policy factors (α_2 and α_4) grow too much in absolute value.

The existence of negative values of α_4 can be understood as a positive effect of residents, increasing the affluence of tourists.

Figure 2 shows the time independent solutions for tourist and residents for a range of α_2 values. Limit points are LP = -0.0020048609 and LP = 0.0107725061.

5 Conclusions

The dynamic interaction between tourists and residents has been modeled through a nonlinear system of ordinary differential equations. The model takes into account the natural growth of both populations and external actions associated with political decisions and reaction actions of tourists and resident populations. Initially, the model has been applied to the city of Barcelona, but it will be necessary to do more future work to calibrate the parameters of this city.

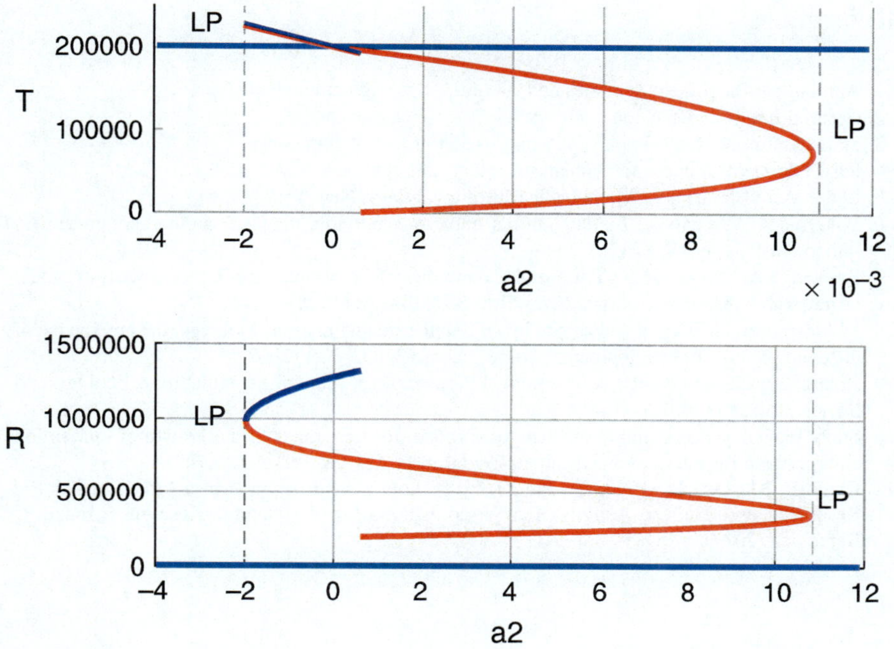

Fig. 2 Time independent solutions for tourists (T) and residents (R) for a range of the parameter α_2. BP stands for Branch Point and LP for Limit Point. Red and blue are for unstable and stable solutions respectively

The proposed equations describe the problem and the expected behavior, but the specific results must be taken with some caution because of the strong dependence of the associated parameters. It is a non-systematic study and the specific results must be understood as one of the possible scenarios. In this scenario, our results show that sustainability, that is, the positive values of the population of tourists and residents, is possible under the appropriate economic and social decisions.

Our bifurcation analysis varying the parameters associated to political and reacting actions have found several stable stationary solutions. The existence of Limit Points indicates that stationary solutions can not be found when policy factors grow beyond that point. There is also a stationary solution corresponding to zero residents.

Acknowledgments This work was partially supported by the Department of Applied Mathematics of the Universitat Politecnica de Catalunya. Gerard Olivar acknowledges Colciencias under project 111974558276 Modelado y Simulación del Metabolismo Urbano de Bogota D.C.

References

1. Ajuntament Barcelona: Statistics and surveys, http://ajuntament.barcelona.cat
2. Gremi d'hotels de Barcelona, http://www.barcelonahotels.org
3. Statista, https://www.statista.com/topics/4156/tourism-in-barcelona/
4. Servei de premsa, http://ajuntament.barcelona.cat/premsa/tag/turisme/
5. Lotka, A.J.: Elements of Mathematical Biology. Dover, New York (1956)
6. Volterra, V.: Variazioni e fluttuazioni del numero d'individui in specie animali conviventi (in Italian), vol. 5, pp. 113 (1926)
7. Brander, J.A., Taylor, M.S.: The simple economics of Easter island: a Ricardo-Malthus model of renewable resource use. Am. Econ. Rev. **88**, 119–138 (1998)
8. D'Alessandro, S.: Non-linear dynamics of population and natural resources: the emergence of different patterns of development. Ecolog. Econ. **62**, 473–481 (2007)
9. Rinaldi S., Casagrandi R.: A Theoretical Approach to Tourism Sustainability. Ecology and Society, Silsden (2002)
10. Sinay L., Sinay L.: A simple mathematical model for the effects of the growth of tourism on environment. In: International Tourism Conference. Alanya, Turkey (2006)
11. Cholo I., Massana I., Olivar G., Prat, J.A., Trullols E.: Bifurcations in a Model of Natural Resources and Human Activity. In: World Conference on Natural Resource Modeling, Barcelona (2017)

Convergence of Dual Infinity Series

Frantisek Klimenda, Josef Soukup, Blanka Skocilasova, Jan Skocilas, and Lenka Rychlikova

Abstract The solution of the system of the partial differential motion equations describing the movement of plate element by Fourier's series is presented in the article. The investigated function is expressed by product of three or four functions of the particular variables. These functions are demanded relation for the calculation of the displacement components, rotation components and stress components. These functions are defined in the form of the dual infinite series. The sum of these functions is necessary to perform by the numerical summarization process—element by element. The convergence of these series has to be proved before, namely in the equations of stresses. The methodology is presented by the calculation of the shear stress for Kirchhoff's models of thin isotropic and orthotropic plates without corrections. This methodology is valid in general for the solution of motion partial differential equations of plate elements by Fourier's method.

Keywords Dual infinity series · Kirchhoff's model · Isotropic thin plate · Orthotropic thin plate

1 Introduction

Many problems of the body mechanics are defined by the system of partial differential equations. The analytical methods are preferred to solve the system. One of the suitable method is application of Fourier's method. This method can be applied for solution of the system of motion partial differential equations of plate elements (thin plates).

F. Klimenda (✉) · J. Soukup · B. Skocilasova · L. Rychlikova
Faculty of Mechanical Engineering, University of J. E. Purkyne in Usti nad Labem, Usti nad Labem, Czech Republic
e-mail: frantisek.klimenda@ujep.cz

J. Skocilas
Faculty of Mechanical Engineering, Czech Technical University in Prague, Praha, Czech Republic

J. Awrejcewicz (ed.), *Perspectives in Dynamical Systems II: Mathematical and Numerical Approaches*, Springer Proceedings in Mathematics & Statistics 363,
https://doi.org/10.1007/978-3-030-77310-6_3

Fourier's method is based on neglecting the mutually dependence of the variables. Solution has been searched in form of product of functions. The number of functions is equal to number of variables and each function corresponds to one variable. In case of the solution of system of partial differential motion equations three or four variables have to be defined by product of three or four functions. The demanded relationships for evaluation of the particular components of the displacement, rotation and stress are expressed in form of double infinite series. The sum of the series has to be determined by sequent numerical summarization process—element by element. Before the evaluation of series, their convergence has to be validated.

2 Materials and Methods

The process of convergence evaluation of double infinity series is shown for case of shear stress solving of Kirchhoff's model for thin isotropic and orthotropic plates without correction. The plate is supported on its perimeter. The load is acting at top face of the plate by alone force with normal direction to the surface or more precisely by continuous load acting on the small area. The equations of the plate deformation is not presented here. It is necessary to determine components of the displacement (u, v, w), velocities ($\dot{u}$, $\dot{v}$, $\dot{w}$) (and stress (σ_x, σ_y, τ_{xy}, τ_{xz}, τ_{yz}) in direction of particular axes in the general point of the thin rectangular elastic plate (dimensions $a \times b \times h$) [1], see Fig. 1.

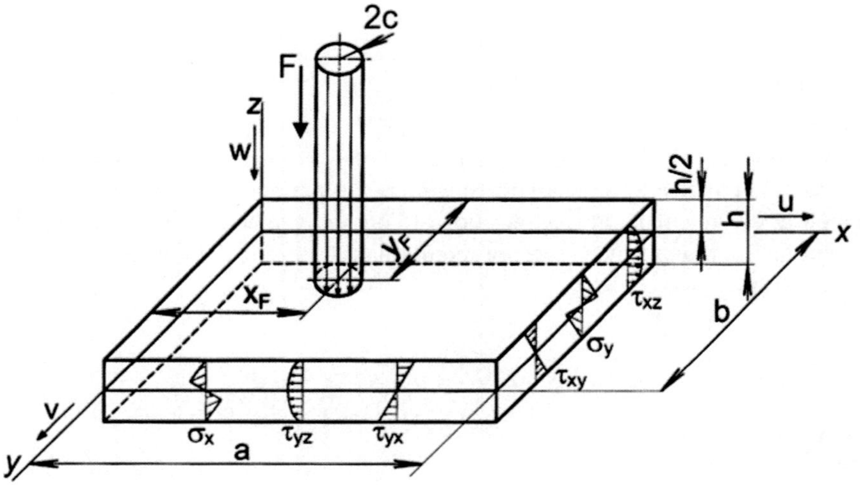

Fig. 1 Thin plate—coordinates, dimensions and stress components

The most important is to determine the shear components of stress τ_{xz}, τ_{yz}, and tensile stress σ_x, σ_y for particular plate models under calculations of thin plate by approximate analytic methods. According to [1, 2], normal stresses in direction x and y of thin plate induced by force F are defined (see Fig. 1)

$$\sigma_x\,(x;\,y;\,z;\,t) = \frac{E_x z}{1-\mu_{yx}\mu_{xy}}\frac{4F}{ab}\sum_{m=1}^{\infty}\sum_{n=1}^{\infty}\frac{p_{mn}}{\Psi_{mn}}\frac{\alpha_n^2+\mu_{yx}\beta_m^2}{\omega_{mn}} sin\alpha_n x sin\beta_m y T(t) \tag{1}$$

$$\sigma_y\,(x;\,y;\,z;\,t) = \frac{E_y z}{1-\mu_{yx}\mu_{xy}}\frac{4F}{ab}\sum_{m=1}^{\infty}\sum_{n=1}^{\infty}\frac{p_{mn}}{\Psi_{mn}}\frac{\beta_m^2+\mu_{xy}{\alpha_n}^2}{\omega_{mn}} sin\alpha_n x sin\beta_m y T(t) \tag{2}$$

where E_x, E_y—Young's tensile modules, F—alone force (Heaviside's jump function with radius c), T—function of harmonic oscillation, a, b, h—plate dimensions, p_{mn}—dimensionless parameter, μ_{yx}, μ_{xy}—Poisson's numbers, ω_{mn}—eigen frequency of oscillation, Ψ_{mn}—parameter (for Kirchhoff's model $\Psi_{mn} = \rho h$), α_n, β_m—parameters

$$\alpha_n = \frac{n\pi}{a} \text{ for } n = 1, 2, 3, \ldots n, \quad \beta_m = \frac{m\pi}{b} \text{ for } m = 1, 2, 3, \ldots m \tag{3}$$

Similarly the components of shear stress are defined

$$\tau_{xz}\,(x;\,y;\,z;\,t) = \left[1-\left(\frac{2z}{h}\right)^2\right]\frac{F\rho h^2}{2ab}\sum_{m=1}^{\infty}\sum_{n=1}^{\infty}\frac{p_{mn}}{\Psi_{mn}}A_x\alpha_n\omega_{mn} \tag{4a}$$

$$cos\alpha_n x sin\beta_m y T(t) \tag{4b}$$

$$\tau_{yz}\,(x;\,y;\,z;\,t) = \left[1-\left(\frac{2z}{h}\right)^2\right]\frac{F\rho h^2}{2ab}\sum_{m=1}^{\infty}\sum_{n=1}^{\infty}\frac{p_{mn}}{\Psi_{mn}}A_y\beta_m\omega_{mn} \tag{5a}$$

$$sin\alpha_n x cos\beta_m y T(t) \tag{5b}$$

where A_x, A_y—parameter depends on stiffness module.

2.1 Isotropic Thin Plate: Kirchhoff's Model Without Corrections

The solution of the partial differential equations is performed by Fourier's method. For this plate, the normal stresses are defined (without solving procedure) in direction x and y

$$\sigma_x\,(x,y,z,t) = \frac{Ez}{1-\mu^2}\cdot\frac{16}{abc}\cdot\frac{F}{\rho h}\sum_{m=1}^{\infty}\sum_{n=1}^{\infty}\frac{J_1\,(\gamma_{mn}c)}{\gamma_{mn}}\cdot\frac{\alpha^2+\mu\beta^2}{\omega_{mn}^2} \tag{6a}$$

$$\sin\alpha x_F.\sin\beta y_F.\sin\alpha x.\sin\beta y.\sin^2\frac{\omega_{mn}}{2}t \tag{6b}$$

$$\sigma_y\,(x,y,z,t) = \frac{Ez}{1-\mu^2}\cdot\frac{16}{abc}\cdot\frac{F}{\rho h}\sum_{m=1}^{\infty}\sum_{n=1}^{\infty}\frac{J_1\,(\gamma_{mn}c)}{\gamma_{mn}}\cdot\frac{\beta^2+\mu\alpha^2}{\omega_{mn}^2} \tag{7a}$$

$$\sin\alpha x_F.\sin\beta y_F.\sin\alpha x.\sin\beta y.\sin^2\frac{\omega_{mn}}{2}t \tag{7b}$$

where $J_1(\gamma_{mn}c)$—Bessel's function of first kind, first order (index) for argument $\gamma_{mn}c$. $\mu = \mu_{xy} = \mu_{yx}$—Poisson's number.

Similarly the components of shear stress are defined

$$\tau_{xz}\,(x,y,z,t) = \left[1-\left(\frac{2z}{h}\right)^2\right]\frac{24F}{abch}\sum_{m=1}^{\infty}\sum_{n=1}^{\infty}\frac{J_1\,(\gamma_{mn}c)}{\gamma_{mn}}\cdot\frac{\alpha_n}{{\alpha_n}^2+{\beta_m}^2}sin\alpha_n x_F. \tag{8a}$$

$$.sin\beta_m y_F.cos\alpha_n x.sin\beta_m y.\sin^2\frac{\omega_{mn}}{2}t \tag{8b}$$

$$\tau_{yz}\,(x,y,z,t) = \left[1-\left(\frac{2z}{h}\right)^2\right]\frac{24F}{abch}\sum_{m=1}^{\infty}\sum_{n=1}^{\infty}\frac{J_1\,(\gamma_{mn}c)}{\gamma_{mn}}\cdot\frac{\beta_m}{{\alpha_n}^2+{\beta_m}^2}sin\alpha_n x_F. \tag{9a}$$

$$.sin\beta_m y_F.sin\alpha_n x.cos\beta_m y.\sin^2\frac{\omega_{mn}}{2}t \tag{9b}$$

2.2 Orthotropic Thin Plate: Kirchhoff's Model Without Corrections

For orthotropic plate the same equations are valid, where $\mu \neq \mu_{xy} \neq \mu_{yx}$. Normal stresses

$$\sigma_x(x, y, z, t) = \frac{E_x z}{1 - \mu_{yx}\mu_{xy}} \cdot \frac{16}{abc} \cdot \frac{F_0}{\rho h} \sum_{m=1}^{\infty} \sum_{n=1}^{\infty} \frac{J_1(\gamma_{mn} c)}{\gamma_{mn}} \cdot \frac{\alpha_n{}^2 + \mu_{yx}\beta_m{}^2}{\omega_{mn}^2} \tag{10a}$$

$$sin\alpha_n x_F.sin\beta_{m,} y_F.sin\alpha_n x.sin\beta_{m,} y.\sin^2\frac{\omega_{mn}}{2}t \tag{10b}$$

$$\sigma_y(x, y, z, t) = \frac{E_x z}{1 - \mu_{yx}\mu_{xy}} \cdot \frac{16}{abc} \cdot \frac{F_0}{\rho h} \sum_{m=1}^{\infty} \sum_{n=1}^{\infty} \frac{J_1(\gamma_{mn} c)}{\gamma_{mn}} \cdot \frac{\beta_m{}^2 + \mu_{xy}\alpha_n{}^2}{\omega_{mn}^2}$$

$$sin\alpha_n x_F.sin\beta_m y_F.sin\alpha_n x.sin\beta_m y.\sin^2\frac{\omega_{mn}}{2}t \tag{11a}$$

$$sin\alpha_n x_F.sin\beta_m y_F.sin\alpha_n x.sin\beta_m y.\sin^2\frac{\omega_{mn}}{2}t \tag{11b}$$

Shear stresses

$$\tau_{xz}(x, y, z, t) = \left[1 - \left(\frac{2z}{h}\right)^2\right] \frac{24}{abc} \cdot \frac{F_0}{\rho h^2} \sum_{m=1}^{\infty} \sum_{n=1}^{\infty} \frac{J_1(\gamma_{mn} c)}{\gamma_{mn}}. \tag{12a}$$

$$\frac{\alpha\left[2D_{xy}\beta_m{}^2 + D_x\left(\alpha_n{}^2 + \mu_{yx}\beta_m{}^2\right)\right]}{\omega_{mn}^2}.sin\alpha_n x_F.sin\beta_m y_F.cos\alpha_n x. \tag{12b}$$

$$sin\beta_m y.\sin^2\frac{\omega_{mn}}{2}t \tag{12c}$$

$$\tau_{yz}(x, y, z, t) = \left[1 - \left(\frac{2z}{h}\right)^2\right] \frac{24}{abc} \cdot \frac{F_0}{\rho h^2} \sum_{m=1}^{\infty} \sum_{n=1}^{\infty} \frac{J_1(\gamma_{mn} c)}{\gamma_{mn}}. \tag{13a}$$

$$\frac{\beta_m \left[2D_{xy}\alpha_n{}^2 + D_y \left(\beta_m{}^2 + \mu_{xy}\alpha_n{}^2\right)\right]}{\omega_{mn}^2}.$$

$$.sin\alpha_n x_F.sin\beta_m y_F.sin\alpha_n x. \tag{13b}$$

$$cos\beta_m y.\sin^2\frac{\omega_{mn}}{2}t \tag{13c}$$

The functions of the stresses (and displacements and velocities) are expressed in form of double infinity series. By each derivation the rate of series convergence is decreasing, eventually series after derivation is not convergent. Therefore the convergence of particular series is necessary validated. The criteria of convergence are rather complicated [3, 4]. Absolute or equiconvergence of series is possible to validate by direct comparison test. The determination of the sum is achieved by addition process element by element. Necessary number of series element, those has to be summarized, depends on rate of convergence and required accuracy of calculation. The maximum number of elements of finite series substituting infinite series is possible to determine from series rest or subset. With respect to slow rate of convergence of some series, e.g. for q_{xz} and q_{yz} (shearing force) and its corresponding components of stress τ_{xy}, τ_{yz}, the different solution of the system of three differential equations has to be find but the solution always lead to double infinity series.

3 Results and Discussion

The convergence has to be validated for particular calculation plate model which is obvious from series comparison. Especially for series in equations for stress components τ_{yz}, and τ_{yz}, or σ_x, and σ_y the convergence has to be validated. If these series converge (it can be assumed as upper bound of series in equations for other variables—displacements, velocities, shear forces, moments, etc.), then simple series also converge with higher rate of convergence, i.e. the lower number of elements can be summarized for required accuracy of calculation.

During validation of convergence it is necessary to apply three basic observations [5] beside common rules for infinite series operations. These observations are presented by followed statements

Statement 1 (comparison criteria)

Assumed series $A = a_1 + a_2 + a_3 + \ldots\ldots$
and series $\bar{A} = \bar{a}_1 + \bar{a}_2 + \bar{a}_3 + \ldots\ldots$

are series with positive elements and $a_i \leq \overline{\overline{a_i}}$ for almost all indexes i. Series $\bar{A}$ is upper bound to series A. From convergence of series $\bar{A}$ arise series A convergence.

Statement 2

If $A = a_1 + a_2 + a_3 + \ldots\ldots$
$B = b_1 + b_2 + b_3 + \ldots\ldots$
then they are series with positive elements, from which series B converge, and also $\frac{a_{n+1}}{a_n} \leq \frac{b_{n+1}}{b_n}$ for almost all n, i.e. for n with exception finite number of elements, then also series A converge.

Statement 3

Assuming series $\sum_{k=1}^{\infty} |a_{ik}|$ converge for each index i. If sum of this series is $s_i = \sum_{k=1}^{\infty} |a_{ik}|$ and series $\sum_{i=1}^{\infty} s_i$ converge, then double series $\sum_{i=1}^{\infty} \sum_{k=1}^{\infty} a_{ik}$ is absolutely convergent.

3.1 *Convergence of Double Infinite Series*

First of all the proof of convergence of double series will be performed. The example is for calculation of component of shear stress (8a) and (8b), i.e. $\tau_{xz} = konst.f(z)$. A for Kirchhoff's model of isotropic plate (without correction)

$$A = \sum_{n=1}^{\infty} \sum_{m=1}^{\infty} \frac{J_1(\gamma_{mn} c)}{\gamma_{mn}} \cdot \frac{\alpha(\alpha^2 + \beta^2)}{\omega_{mn}^2}.\sin\alpha x_F.\sin\beta y_F.\cos\alpha x.\sin\beta y. \tag{14a}$$

$$\sin^2 \frac{\omega_{mn}}{2} t \tag{14b}$$

It is valid

$$J_1(\gamma_{mn} c) < 1 \text{ and } \lim J_1(\gamma_{mn} c) = 0 \tag{15}$$

for $m \to \infty$ and $n \to \infty$.

And also

$$\left|\sin\beta y_F.\sin\beta y.\sin^2 \frac{\omega_{mn}}{2} t\right| \leq 1$$

For arguments $\gamma_{mn} = \sqrt{\alpha^2 + \beta^2}$ is valid

$$a_{mn} = \frac{\alpha\left(\alpha^2+\beta^2\right)}{\sqrt{\alpha^2+\beta^2}\left[\alpha^2+\beta^2\right]} = \frac{1}{\left(\alpha^2+\beta^2\right).\sqrt{1+\left(\frac{\beta}{\alpha}\right)^2}} \tag{16}$$

After arrangement

$$a_{mn} = \left(\frac{b}{\pi}\right)^2 \frac{1}{\sqrt{1+\left(\frac{am}{bn}\right)^2}}.\frac{1}{m^2+\left(\frac{b}{a}n\right)^2} \le \left(\frac{b}{\pi}\right)^2 \frac{1}{\sqrt{1+\left(\frac{a}{b}.\frac{1}{n}\right)^2}} \frac{1}{m^2+\left(\frac{b}{a}n\right)^2} \tag{17}$$

Based on these raw assumptions it is possible to choose upper bound to double series A (14a) and (14b) in form

$$\left(\frac{b}{\pi}\right)^2 \sum\nolimits_{n=1}^{\infty}\sum\nolimits_{m=1}^{\infty} \frac{1}{\sqrt{1+\left(\frac{a}{b}.\frac{1}{n}\right)^2}}.\frac{\sin n\frac{\pi x_F}{a}.\cos n\frac{\pi x}{a}}{m^2+\left(\frac{b}{a}n\right)^2} \tag{18}$$

After these steps convergence of double series has to be proved by the application of statement 1 and 3

$$\overline{A} = \sum\nolimits_{n=1}^{\infty} \frac{\sin n\frac{\pi x_F}{a}.\cos n\frac{\pi x}{a}}{\sqrt{1+\left(\frac{a}{b}.\frac{1}{n}\right)^2}} \sum\nolimits_{m=1}^{\infty} \left|\frac{1}{m^2+\left(\frac{b}{a}n\right)^2}\right| \tag{19}$$

Series for m elements is absolutely convergent for every n and its summary s_n is [6]

$$s_n = \frac{\pi a}{2bn}\coth\left(\pi\frac{b}{a}n\right) - \frac{1}{2\left(\frac{b}{a}n\right)^2} \tag{20}$$

After arrangement

$$s_n = \frac{1}{2}\left[\frac{\pi}{\frac{b}{a}n}.\frac{1+\exp\left(-2\pi\frac{b}{a}n\right)}{1-\exp\left(-2\pi\frac{b}{a}n\right)} - \frac{1}{\left(\frac{b}{a}n\right)^2}\right] \tag{21}$$

Next, the series convergence has to be validate

$$\overline{A} = \sum\nolimits_{n=1}^{\infty} \frac{s_n}{\sqrt{1+\left(\frac{a}{b}.\frac{1}{n}\right)^2}}.\sin n\frac{\pi x_F}{a}.\cos n\frac{\pi x}{a} \tag{22}$$

After arrangement

$$\overline{A}=\frac{1}{2}\left[\frac{a}{b}\pi\sum_{b=1}^{\infty}\frac{\sin n\frac{\pi x_F}{a}.\cos n\frac{\pi x}{a}}{n\sqrt{1+\left(\frac{a}{b}.\frac{1}{n}\right)^2}}\cdot\frac{1+\exp\left(-2\pi\frac{b}{a}n\right)}{1-\exp\left(-2\pi\frac{b}{a}n\right)}-\left(\frac{a}{b}\right)^2\sum_{n=1}^{\infty}\frac{\sin n\frac{\pi x_F}{a}.\cos n\frac{\pi x}{a}}{n^2\sqrt{1+\left(\frac{a}{b}.\frac{1}{n}\right)^2}}\right] \tag{23}$$

Both series converge, which can be proven by suitable selection of convergence upper bound series. With respect to

$$\frac{1+\exp\left(-2\pi\frac{b}{a}n\right)}{\left(1-\exp\left(-2\pi\frac{b}{a}n\right)\right)\sqrt{1+\left(\frac{a}{b}.\frac{1}{n}\right)^2}}\leq\frac{1+\exp\left(-2\pi\frac{b}{a}\right)}{1-\exp\left(-2\pi\frac{b}{a}\right)}=k_1 \tag{24}$$

And series is

$$\bar{A}_1=\sum_{n=1}^{\infty}\frac{\sin n\frac{\pi x_F}{a}.\cos n\frac{\pi x}{a}}{n} \tag{25}$$

which converge and for $0 < x < a$ the sum is [6]

$$\bar{A}_1=\sum_{n=1}^{\infty}\frac{\sin n\frac{\pi x_F}{a}.\cos n\frac{\pi x}{a}}{n}=-\frac{1}{2}\quad\text{for } 0<x_F<x \tag{26}$$

$$\bar{A}_1=\sum_{n=1}^{\infty}\frac{\sin n\frac{\pi x_F}{a}.\cos n\frac{\pi x}{a}}{n}=\frac{\pi}{4}\left(1-2\frac{x}{a}\right)\quad\text{for } x_F=x \tag{27}$$

$$\bar{A}_1=\sum_{n=1}^{\infty}\frac{\sin n\frac{\pi x_F}{a}.\cos n\frac{\pi x}{a}}{n}=\frac{\pi}{2}\left(1-\frac{x_F}{a}\right)\quad\text{for } 0<x_F<a \tag{28}$$

After multiplication of this series ($\bar{A}_1$) k1 this series is upper bound to first series (8a) and (8b) and therefore this series also converge.

Because it is valid

$$\frac{\left|\sin n\frac{\pi x_F}{a}.\cos n\frac{\pi x}{a}\right|}{\sqrt{1+\left(\frac{a}{b}.\frac{1}{n}\right)^2}}\leq 1=k_2 \tag{29}$$

The series is absolutely converging $\overline{A_2}=\sum_{n=1}^{\infty}\frac{1}{n^2}=\frac{\pi^2}{6}$ multiplied by constant k_2, because it is upper bound to second series (8a) and (8b) and therefore the series absolutely converge as well.

The proof of series convergence $\bar{A}_1$ and $\bar{A}_2$ thereby $\bar{A}$ the convergence of double series A is also proven by 3 statement in relationship to τ_{xz} for Kirchhoff's model. Its sum is possible to determine by consequent summarization of series elements. For sum of series A the followed relation has to be valid

$$A < \left(\frac{b}{\pi}\right)^2 \cdot \frac{1}{2} \cdot \frac{a}{b} \left[\pi k_1 \overline{A_1} - \frac{a}{b} . k_2 \overline{A_2}\right] \tag{30}$$

The performed selection of the upper bound to particular series is motivated by the simplest establishment of convergence. For better estimation of the series sum, it is possible to select "more close" upper bounds. The calculation procedure will be more complicated and extensive.

For Kirchhoff's model of orthotropic plate, the relevant series with relation to calculation of shear stress component $\tau_{xz} = konst. f(z)$. A

$$A = \sum\nolimits_{n=1}^{\infty} \sum\nolimits_{m=1}^{\infty} \frac{J_1(\gamma_{mn} c)}{\gamma_{mn}} \cdot \frac{\alpha\left[\alpha^2 D_x + \beta^2 . D_2\right]}{\left[D_x \alpha^4 + D_1 \alpha^2 \beta^2 + D_y \beta^4\right]} . \sin\alpha x_F . \sin\beta y_F . \cos\alpha x . \tag{31a}$$

$$. \sin\beta y . \sin^2 \frac{\omega_{mn}}{2} t \tag{31b}$$

which corresponds to series (6a), (6b), (7a), (7b). For series (31a) and (31b) where $D_1 = D_x \mu_{xy} + D_y \mu_{yx}$ and $D_2 = D_x \mu_{xy} + 2D_{xy}$ is possible to arrange element

$$a_{mn} = \frac{\alpha\left[\alpha^2 D_x + \beta^2 D_2\right]}{\sqrt{\alpha^2 + \beta^2}\left[D_x \alpha^4 + D_1 \alpha^2 \beta^2 + D_y \beta^4\right]} \tag{32}$$

into the form

$$a_{mn} = \frac{D_x}{\sqrt{1 + \left(\frac{am}{bn}\right)^2}\left[D_x \left(\frac{bn}{am}\right)^2 + D_1 + D_y \left(\frac{am}{bn}\right)^2\right] \left(\frac{\pi m}{b}\right)^2} + \tag{33a}$$

$$+ \frac{D_2}{\sqrt{1 + \left(\frac{am}{bn}\right)^2}\left[D_x \left(\frac{bn}{am}\right)^2 + D_1 + D_y \left(\frac{am}{bn}\right)^2\right] \left(\frac{\pi n}{a}\right)^2} \tag{33b}$$

And after next arrangement it is possible to select for upper bound

$$a_{mn} \leq \frac{D_x}{D_1 \left(\frac{\pi}{b}\right)^2 m^2 + D_x \left(\frac{\pi}{a}\right) n^2} + \frac{D_2}{D_1 \left(\frac{\pi}{a}\right)^2 n^2 + D_y \left(\frac{\pi}{b}\right) m^2} \tag{34}$$

Then it is possible to select upper bound to series (31a) and (31b) in form

$$\overline{A}=\sum_{n=1}^{\infty}\sum_{m=1}^{\infty}\left(\frac{b}{\pi}\right)^{2}\left[\frac{D_x}{D_1}\cdot\frac{1}{m^2+(p_1n)^2}+\frac{D_2}{D_y}\frac{1}{m^2+(p_2n)^2}\right].\sin n\frac{\pi x_F}{a}.\cos n\frac{\pi x}{a} \tag{35}$$

where

$$D_x = \frac{E_x h^3}{12\left(1-\mu_{yx}\mu_{xy}\right)} \quad D_y = \frac{E_y h^3}{12\left(1-\mu_{yx}\mu_{xy}\right)} \quad D_{xy} = G_{xy}\frac{h^3}{12} \tag{36}$$

$$p_1 = \frac{b}{a}.\sqrt{\frac{D_x}{D_1}} \quad p_2 = \frac{b}{a}.\sqrt{\frac{D_1}{D_y}} \tag{37}$$

G_{xy}—shear modulus in plane *xy*, D_{xy}—stiffness modulus in torque, D_x—stiffness modulus in direction *x*, D_y—stiffness modulus in direction *y*.

Next process of proof for series (35) is similar to series (18) for isotropic plate.

The validation of series convergence for shear stress components τ_{yz} and τ_{xy} and for normal stress components σ_x and σ_y is possible to perform by the same procedure.

4 Conclusion

Application of Fourier's method on solution of motion partial differential equations was presented. The solution for components of displacements, velocities and stresses are derived for oscillation of the thin isotropic and orthotropic plate excited by alone force. The limitation and calculation procedure of solving the convergence problems using Fourier's method were proposed with application to example of isotropic and orthotropic thin plate described by Kirchhoff's model.

Acknowledgments This work was supported by internal grant of Jan Evangelista Purkyně in Ústí nad Labem, Faculty of Mechanical Engineering no. UJEP-IGS-2018-48-002-1 and no. UJEP-SGS-2018-48-002-2.

References

1. Volek, J.: Nestacionární napjatost tenké desky vyvolaná příčnou osamělou silou. Bulletin vědeckých, výzkumných a pedagogických prací ústavu za rok 2000. Ústav techniky a řízení výroby, Univerzita J. E. Purkyně, Ustí nad Labem (2000)
2. Leitmann, M.J.: The linear theory of viscoelasticity. In: Encyklopedia of Physics. Springer, Berlin (1973)

3. Petr, K.: Počet diferenciální. SČMF, Praha (1923)
4. Tolstov, G.: Fourniersreihen, Berlin (1955)
5. Rektorys, K.: Přehled užité matematiky. SNTL, Praha (1968)
6. Prudnikov, A.P., Bryčkov, J.A., Maričev, O.I.: Integrali i rjady. Elementarnyje funkcii. Nauka, Moskva (1981)

Full Spectrum Analysis for Studying the Backward Whirl in Accelerated Rotor Systems

Mohammad A. Al-Shudeifat, Oleg Shiryayev, Tariq Alzarooni, and Chandrasekhar Nataraj

Abstract The backward whirl (BW) phenomena in intact and cracked rotor systems that exhibit recurrent acceleration and deceleration during startup and coast down operations has not been well-studied in the literature. However, for startup and coast down operations during which a frequent passage through critical forward whirl (FW) speeds takes place, the BW orbits are found to be immediately captured after the passage through these critical FW rotational speeds. The zones of BW orbits are observed to be significantly affected by the appearance of crack damages that are accompanied with isotropic or anisotropic bearings at the shaft supports. The finite element model of the cracked rotor-bearing-disk system is employed here to obtain the linear-time-variant (LTV) equations of motion for the numerical simulation. The obtained LTV mathematical model represents a nonlinear dynamical model of the system. Consequently, the full spectrum analysis (FSA) is successfully employed here to the numerical simulation and the experimental whirl responses of the considered system to confirm the existence of the BW zones of shaft rotational speeds that exist after the passage through the critical and subcritical FW whirl rotational speeds. The obtained results for the intact and cracked rotor systems with anisotropic bearings verify the robustness of the FSA as a powerful tool of capturing the BW zones in cracked rotor systems, especially in the experimental whirl response.

Keywords Cracked rotor · Backward whirl · Full spectrum analysis

---

M. A. Al-Shudeifat (✉) · T. Alzarooni
Department of Aerospace Engineering, Khalifa University of Science and Technology, Abu Dhabi, UAE
e-mail: mohd.shudeifat@ku.ac.ae; tariq.alzarooni@ku.ac.ae

O. Shiryayev
Department of Mechanical Engineering, University of Alaska Anchorage, Anchorage, AK, USA
e-mail: oshiryayev@alaska.edu

C. Nataraj
Mechanical Engineering, Villanova University, Villanova, PA, USA
e-mail: c.nataraj@villanova.edu

J. Awrejcewicz (ed.), *Perspectives in Dynamical Systems II: Mathematical and Numerical Approaches*, Springer Proceedings in Mathematics & Statistics 363,
https://doi.org/10.1007/978-3-030-77310-6_4

1 Introduction

The study of backward whirl (BW) phenomena in a rotary system could help assessing the rotor condition in terms of predicting premature failures. Such phenomena could be found linked to poor rotor conditions including crack initiation and propagation, bearings tears and wears, shaft rubbing and so on. Consequently, it is of interest to investigate BW-based features that could be utilized in a damage detection technique for enhancing vibration health monitoring of miscellaneous rotary systems. The effect of propagation of fatigue cracks and bearing damages on BW excitation in accelerated rotor systems that exhibits recurrent passage through their critical forward whirl rotational speeds during startup and coast down operations has not been well-established in literature. In [1–3], it was numerically and theoretically found that the propagation of the crack in rotor systems could excite backward whirl orbits during the passage through critical forward whirl rotational speeds. In [4], the effect of crack depth of a rotor-bearing-disk system on FW and BW amplitudes and whirl orbit shapes were investigated through the whirl response obtained by the harmonic balance solution using open and breathing crack models followed by experimental verification. The finite element (FE) model was employed in [5] whereby whirl orbits near critical and sub-critical speed ranges of the rotor were discussed. It was shown that there exists some speed range near the critical speed, where the temporary whirl direction reversal to backward and phase shift were noticed.

The BW response was also investigated in [6] for a cracked shaft using the FE model. It was reported that BW zones are easily captured for higher bearing clearance value regardless of crack condition, which was attributed to bearing film cross-coupled stiffness. The BW was studied in [7] whereby FE model was also employed to analyze the dynamic response of functionally graded (FG) shaft with multiple cracks. Results show that besides being affected by crack locations, orientations and size, the extent of percentage of reductions in fundamental frequencies and critical speeds are also influenced by the power-law gradient index of the FG shaft. The BW has been experimentally investigated in [8] using double-disk and multi-bearing apparatus on which vibration displacement data were acquired. It was reported that BW orbits have been captured when rubbing and misalignment forces were present.

Numerical and experimental analysis were used to analyze the existence of FW and BW zones during rotor's run-up and coast down operation using solely an open crack model in [9–11]. It was noticed that BW zones appear after passing the critical FW speed at low vibration amplitudes between local transient peaks of FW amplitudes. In [11], a 4-DOF rotor system was studied by incorporating gyroscopic effect into the classical Jeffcott Rotor system model. However, the backward whirl zones were not captured in the numerical simulation response of this model even though they have been captured in the experimental whirl response.

The full spectrum transform analysis was utilized by a few researchers of rotary systems for fault diagnostic purpose [11–15]. More detailed description of FSA concerning the benefits, the methodology, and signal correlations can be reviewed in [12, 13]. In a nutshell, the FSA plot displays the correlation between the vibration displacement data from the lateral X (Real) and vertical Y (Imaginary) components of the rotor or casing responses. These displacement components can be obtained either by using theoretical analysis or measurement hardware (i.e., proximity probes).

A few examples of analysis of failure scenarios have already been outlined in [12]. Experimental-based FSA was used in [14] to investigate rotor-bearing system failure including unbalance, crack, rotor-stator rub, and misalignment at sub-critical rotational speeds. The authors reflected that FSA is favorable considering the limitation of conventional Fourier Spectrum (i.e., FFT) in addressing the whirl nature (i.e., forward/backward whirl) of rotor faults. It was found out that the spectra of the cracked rotor or the rotor with rubbing were drastically dominated by FW frequencies; whereas for misalignment, whether it was parallel or angular type, the spectrum was strongly dominated by harmonics corresponding to BW frequencies especially at higher components. Furthermore, experimental-based FSA was also compared against classical theoretical analysis results of overhung rotor-disk system for validation purpose [15].

The FSA will be investigated here considering finite element model rather than the four degree-of-freedom model in [11] of the cracked rotor-bearing-disk system. By using the FE model rather than the 4-dofs model, anisotropic bearings can be incorporated with the cracked rotor in the numerical simulation. Subsequently, the FSA is employed here to analyze the whirl response obtained numerically from the FE model and the corresponding experimental whirl response to further confirm the existence of BW zones of rotational speeds in the transient operation regime.

2 Cracked Rotor Modeling

The open transverse crack in the shaft cross-section is shown in Fig. 1, where the depth of the crack in the radial direction is represented by h. At the beginning of shaft rotation, the crack opening orientation is assumed to be at zero angle with respect to the stationary X axis as shown in Fig. 1a. The unbalance force vector angle is measured with respect to the crack opening direction and represented by β.

The area moments of inertia I_x and I_y of the cracked shaft cross-section in Fig. 1 with respect to the rotating coordinates x and y are, respectively, given as in [1, 2]:

$$I_x = \frac{\pi R^4}{4} - \frac{R^4}{12}\left((1-\mu)\left(2\mu^2 - 4\mu - 3\right)\gamma + 3\sin^{-1}\gamma\right) \quad (1)$$

$$I_y = \frac{\pi R^4}{4} + \frac{R^4}{4}\left((1-\mu)\left(2\mu^2 - 4\mu + 1\right)\gamma + \sin^{-1}(1-\mu)\right) \quad (2)$$

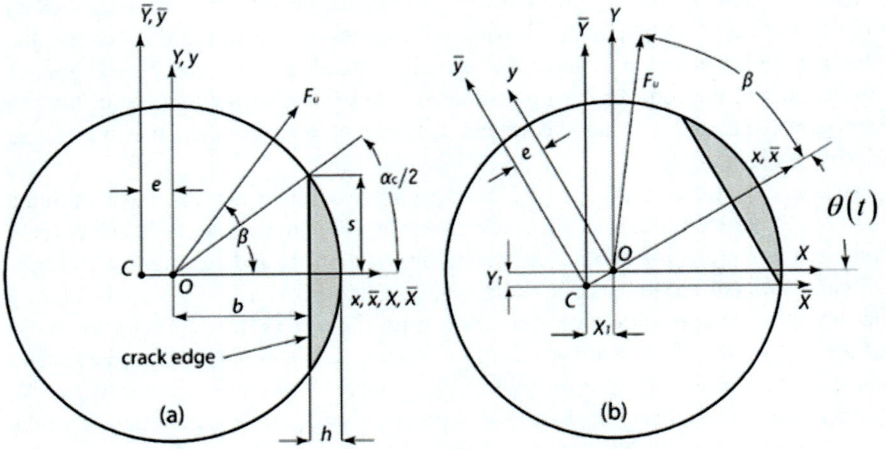

Fig. 1 Schematics of the crack in the shaft cross-section: (**a**) before rotation, and (**b**) after rotation by angle $\theta(t)$

where, $\mu = h/R$ is the ratio of crack depth h to the shaft radius R, and$\gamma = \sqrt{\mu\,(2-\mu)}$. The uncracked area A_{ce} in the cross-section and its centroid location e on the x-axis are calculated respectively as

$$A_{ce} = R^2\left(\pi - \cos^{-1}(1-\mu) + (1-\mu)\,\gamma\right) \tag{3}$$

$$e = \frac{2R^3}{3A_{ce}}[\mu\,(2-\mu)]^{\frac{3}{2}}. \tag{4}$$

If the open crack appears n the jth element, the stiffness matrix in the rotating centroidal coordinates axes $\overline{x}$ and $\overline{y}$ is given as in [1, 2]:

$$\mathbf{k}_R^j = \frac{E}{l^3}\begin{bmatrix} 12I_{\overline{y}} & 0 & 0 & 6lI_{\overline{y}} & -12I_{\overline{y}} & 0 & 0 & 6lI_{\overline{y}} \\ 0 & 12I_{\overline{x}} & -6lI_{\overline{x}} & 0 & 0 & -12I_{\overline{x}} & -6lI_{\overline{x}} & 0 \\ 0 & -6lI_{\overline{x}} & 4l^2I_{\overline{x}} & 0 & 0 & 6lI_{\overline{x}} & 2l^2I_{\overline{x}} & 0 \\ 6lI_{\overline{y}} & 0 & 0 & 4l^2I_{\overline{y}} & -6lI_{\overline{y}} & 0 & 0 & 2l^2I_{\overline{y}} \\ -12I_{\overline{y}} & 0 & 0 & -6lI_{\overline{y}} & 12I_{\overline{y}} & 0 & 0 & -6lI_{\overline{y}} \\ 0 & -12I_{\overline{x}} & 6lI_{\overline{x}} & 0 & 0 & 12I_{\overline{x}} & 6lI_{\overline{x}} & 0 \\ 0 & -6lI_{\overline{x}} & 2l^2I_{\overline{x}} & 0 & 0 & 6lI_{\overline{x}} & 4l^2I_{\overline{x}} & 0 \\ 6lI_{\overline{y}} & 0 & 0 & 2l^2I_{\overline{y}} & -6lI_{\overline{y}} & 0 & 0 & 4l^2I_{\overline{y}} \end{bmatrix} \tag{5}$$

where $I_{\overline{x}} = I_x$, $I_{\overline{y}} = I_y - A_{ce}e^2$, E is the elastic modulus of the shaft material, and l is the element length. Accordingly, the stiffness matrix in the fixed coordinates $\overline{X}$

and $\overline{Y}$ is obtained by the following transformation:

$$\mathbf{k}_F^j = \boldsymbol{\Psi}\, \mathbf{k}_R^j \boldsymbol{\Psi}^T \tag{6}$$

where $\boldsymbol{\Psi}$ is the 8×8 transformation matrix given in [1, 2]. Therefore, the cracked element stiffness matrix $\mathbf{k}_F^j$ in the fixed centroidal coordinates is obtained as

$$\mathbf{k}_F^j = \mathbf{k}_1^j + \mathbf{k}_2^j \cos(2\theta(t)) + \mathbf{k}_3^j \sin(2\theta(t)) \tag{7}$$

where $\mathbf{k}_1^j$, $\mathbf{k}_2^j$, and $\mathbf{k}_3^j$have been derived in [1, 2, 9], and $\theta(t)$ is the rotation angle. Accordingly, the FE equations of motion of the cracked rotor system in the fixed X and Y coordinates are expressed as:

$$\mathbf{M}\ddot{\mathbf{q}}(t) + \hat{\mathbf{C}}\dot{\mathbf{q}}(t) + (\mathbf{K}_1 + \mathbf{K}_2 \cos(2\theta(t)) + \mathbf{K}_3 \sin(2\theta(t)))\, \mathbf{q}(t) = \mathbf{F}_1 \cos(\theta(t)) + \\ + \mathbf{F}_2 \sin(\theta(t)) + \mathbf{F}_g \tag{8}$$

where $\mathbf{M}$ is the mass matrix, $\mathbf{K}_1$ is $4(N + 1) \times 4(N + 1)$ stiffness matrix in which $\mathbf{k}_1^j$ is merged. In $\mathbf{K}_2$ and $\mathbf{K}_3$, the entries are zeros except for the cracked element where $\mathbf{k}_2^j$ is merged in $\mathbf{K}_2$ and $\mathbf{k}_3^j$ is merged in $\mathbf{K}_3$. The vectors $\mathbf{F}_1$ and $\mathbf{F}_2$ represent the unbalance force amplitudes and $\mathbf{F}_g$ represents the gravity force vector. The sum of gyroscopic and damping matrices is represented by $\hat{\mathbf{C}} = \Omega\mathbf{G} + \mathbf{C}$. The translational and rotational oscillations are expressed by the vector $\mathbf{q}(\mathrm{t}) = \left[\boldsymbol{q}_1^T\, \boldsymbol{q}_2^T \ldots \boldsymbol{q}_i^T \ldots \boldsymbol{q}_{N+1}^T\right]^T$, where $\boldsymbol{q}_i^T(t) = \left[u_i\ v_i\ \varphi_i^x\ \varphi_i^y\right]$ is the vector of the ith node translational and rotational coordinates.

Incorporating a constant angular acceleration α during the transient startup and coast down operations of the considered rotor system converts the equations of motion and their corresponding state-space representation into linear-time-varying (LTV) systems. Therefore, running the rotor system from standstill at constant angular acceleration alters the equations of motion where the angle of rotation and the angular rotational speed becomes $\theta(t) = \alpha t^2/2$ and $\Omega(t) = \alpha t$, respectively, for $\Omega(0) = 0$ and $\theta(0) = 0$. Accordingly, the gyroscopic matrix is rewritten as $\overline{\mathbf{G}}(t) = \alpha t\mathbf{G}$ and the unbalanced force vector components at the *i-th* node are rewritten as

$$f_x^i(t) = m\varepsilon\alpha^2 t^2 \cos\left(\alpha t^2/2\right) - m\varepsilon\alpha \sin\left(\alpha t^2/2\right)$$

$$f_y^i(t) = m\varepsilon\alpha^2 t^2 \sin\left(\alpha t^2/2\right) + m\varepsilon\alpha \cos\left(\alpha t^2/2\right) \tag{9}$$

where the $m\varepsilon$ is the product of the unbalance mass m and the unbalance mass eccentricity ε. The numerical integration whirl response of the above LTV equations of motion is obtained and compared with the experimental whirl response. Accordingly, the FSA is applied to both numerical simulation and experimental whirl responses to identify the zones of rotational speeds at which BW orbits are captured.

3 Experimental Setup and Model Parameters

The double-disk rotor-bearing configuration in Fig. 2 of which the physical parameters are listed in Table 1 is considered here for obtaining the numerical simulation and the experimental whirl responses. The system is divided into 6 elements to obtain the numerical simulation whirl response near the location of the right-side bearing by the numerical integration of the LTV equations of motion shown in Eq. (8).

The physical damping is obtained from $\mathbf{C} = \gamma\mathbf{M} + \xi\mathbf{K}$ where $\gamma = 5 \times 10^{-4}$ s^{-1} and $\xi = 5 \times 10^{-4}$ s. In addition, anisotropic bearings with stiffness values $k_{xx} = 5 \times 10^{6}$ N/m and $k_{yy} = 7 \times 10^{7}$ N/m are incorporated in the numerical simulation. The experimental whirl amplitudes are collected near the right bearing using two perpendicular proximity probes. For both numerical and experimental

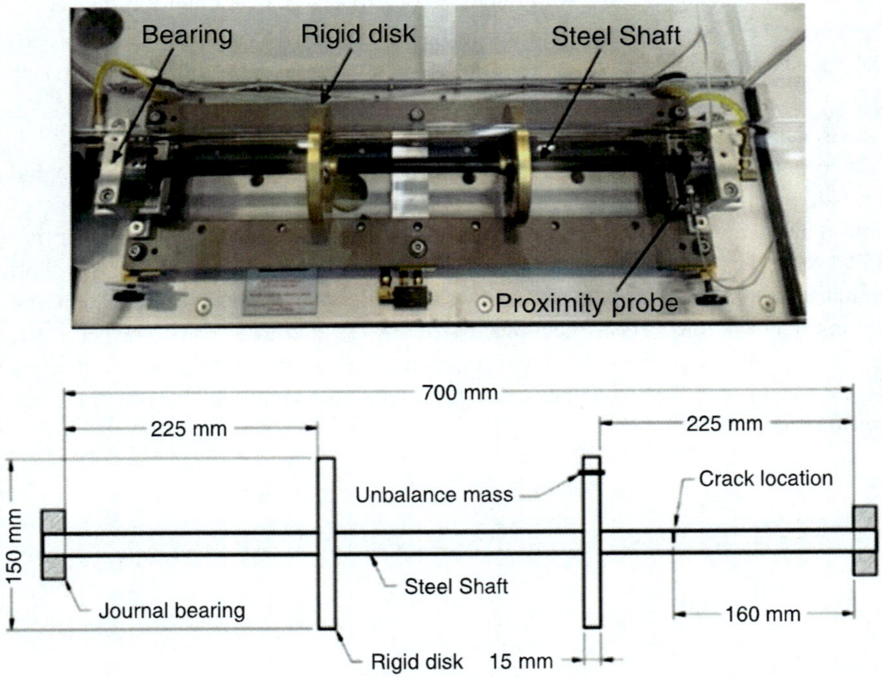

Fig. 2 Rotor-disk setup for experimental testing in for double-disk configuration (DD)

Table 1 Physical parameters of the considered rotor-disk system

Description	Value	Description	Value
Length of the rotor, L	0.7 m	Disk outer radius, R_o	75×10^{-3} m
Radius of the rotor, R	19.06×10^{-3} m	Disk inner radius, R_i	19×10^{-3} m
Density of rotor, ρ_s	7850 kg/m^3	Density of disk, ρ_d	2700 kg/m^3
Modulus of elasticity, E	2.1×10^{11} N/m^2	Mass of the disk, m_d	0.663 kg
Bearing stiffness (k_{xx}, k_{yy})	5×10^6, 7×10^7 N/m	Mass unbalance, $m_e d$	2×10^{-5} kg m
Bearing damping (c_{xx}, c_{yy})	5×10^2 N s/m	Mass unbalance angle, β	Varying

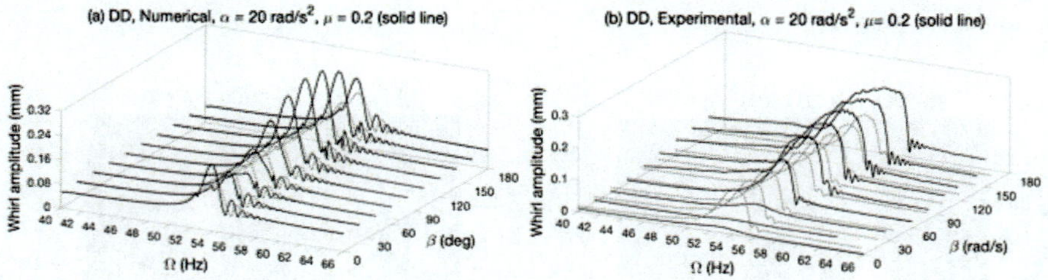

Fig. 3 Numerical and experimental whirl amplitudes in (**a**) and (**b**), respectively, versus shaft rotation speeds and unbalance force vector orientations for the cracked DD configuration

results, the envelope of the resultant whirl amplitudes $z = \sqrt{u^2 + v^2}$ of the horizontal and vertical displacements is plotted for comparisons.

The zones of BW shaft rotational speeds have been found in [9], to immediately appear after passing the first critical FW speed in the neighborhood of a local minimum of whirl amplitudes. The effect of the unbalance force vector angle β on these BW zones and on the peak FW amplitudes is illustrated in Fig. 3 at constant angular acceleration rate of $\alpha = 20$ rad/s^2. The zones BW rotational speeds are observed to nearly disappear for some range of unbalance force vector angles.

In Fig. 4 the BW zones are identified for selected values of angular acceleration rates, unbalance force angle and crack depth where several zones of BW rotational speeds are captured at higher angular acceleration rate of $\alpha = 50$ rad/s^2. Therefore, the number of BW zones is observed to be sensitive to the angular acceleration rate and the unbalance force angle orientation for the cracked rotor systems.

The FSA is applied to confirm the BW whirl zones in the numerical and experimental results for the considered intact and the corresponding cracked DD rotor configurations. The FSA plots are shown in Fig. 5a, b for the numerical whirl results and in Fig. 6a, b for the corresponding experimental whirl results. In these figures, the FSA plots clearly identify and confirm the existence of BW zones of rotational speeds in the intact and cracked DD configurations. The intensity and the extent of the BW zone/zones are observed to be affected by the appearance of the crack as shown. Moreover, an excellent agreement between the FSA of the numerical simulation whirl response and that of the experimental whirl response is clearly observed in Figs. 5 and 6 in identifying the BW zone of rotational speeds

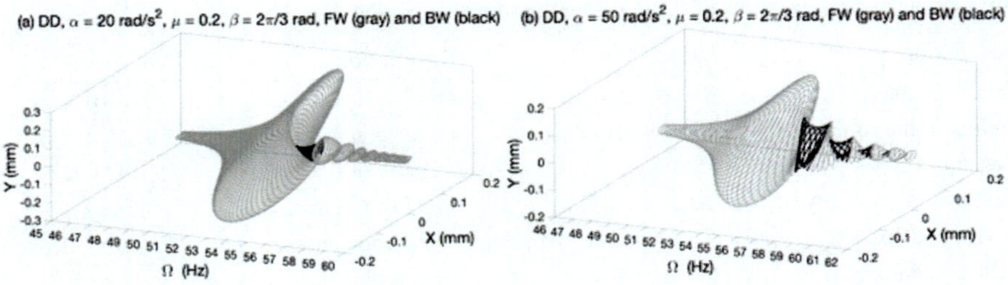

Fig. 4 Numerical whirl amplitudes of the cracked DD configuration in (**a**) for $\alpha = 20$ rad/s^2 and in (**b**) for $\alpha = 50$ rad/s^2

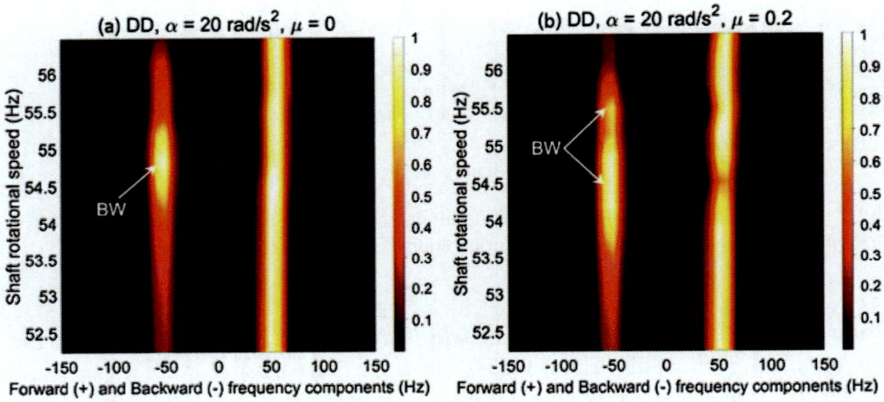

Fig. 5 FSA of the numerical whirl responses of the intact DD system in (**a**) and the cracked one in (**b**) at $\beta = 2\pi/3$ rad

near $\Omega \approx 54.5$ Hz. In Fig. 7, the effect of higher acceleration rate on the extent of the BW zone is also observed in the experimental results.

The BW zones are also captured by the FSA after the passage through the subcritical ($\Omega/2$) rotational speeds. The corresponding FSA plot for the BW zones of rotational speeds after the passage through the subcritical rotational speed is shown in Fig. 8 for low and high acceleration rates. Consequently, the FSA is also found to be very sensitive and efficient tool in capturing BW zones of rotational speeds after the passage through the critical forward whirl speeds and their corresponding subcritical frequency components.

4 Concluding Remarks

The FSA method is applied here to confirm the existence of the new backward whirl phenomena that immediately appear after the passage through the critical forward whirl rotational speed in accelerated intact and cracked rotor-disk systems.

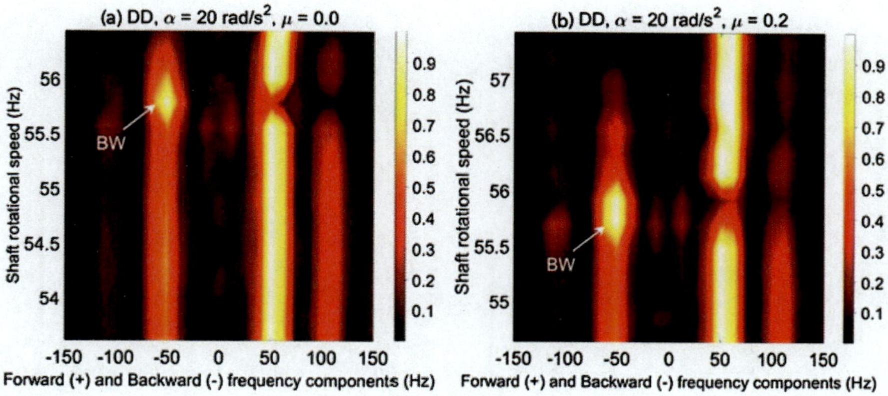

Fig. 6 FSA of the experimental whirl responses of the intact DD system in (**a**) and the cracked one in (**b**) at $\beta = 2\pi/3$ rad

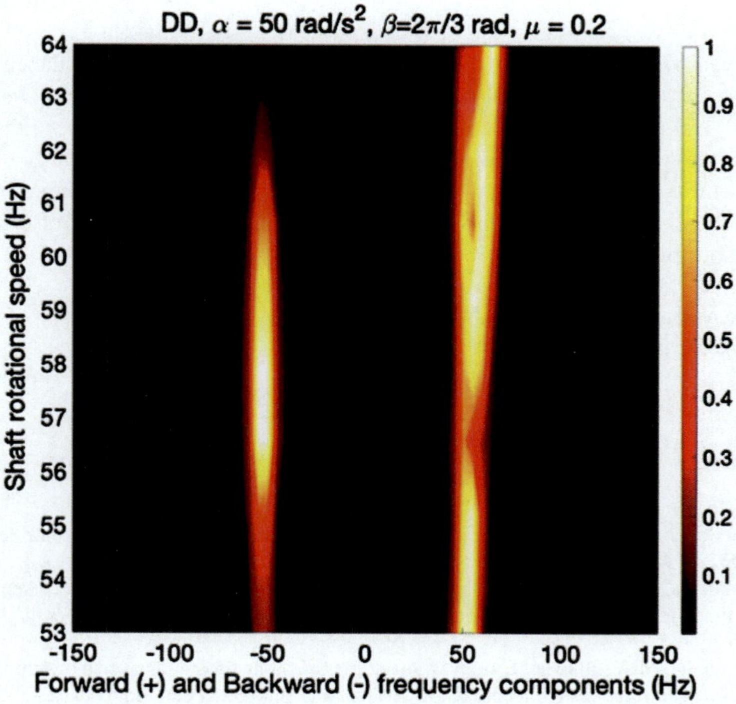

Fig. 7 FSA of the experimental whirl response of the cracked DD system for a high angular acceleration rate

The FSA method is applied to the numerical and experimental whirl responses where the effect of the angular acceleration of the shaft and the unbalance force vector orientation were also investigated. The extent and the number of BW zones

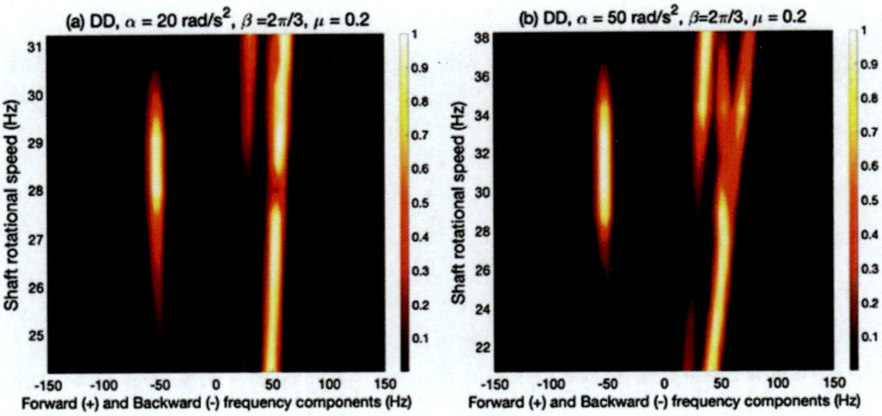

Fig. 8 FSA of numerical whirl response of the cracked DD configuration in (**a**) for $\alpha = 20$ rad/s^2 and in (**b**) for $\alpha = 50$ rad/s^2

is observed to be sensitive to the angular acceleration rate, the unbalance force angle orientation, and the crack depth. In addition, an excellent agreement between the numerical simulation predictions and the experimental validation results in the FSA and the whirl response plots has been observed. Therefore, this study verifies the robustness of the FSA method in confirming the new BW phenomena in the numerical and experimental whirl responses of the intact and cracked rotor system with anisotropic bearings.

Acknowledgments The authors are grateful for the support provided for this project by the Khalifa University of Science and Technology.

References

1. Al-Shudeifat, M.A.: On the finite element modeling of an asymmetric cracked rotor. J. Sound Vib. **332**, 2795–2807 (2013). https://doi.org/10.1016/j.jsv.2012.12.026
2. Al-Shudeifat, M.A.: Stability analysis and backward whirl investigation of cracked rotors with time-varying stiffness. J. Sound Vib. **348**, 365–380 (2015). https://doi.org/10.1016/j.jsv.2015.03.007
3. Al-Shudeifat, M.A., Butcher, E.A.: New breathing functions for the transverse breathing crack of the cracked rotor system: approach for critical and subcritical harmonic analysis. J. Sound Vib. **330**, 526–544 (2011). https://doi.org/10.1016/j.jsv.2010.08.022
4. Al-Shudeifat, M.A., Butcher, E.A., Stern, C.R.: General harmonic balance solution of a cracked rotor-bearing-disk system for harmonic and sub-harmonic analysis: analytical and experimental approach. Int. J. Eng. Sci. **48**, 921–935 (2010). https://doi.org/10.1016/j.ijengsci.2010.05.012
5. Jun, O.S., Gadala, M.S.: Dynamic behavior analysis of cracked rotor. J. Sound Vib. **309**, 210–245 (2008). https://doi.org/10.1016/j.jsv.2007.06.065

6. Sekhar, A.S., Prabhu, B.S.: Vibration and stress fluctuation in cracked shaft. J. Sound Vib. **169**, 655–667 (1994). https://doi.org/10.1006/jsvi.1994.1039
7. Gayen, D., Chakraborty, D., Tiwari, R.: Finite element analysis for a functionally graded rotating shaft with multiple breathing cracks. Int. J. Mech. Sci. **134**, 411–423 (2017). https://doi.org/10.1016/j.ijmecsci.2017.10.027
8. Nembhard, A.D., Sinha, J.K., Yunusa-Kaltungo, A.: Experimental observations in the shaft orbits of relatively flexible machines with different rotor related faults. Meas. J. Int. Meas. Confed. **75**, 320–337 (2015). https://doi.org/10.1016/j.measurement.2015.08.007
9. Al-Shudeifat, M.A.: New backward whirl phenomena in intact and cracked rotor systems. J. Sound Vib. **443**, 124–138 (2019). https://doi.org/10.1016/j.jsv.2018.11.038
10. Al-Shudeifat, M.A., Al Hosani, H., Saeed, A.S., Balawi, S.: Effect of unbalance force vector orientation on the whirl response of cracked rotors. ASME J. Vib. Acoust. **141**, 021001–021001 (2019). https://doi.org/10.1115/1.4041462
11. Alhammadi, F.K., Al-Shudeifat, M.A., Shiryayev, O.: Effect of angular acceleration and unbalance force orientation on the backward whirl in cracked rotors. In: Proceedings of the International Mechanical Engineering Congress and Exposition IMECE 2018, IMECE2018-87476 (2018). https://doi.org/10.1115/IMECE2018-87476
12. Goldman, P., Muszynska, A.: Application of full spectrum to rotating machinery diagnostics. Orbit. **20**(1), 17–21 (1999). http://amconsulting.intellibit.com/499agnes.pdf
13. Tůma, J., Biloš, J.: Full spectrum analysis in journal bearing diagnostics. In: Proceedings of the International Carpathian Control Conference ICCC2004 (2004). http://homel.vsb.cz/~tum52/publications/TumaPaper_ICCC2004.pdf
14. Patel, T.H., Darpe, A.K.: Application of full spectrum analysis for rotor fault diagnosis. In: Gupta, K. (ed.) Proceedings of IUTAM Symposium on Emerging Trends in Rotor Dynamics, pp. 535–545. Springer, Dordrecht (2011)
15. Wu, X., Naugle, C., Meagher, J.: A full spectrum analysis methodology applied to an anisotropic overhung rotor. J. Appl. Mech. Eng. **5** (2016). https://doi.org/10.4172/2168-9873.1000232

Switched Reluctance Motor Dynamic Eccentricity Modelling

Jakub Lorencki

Abstract SRM is a brushless electric motor built of iron. It is used in places where durability and efficiency are important. This is related to its increased resistance to damage than in other electric motors. Due to these applications, proper diagnostics of such a motor is a very important factor. Like any other electric motor it can be susceptible to various mechanical and electrical damages. One of the most common faults is dynamic eccentricity which occurs when the center of the rotor is not at the center of rotation and minimum air gap revolves with the rotor. This phenomenon will be simulated using the finite element method of the FEMM software. And then the data from this method is used in the Matlab program for dynamic simulations. Then it will be possible to see how this mechanical fault affects the motor's performance. This study can also compared with experimental research on a test stand that had been performed before.

Keywords SRM · Eccentricity · Motor diagnostics

1 Introduction

In recent years, significant environmental degradation can be observed as a result of such factors as: global warming, greenhouse effect, depletion of fossil fuels, increase in their prices and increasing amount of carbon dioxide in the atmosphere [1]. These phenomena indirectly motivated the industry to return to the electric drive in many applications, especially for electric and hybrid vehicles [2, 3].

The electric drive located there must face a number of requirements. One of the motor type that can be applied there is the switched reluctance motor (SRM) (Fig. 1). Despite the weaker energy properties compared to the more popular electric motors, in particular for the permanent magnet motor (BLDC), SRM has some advantages,

J. Lorencki (✉)
Warsaw University of Technology, Warszawa, Poland
e-mail: jakub.lorencki@pw.edu.pl

J. Awrejcewicz (ed.), *Perspectives in Dynamical Systems II: Mathematical and Numerical Approaches*, Springer Proceedings in Mathematics & Statistics 363,
https://doi.org/10.1007/978-3-030-77310-6_5

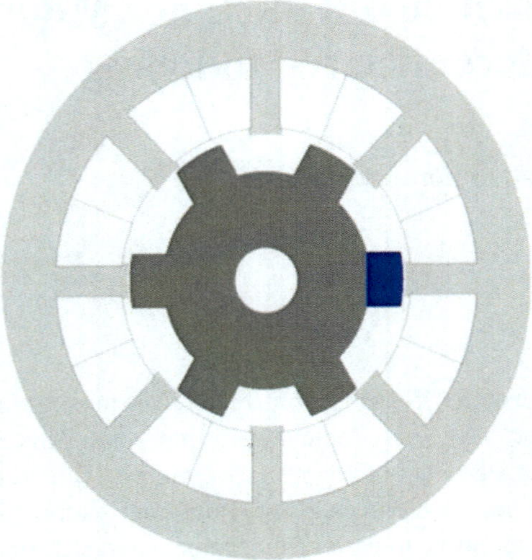

Fig. 1 Illustration of cross-section of Switched Reluctance Motor

e.g. in the form of greater availability of rare-earth materials (especially outside China) from which electrotechnical plates are made for it (i.e. different iron alloys). It is also cheaper to produce SRM and its safety of assembly in comparison to the high mass and attraction force of rare earth magnets [4, 5]. The lack of magnets also eliminates the risk of demagnetizing the material by electromagnetic pulses or high temperatures, therefore SRM a application can have a great success in the army or other specialization, where reliability is extremely important, i.e. in such locations where the replacement of the machine, and therefore long-term shutdown of the drive can cause significant damage and costs. Due to its extremely durable construction (lack of windings on the rotor and salient poles on the stator and rotor), the motor can reach rotational speeds of over 10,000 rpm without the danger of rotor damage by centrifugal force.

2 Purpose of Research

The SRM, like other types of electric motors, is susceptible to various types of damages during its operation. These damages can be divided generally into electrical and mechanical ones. This work deals only with mechanical damages.

The most frequent mechanical damages that may occur in electric motors are: eccentricity (misalignment) static and dynamic, imbalance, bearing damage, mechanical looseness or shaft deflection [6, 7].

2.1 *Mechanical Damages of Electrical Motor*

Each type of mechanical defect causes negative effects in the operation of the electric motor and therefore the entire drive, such as higher vibrations, and therefore noise, higher torque pulsation, energy losses and others [8, 9].

The purpose of this work is to indicate whether the introduction of static eccentricity introduces losses at the motor torque. This is a prelude to further, more complex analyzes in motor modeling.

2.2 *Eccentricity*

Eccentricity is a very common damage in electric motors and to a lesser or greater extent it occurs in every motor. It is characterized by a non-homogeneous irregularity of the air gap between the rotor and the stator. This phenomenon can be observed in the current spectrum as a harmonic caused by the changing inductance causing a non-uniform flux in the air gap. In the worst case, eccentricity can lead to rubbing the rotor with the stator, and therefore to permanent damage of them.

There are two types of eccentricity: static and dynamic (Fig. 2). Static eccentricity occurs when the axis of rotation of the rotor identical with its geometrical axis does not coincide with the geometric axis of the stator. Dynamic eccentricity occurs when the axis of rotation of the rotor identical with the geometric axis of the stator does not coincide with the geometric axis of the rotor [10]. Due to the unevenness of the air gap, the magnetic flux changes causing irregular current values visible in the spectrum. In real conditions static and dynamic eccentricity occur simultaneously. There may be many reasons, such as defects in the rotor design (e.g. imbalance, deformation of round geometry), missing or damaged structural elements (e.g. screws), incorrect assembly or damaged bearings [11].

3 Finite Element Magnetic Method for Magnetostatic Analysis

In order to simulate dynamic eccentricity of the 8/6 SRM a Finite Element Method Magnetics (FEMM) was used. It is a software that has 2D interface and static analysis is possible.

A model of the motor (its cross-section) was designed by a FEMM package. It was inspired by the real SRM motor that was previously used for the experiments on the test bench. The iron sheet that was used for the stator and rotor is M-19 steel and its B-H magnetization curve is depicted in Fig. 3.

Number of turns per phase is equal to 70 and the magnitude of the current is of 20 A.

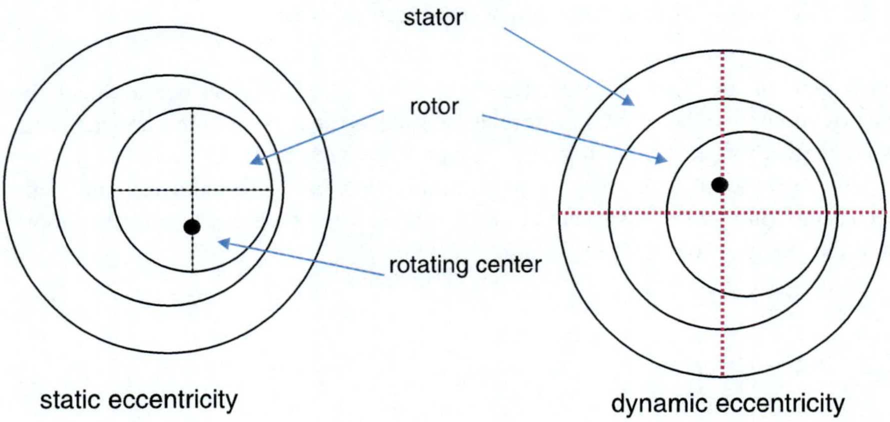

Fig. 2 Illustration of static dynamic eccentricity

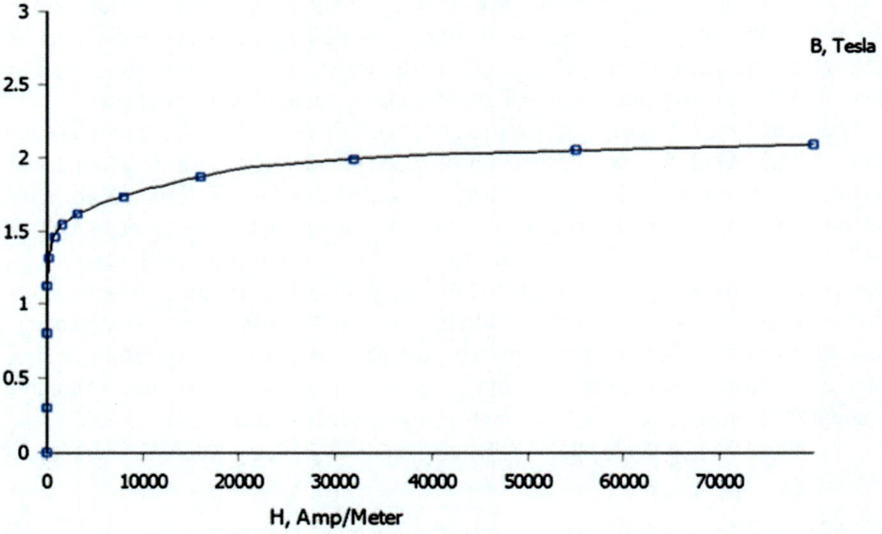

Fig. 3 Magnetization curve for iron sheet of M-19 steel

The other parameters of this motor are: number of phases = 4; stator poles = 6; rotor poles = 8; Coil turns per winding = 70; motor length = 22.5 mm; stator outer diameter = 40.2 mm, air gap length = 0.105 mm, rotor outer diameter = 20.79 mm, shaft diameter = 4.73 mm, stator inner diameter = 21 mm, height of tooth = 5.5 mm, width of tooth tip = 3.5 mm, height of tooth foot = 0.5 mm, width of tooth = 3 mm, rotor inner diameter = 4.73 mm, width of lobe = 3 mm, height of lobe = 2.51 mm. Static torque and magnetizing curve are measured every 5° angles from 0° to 900°. The results are shown in Figs. 4 and 5.

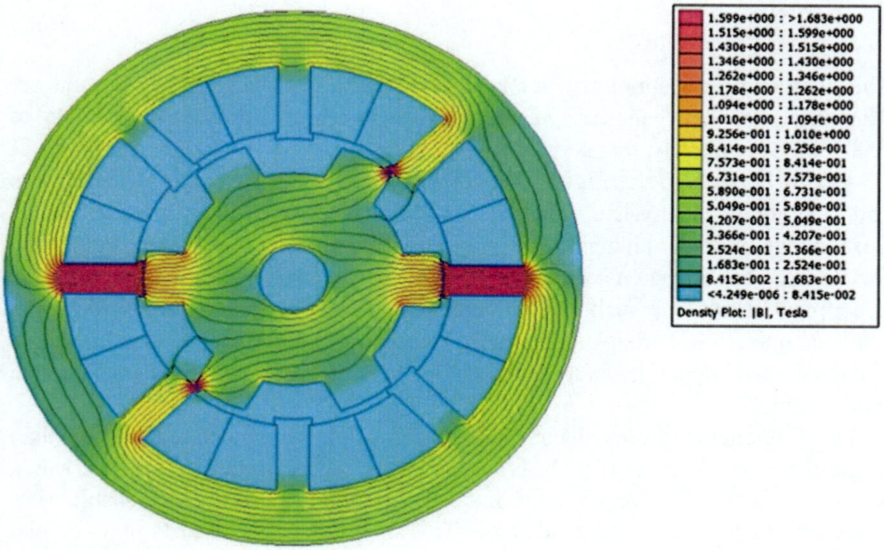

Fig. 4 Flux distribution density and density shadow from a healthy motor (0.105 mm air-gap)

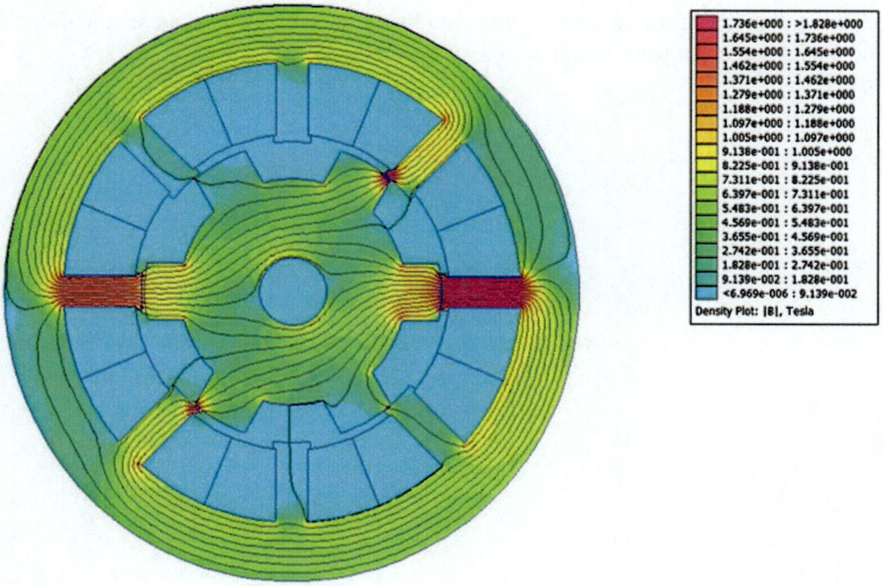

Fig. 5 Flux distribution density and density shadow from 95% eccentric air-gap (0.005 mm of air-gap)

4 Results

The study compares a motor with a healthy state and static eccentricity introduced. The difference between them is the air gap has been reduced from 0.105 to 0.005 mm respectively, that is by 95%.

As can be seen in the Fig. 6, despite such a significant change in the air gap and, consequently, physical quantities in the motor and dynamic parameters, its flux linkage does not change much and it is difficult to state clearly whether the flux linkage has increased or decreased value in eccentric state. In any case, this experiment shows that such a big modification in the motor's geometry does not affect its operation. Perhaps it results from the fact that it is a small-sized machine and maybe with larger geometry and larger powers, such differences would be much more visible.

The differences between electromagnetic torque plots are much more visible than in previous comparison (Fig. 7). Not only the eccentric torque has higher value, also the differences between phases in this torque plot are also distinguishable, very interesting is the peak in the third phase. This shows that non-linearity plays crucial role in this type of phenomenon.

Lastly in Figs. 8 and 9 were presented 3D plots. They have been concentrated five times using the Matlab interpolation function. In these graphs, a constant value of current and angle of rotation was given, and flux and torque values were calculated. The charts are given for the correct operation of the motor.

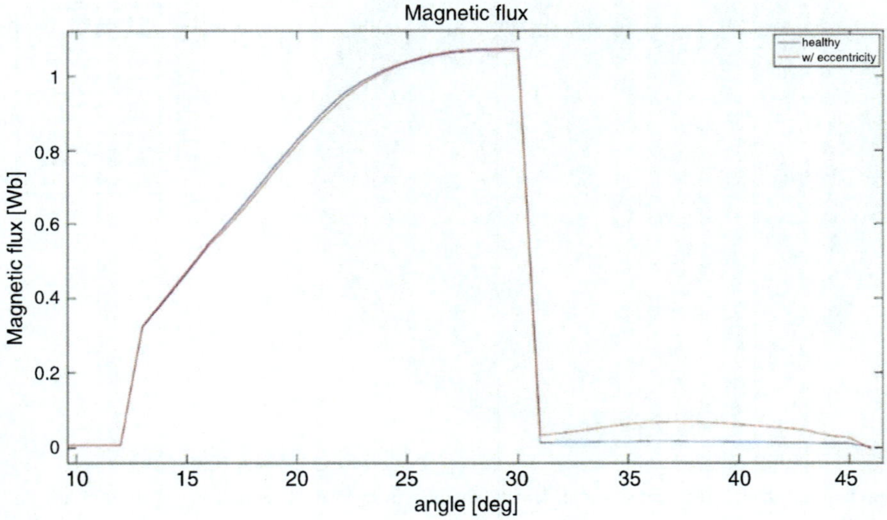

Fig. 6 Flux linkage difference between eccentric and non-eccentric motor

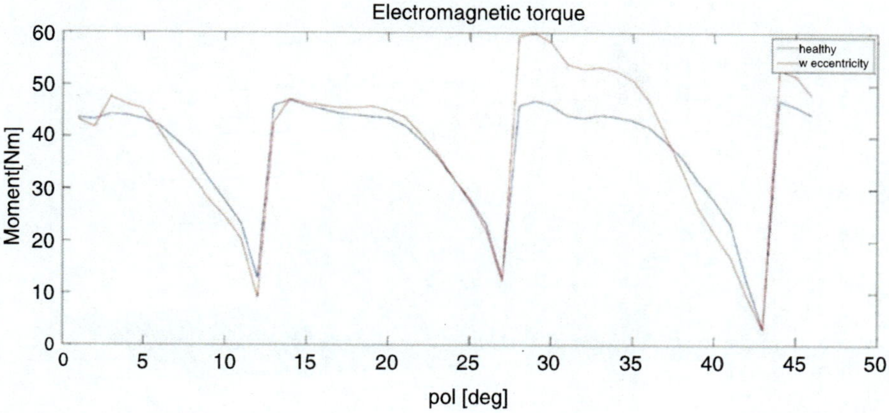

Fig. 7 Static torque difference between eccentric and non-eccentric motor

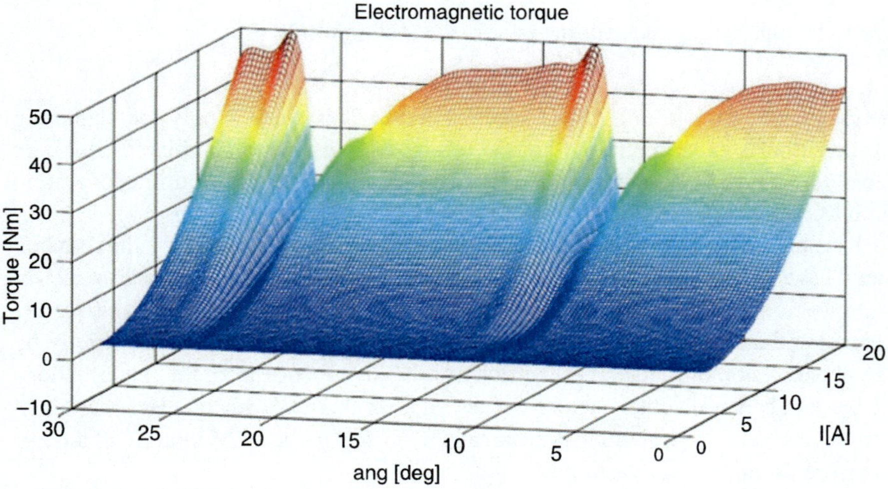

Fig. 8 Electromagnetic torque (interpolated) plotted on current and angle axes

The regularity with the real functioning and phenomena in electric motors an be observed here. With misalignment, the graphs are linear, with coaxial, when saturation occurs, the graphs are non-linear.

5 Conclusions

By using the FEMM package, the author wanted to initiate the process of modeling mechanical damages in the reluctance motor in static mode. Despite its simplicity, this package offers many possibilities for modeling and comparing different mag-

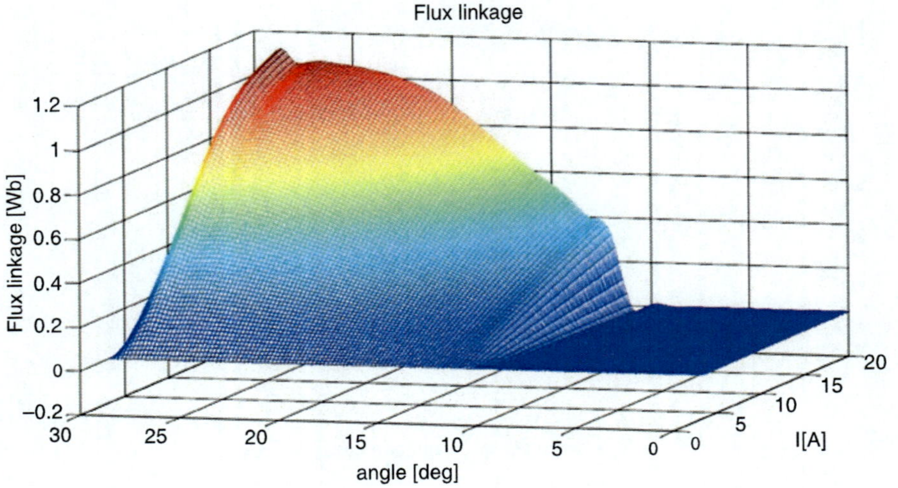

Fig. 9 Flux linkage (interpolated) plotted on current and angle axes

netic parameters, e.g. in an SRM electric motor. This work is only the beginning of the author's experiments, and in the future it is possible to combine the look-up table with FEMM and incorporate it into a dynamic model realized in cooperation with Matlab.

It would be a good comparison as the experiments once achieved on the research stand have to be modeled in the FEMM environment. It would perhaps also answer the question whether the phenomena of disturbance of physical quantities, i.e. current and acceleration (vibration) in the damaged motor were caused strictly by the phenomenon of dynamic eccentricity or the disturbance from the imperfections of the test bench or other factors connected with this experiment. Also, certainly, further accurate research related to damage modeling in the SRM motor will answer many bothering and interesting issues.

References

1. Ehsani, M., Gao, Y., et al.: Modern Electric, Hybrid Electric, and Fuel Cell Vehicles. CRC Press, London (2018)
2. Biernat, A.: Analiza sygnałów diagnostycznych maszyn elektrycznych. Oficyna Wydawnicza Politechniki Warszawskiej, Warszawa (2015)
3. Hööka, M., Tang, X.: Depletion of fossil fuels and anthropogenic climate change: a review. Energy Policy. **52**, 797 (2013)
4. Sperling, D., Deborah, G.: Two Billion Cars: Driving Toward Sustainability. Oxford University Press, New York (2009)
5. Lorencki, J., Tokarzewski, J.: Osiągi pojazdów o napędzie niekonwencjonalnym. Zeszyty Naukowe Instytutu Pojazdów. **5**, 81 (2010)

6. Rarity of Rare Earths Prompts a Renaissance in Motor Design. http://www.dpaonthenet.net/article/57042/Rarity-of-rare-earths-prompts-a-renaissance-in-motor-design.aspx. 20 Sept 2016
7. How Beijing Cornered the Rare Earths Market. http://www.foreignaffairs.com/articles/137602/damien-ma/china-digs-it. 20 Sept 2016
8. Toliyat, H.: Electric Machines: Modeling, Condition Monitoring, and Fault Diagnosis. CRC Press, London (2012)
9. Vibration Diagnostic Guide—SKF Reliability Systems (2000). www.skfreliability.com. 20 Sept 2016
10. Lorencki, J., Radkowski, S.: Procedury i Analiza usterek w silnikach synchronicznych z magnesami trwałymi. Zeszyty Naukowe Instytutu Pojazdów. **2**, 98 (2014)
11. Lorencki, J., Radkowski, S.: Elektryczna diagnostyka asymetrii wirnika w pojazdach hybrydowych **4**, 95 (2013)

Harmonic Transfer Path Analysis of a Wine Refrigerator

Wolfgang Alois Hörtnagel, Stefan Plagg, and Fadi Dohnal

Abstract Transfer path analysis (TPA) is increasingly being applied in the industry when it comes to a new product generation of lightweight and therefore highly flexible structures. TPA helps identifying critical locations and components of the overall structure that contribute to specific vibration observations. Typically TPA needs to be balanced between needed accuracy and time efficiency/cost. Several TPA methodologies exist and need to be adapted to the specific system under consideration. We develop a robust algorithm for the estimation of the frequency response functions of a complex, flexible structure like a commercial wine refrigerator. This algorithm leads to an improved TPA of the overall system and can help optimizing future designs by defining desirable characteristics of critical locations.

Keywords Transfer path analysis · Regularization

W. A. Hörtnagel (✉)
UMIT – Private University for Health Sciences, Medical Informatics and Technology, Division for Mechatronics Lienz, Lienz, Austria

Liebherr-Hausgeräte Lienz, Lienz, Austria
e-mail: wolfgang.hoertnagel@umit.at
https://www.umit.at/melz

S. Plagg
Liebherr-Hausgeräte Lienz, Lienz, Austria

F. Dohnal
UMIT – Private University for Health Sciences, Medical Informatics and Technology, Division for Mechatronics Lienz, Lienz, Austria

J. Awrejcewicz (ed.), *Perspectives in Dynamical Systems II: Mathematical and Numerical Approaches*, Springer Proceedings in Mathematics & Statistics 363,
https://doi.org/10.1007/978-3-030-77310-6_6

1 Introduction

Transfer path analysis (TPA) has been widely used for analyzing the transfer characteristics of vibration and noise for example in [1]. Van der Seijs et al. [2] have reviewed a wide range of TPA techniques and classified them into three categories, namely classical, component-based and transmissiblity-based TPA. TPA is particularly useful when the actual vibrating mechanisms are too complex to model or measure directly, as it allows the representation of a source by forces and vibrations acting at the interfaces of the passive side. In this way the source excitations can be separated from the structural/acoustic transfer characteristics, allowing us to identify and to troubleshoot the dominant paths of vibration transmission. Due to environmental noise and unavoidable measurement errors, the problem of estimating the path distributions often becomes ill-posed. This paper proposes a multi-stage approach for this issue that could improve both, the smoothness as well as the accuracy of the estimation.

A wine cabinet as depicted in Fig. 1 was examined when conducting the necessary experimental measurements. The long term effect of storage conditions on wine has been previously researched. Notable is the experiment of Chung et al. [3] whose group conducted a chemical analysis after subjecting wine to various long term vibrations during an 18-month storage. The results showed that some physico-chemical properties continously change during storage and vibration can result in significant changes in the composition of wine. Minimizing the vibration assists retaining the wine aroma and is therefore recommended for optimal storage of wine. Tao et al. [4] came to the same conclusion, that for the production of high-quality wine, vibration of bottles should be avoided.

2 Bias of Power Estimations

One of the fundamental questions when discussing estimation techniques is whether the expected value of the estimate matches the properties of the measured quantity. Estimation biases can be introduced by noise and have to be taken into consideration.

2.1 *Power Spectral Density*

The spectral properties of signals can be characterised by various functions like the Fourier transform. This paper focuses on power spectral densities as a descriptor. To illustrate the basic concepts, we introduce two signals $x(t)$ and $y(t)$ and additive noise labelled as $m(t)$ and $n(t)$, respectively,

Fig. 1 Wine cabinets need to offer protection against UV light and vibrations to prevent outside factors from degrading the wine's quality. Optimal storage conditions are prerequisite for long-term storage

$$\tilde{x}(t) = x(t) + m(t) \qquad \tilde{y}(t) = y(t) + n(t) \tag{1a}$$

$$\tilde{X}(\Omega) = X(\Omega) + M(\Omega) \qquad \tilde{Y}(\Omega) = Y(\Omega) + N(\Omega) \tag{1b}$$

$$S_{xx}(\Omega) = \lim_{T\to\infty} \frac{1}{2T} \left\{X_T^*(\Omega)X_T(\Omega)\right\} \quad S_{xy}(\Omega) = \lim_{T\to\infty} \frac{1}{2T} \left\{X_T^*(\Omega)Y_T(\Omega)\right\} \tag{1c}$$

Herein, tilde denotes the noisy signals and X and Y the Fourier transforms of the underlying time signals. Ω is the Fourier frequency and T the time interval of the measurement. The auto power spectral density (APSD) S_{xx} of a time signal x indicates how its power is distributed over the frequency range Ω. It is a real function and does not contain phase information. One cannot backcalculate the original signal from its APSD. The cross power spectral density (CPSD) S_{xy} is calculated similarly and is proportional to the product of two signals x and y. It is complex-valued and dependent on the relative phase difference between the two signals.

Under the assumption of stochastically independent and zero-mean noise, the auto power spectral density of a noisy signal is slightly bigger than the APSD of a noise-free signal

$$S_{\tilde{x}\tilde{x}}(\Omega) = S_{xx}(\Omega) + S_{mm}(\Omega) \geq S_{xx}(\Omega) \tag{2}$$

The CPSD is unbiased under our assumptions.

2.2 H_x-Estimators

The simplest estimation method calculates the transfer function between x and y as the ratio of its Fourier transforms. This definition is usually labelled H_0-estimator, see [5]. Repeated measurements can be averaged but this method will lead to a biased result that depends on the properties of the overlaying noise within the measured signals

$$H_0(\Omega) = \frac{\tilde{Y}(\Omega)}{\tilde{X}(\Omega)} \approx \frac{\sum_i \left[\tilde{Y}_i(\Omega)\right]}{\sum_i \left[\tilde{X}_i(\Omega)\right]} \tag{3}$$

The expected relative error of this estimator can be calculated by introducing additive noise to the noise-free signals of input x and output y according to Eqs. 1. The estimation can be slightly higher or lower than the true value. The total relative error can be approximated by the difference betweem the relative output noise and the relative input noise

$$\frac{\Delta H_0(\Omega)}{H(\Omega)} = \frac{H_0 - H}{H} = \frac{N(\Omega)/Y(\Omega) - M(\Omega)/X(\Omega)}{1 + M(\Omega)/X(\Omega)} \approx \frac{N(\Omega)}{Y(\Omega)} - \frac{M(\Omega)}{X(\Omega)} \tag{4}$$

Note that all quantities represent complex functions which may become negative at some phase values. The subtly different approach of the H_1-estimator utilizes the CPSD and the APSD. The CPSD $S_{\tilde{x}\tilde{y}}$ of two signals can be approximated by averaging the product of the Fourier transforms X and Y within a period of stable or broadband excitation

$$H_1(\Omega) = \frac{S_{\tilde{x}\tilde{y}}(\Omega)}{S_{\tilde{x}\tilde{x}}(\Omega)} \approx \frac{\sum_i \left[\tilde{X}_i^*(\Omega)\tilde{Y}_i(\Omega)\right]}{\sum_i \left[\tilde{X}_i^*(\Omega)\tilde{X}_i(\Omega)\right]} \tag{5a}$$

$$\frac{\Delta H_1(\Omega)}{H(\Omega)} = \frac{H_1 - H}{H} = -\frac{S_{mm}(\Omega)/S_{xx}(\Omega)}{1 + S_{mm}(\Omega)/S_{xx}(\Omega)} \approx -\frac{S_{mm}(\Omega)}{S_{xx}(\Omega)} \tag{5b}$$

The expected relative error of an H_1-estimator is real-valued and negative. The resulting estimate will be of smaller magnitude but in phase with the original system. While the H_1 estimator focuses on the input signal x, the H_2-estimator focusses on the output signal y. The H_2-estimator uses the APSD $S_{\tilde{y}\tilde{y}}$ and the CPSD $S_{\tilde{y}\tilde{x}}$ to obtain an approximation that consistently overestimates the true value

$$H_2(\Omega) = \frac{S_{\tilde{y}\tilde{y}}(\Omega)}{S_{\tilde{y}\tilde{x}}(\Omega)} \approx \frac{\sum_i \left[\tilde{Y}_i^*(\Omega)\tilde{Y}_i(\Omega)\right]}{\sum_i \left[\tilde{Y}_i^*(\Omega)\tilde{X}_i(\Omega)\right]} \tag{6a}$$

$$\frac{\Delta H_2(\Omega)}{H(\Omega)} = \frac{H_2 - H}{H} \approx \frac{S_{nn}(\Omega)}{S_{yy}(\Omega)} \tag{6b}$$

The expected relative error of an H_2-estimator is real-valued and positve. The resulting estimate will be of higher magnitude but in phase with the original system. Using these two estimates as boundaries, we can assume that the true absolute value of the transfer function H will be higher than H_1 but lower than H_2 (see Eq. 7)

$$|H_1(\Omega)| \leq |H(\Omega)| \leq |H_2(\Omega)| \tag{7}$$

By averaging the two previous estimates H_1 and H_2, the resulting heuristic H_3 estimator respects the two boundaries by placing the estimate in the middle

$$H_{3a}(\Omega) = \frac{1}{2}[H_1(\Omega) + H_2(\Omega)] \tag{8a}$$

$$\frac{\Delta H_{3a}(\Omega)}{H(\Omega)} = \frac{H_{3a} - H}{H} \approx \frac{1}{2}\left[\frac{S_{nn}(\Omega)}{S_{yy}(\Omega)} - \frac{S_{mm}(\Omega)}{S_{xx}(\Omega)}\right] \tag{8b}$$

The expected relative error of an H_3-estimator is the average of the errors of the H_1- and the H_2-estimators. In frequency ranges where one of the previous estimators has a much better estimate than the other the H_3 error will be halve as big as the worse estimator. In the frequency range of resonance, this estimator will perform worse than a H_2-estimator would, whereas in regions of antiresonance its performance would be worse than that of the H_1-estimator. A best of both worlds result can be achieved by the use of an H_4-estimator that is calculated as a weighted average between H_1 and H_2 (see Eqs. 9). The weight-function κ is used to indicate which of the two estimators H_1 or H_2 will have a more reliable result. In ranges of resonance where $\kappa \geq 0.5$ the H_4-estimator will be closer to the H_2 result. In antiresonance the H_1 estimate will be preferred

$$\kappa(\Omega) = \frac{|H_3(\Omega)|}{\max_{\Omega}\{|H_3(\Omega)|\}} \tag{9a}$$

$$H_4(\Omega) = [1 - \kappa(\Omega)]\, H_1(\Omega) + \kappa(\Omega) H_2(\Omega) \tag{9b}$$

All higher ordered H_x-estimators have relative errors that are real valued whereas the H_0 estimator has a complex valued relative error. Thus the phase information of the more complex H_x-estimators is more reliable than the results of the H_0 approach. All previous measurements show that the H_4-estimator approximates the true transfer function better than the H_1, H_2 and H_3 variants.

3 Tikhonov Regularization

A transfer path analysis considers the interference between multiple chosen paths. The problem of estimating the multiple-input-multiple-output dependencies is one of the core issues of a TPA

$$\mathbf{X}\mathbf{T}^T = \mathbf{Y} \tag{10}$$

Noise can significantly influence the solutions of Eq. 10 and may produce spurious solutions without physical meanings. Various regularization algorithms can be used to overcome ill-conditioned problems and improve the stability of solutions. The Tikhonov regularization in [6] considers both, fitting degrees and stability of solutions and is thus a common choice in this context.

The key in this process is the introduction of the regularization parameter λ. By minimizing the Euclidean norm

$$\min_{\mathbf{T}} \left\{ \|\mathbf{X}\mathbf{T}^T - \mathbf{Y}\|^2 + \lambda^2 \|\mathbf{T}^T\|^2 \right\} \Leftrightarrow \left(\mathbf{X}^T\mathbf{X} + \lambda^2\mathbf{I}\right)\mathbf{T}^T = \mathbf{X}^T\mathbf{Y} \tag{11}$$

the transmissiblity function matrix can be identified according to

$$\mathbf{T}_\lambda^T = \left(\mathbf{X}^T\mathbf{X} + \lambda^2\mathbf{I}\right)^{-1}\mathbf{X}^T\mathbf{Y} \tag{12}$$

Utilizing the singular value decomposition (SVD) on $\mathbf{X}$, we can obtain the parameters $\boldsymbol{\sigma}_j$, $\mathbf{u}_j$ and $\mathbf{v}_j$

$$\mathbf{X} = \mathbf{U}\boldsymbol{\Sigma}\mathbf{V}^T = \sum_{j=1}^{m} \boldsymbol{\sigma}_j \mathbf{u}_j \mathbf{v}_j^T \tag{13}$$

The generalized cross validation (GCV) can then be used to select optimal parameters λ for Tikhonov regularization

$$\mathrm{GCV}\,(\lambda) = \min_{\lambda} \left\{ \frac{\|\mathbf{X}\mathbf{T}_{\lambda}^{T} - \mathbf{Y}\|^{2}}{m - n + \sum_{j=1}^{m} \left(\frac{\lambda^{2}}{\lambda^{2} + \sigma_{j}^{2}} \right)} \right\} \tag{14}$$

These optimal values of λ can then be inserted into

$$\mathbf{T}_{\lambda}^{T} = \sum_{j=1}^{n} \frac{\sigma_{j}^{2}}{\lambda^{2} + \sigma_{j}^{2}} \frac{\mathbf{u}_{j}^{T}\mathbf{Y}}{\sigma_{j}} \mathbf{v}_{j} \tag{15}$$

With our previously calculated SVD-parameters, the estimate of the transmissiblity function matrix $\mathbf{T}_{\lambda}^{T}$ is fully defined.

4 Multi-Stage Estimators

4.1 Combinatory Approach

Based on the previous section, the heuristic idea of combining the two concepts seems promising. The basic Tikhonov regularization calculates the optimal transmissibility matrix from $\mathbf{X}$ to $\mathbf{Y}$. This concept is similar to the H_0-estimator introduced in Eq. 3. Exploiting this analogy for the calculation of the transmissibility from $\mathbf{S}_{\tilde{x}\tilde{x}}$ to $\mathbf{S}_{\tilde{x}\tilde{y}}$, an effective H_1-estimator is the result of

$$\mathbf{T}_{\lambda,1}^{T} = \left(\mathbf{S}_{\tilde{x}\tilde{x}}^{T} \mathbf{S}_{\tilde{x}\tilde{x}} + \lambda^{2}\mathbf{I} \right)^{-1} \mathbf{S}_{\tilde{x}\tilde{x}}^{T} \mathbf{S}_{\tilde{x}\tilde{y}} \tag{16}$$

H_2- to H_4-estimators are constructed according to the same rules

$$\mathbf{T}_{\lambda,2}^{T} = \left(\mathbf{S}_{\tilde{y}\tilde{x}}^{T} \mathbf{S}_{\tilde{y}\tilde{x}} + \lambda^{2}\mathbf{I} \right)^{-1} \mathbf{S}_{\tilde{y}\tilde{x}}^{T} \mathbf{S}_{\tilde{y}\tilde{y}} \tag{17a}$$

$$\mathbf{T}_{\lambda,3}^{T} = \frac{1}{2} \left(\mathbf{T}_{\lambda,1}^{T} + \mathbf{T}_{\lambda,2}^{T} \right) \tag{17b}$$

$$\mathbf{T}_{\lambda,4}^{T} = [1 - \kappa(\Omega)]\, \mathbf{T}_{\lambda,1}^{T} + \kappa(\Omega) \mathbf{T}_{\lambda,2}^{T} \tag{17c}$$

4.2 Staging

Performing a real measurement, usually not all measurement points are recorded simultaneously. However independent measurements can be affected by different environmental effects. A noise-peak that is present in one measurement but not the other can lead to estimation error. This paper proposes the calculation of part-transmissibility functions that specify the relation between the point of interest and a constantly measured reference point which is fixed for all measurements. This single-input-single-output relation can be established comfortably by the H_4-estimator. Using these part-transmissibility functions as a basis for further path-calculations reduces the impact of environmental noise in the estimation of the transfer paths.

4.3 Summary

The Eqs. 17 summarize the proposed procedure. Performing the Tikhonov regularization while substituting $\mathbf{X}$ and $\mathbf{Y}$ with the corresponding auto or cross power spectral density leads to multi-stage estimators. In addition to the pure transmissibility calculation, it is recommended to perform an intermediate step of calculating part-transmissibility functions that establish a relation between an optional measurement point and a fixed reference measurement point. Using these part-transmissibility functions as a basis for the path-estimations leads to further improvements.

First tests of this proposed estimator have resulted in improved performance. Figure 2 shows a decidedly smoother estimate with pronounced peaks. The smoothness of this transmissiblity function helps greatly in visualizing the transfer functions in false-colours which is common for comparing multiple path contributions in TPA. In addition to the smoother estimation result, the tests have shown a more robust result of the GCV procedure in Eq. 14 when calculating the regularization parameter λ.

5 Conclusions

We developed a robust algorithm for the estimation of the frequency response functions of a complex, flexible structure like a commercial wine refrigerator. This algorithm leads to an improved TPA of the overall system and can help optimizing future designs by defining desirable characteristics of critical locations. In this paper a novel multi-stage approach to transfer path estimation is proposed. The procedure uses a modified Tikhonov regularization for calculating the transmissibility function matrices for a transfer path analysis. First tests have shown improved smoothness and accuracy in the results compared to the commonly applied approach. The

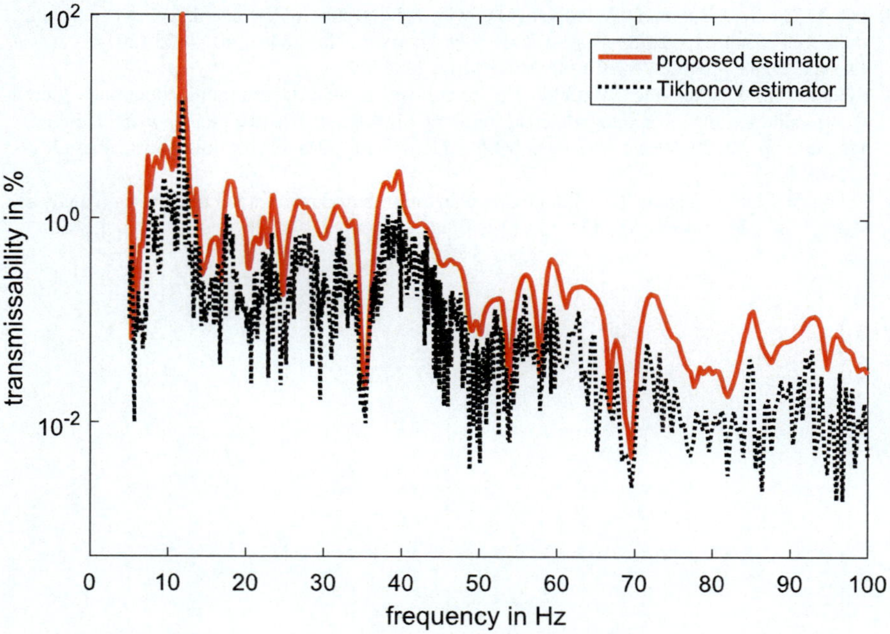

Fig. 2 The proposed estimation process results in a smoother transfer function. This is especially beneficial when performing a transfer path analysis and visualizing the paths in false-colours

proposed algorithm seems to also improve the stability of the estimation process by making the selection of the regularization parameter more robust. Further studies are necessary for validating mathematically the proposed heuristical approach.

Acknowledgments This work was supported by the Österreichische Forschungsfördergesellschaft mbH (FFG) (No. 18819992) and Liebherr-Hausgeräte Lienz.

References

1. Gajdatsy, P., Janssens, K., Desmet, W., der Auweraer, H.V.: Application of the transmissibility concept in transfer path analysis. Mech. Syst. Sig. Process. **24**(7), 1963–1976 (2010). https://doi.org/10.1016/j.ymssp.2010.05.008
2. van der Seijs, M.V., de Klerk, D., Rixen, D.J.: General framework for transfer path analysis: history, theory and classification of techniques. Mech. Syst. Sig. Process. **68–69**, 217–244 (2016). https://doi.org/10.1016/j.ymssp.2015.08.004
3. Chung, H.J., Son, J.H., Park, E.Y., Kim, E.J., Lim, S.T.: Effect of vibration and storage on some physico-chemical properties of a commercial red wine. J. Food Compos. Analy. **21**(8), 655–659 (2008). https://doi.org/10.1016/j.jfca.2008.07.004. Wine: nutrients, bioactive non-nutrients and more

4. Tao, Y., García, J.F., Sun, D.W.: Advances in wine aging technologies for enhancing wine quality and accelerating wine aging process. Crit. Rev. Food Sci. Nutr. **54**(6), 817–835 (2014). https://doi.org/10.1080/10408398.2011.609949. PMID: 24345051
5. Kühlert, H., Markert, R., Witfeld, H.: Identifikation von Übertragungsfunktionen durch Mehrpunkterregung mit breitbandigen korrelierten Signalen. In: Schwingungen in der Fahrzeugdynamik, pp. 42–79. Vieweg+Teubner Verlag, Germany (1991). https://doi.org/10.1007/978-3-322-88807-5_3
6. Cheng, W., Lu, Y., Zhang, Z.: Tikhonov regularization-based operational transfer path analysis. Mech. Syst. Sig. Process. **75**, 494–514 (2016). https://doi.org/10.1016/j.ymssp.2015.12.025

Risk Related Prediction for Recurrent Stroke and Post-stroke Epilepsy Using Fractional Fourier Transform Analysis of EEG Signals

Eva-H. Dulf and Clara-M. Ionescu

Abstract Stroke is a medical condition which can easily affect the quality of life, depending on how extended the stroke is and what regions of the brain are involved. According to the most recent data cited in WHO, Romania is in top three of the countries with increased frequency of stroke and has the second place for having the most deaths and disabilities caused by stroke. Actually, stroke is the second death cause in Romania after cardiac arrest. Today, there are various prevention methods concerning stroke. The hypothesis of the research context is that EEG signal can provide useful information on risk related prediction for recurrent stroke and post-stroke epilepsy. Knowing that there is a certain risk on developing secondary epilepsy after stroke, based on the EEG rhythms, may help in prevention and maybe in reconsidering a new approach in the treatment of this pathology. On the other hand Fractional Fourier Transform (FrFT), a generalization of conventional Fourier Transform, is used with success in many applications like detection of signals in noise, image compression, reduction of side lobe levels using convolutional windows, time-frequency analysis, etc. It can be used in more effective manner compared to Fourier transform with additional degrees of freedom. That was the motivation to analyze the spectra of each component of the EEG signals using FrFT in order to predict recurrent stroke and post-stroke epilepsy incidence. The results prove the efficiency of the method.

Keywords EEG signal analysis · Fractional Fourier Transform · Recurrent stroke · Epilepsy

E.-H. Dulf (✉)
Technical University of Cluj-Napoca, Cluj Napoca, Romania

Obuda University, Physiological Controls Research Center, Budapest, Hungary
e-mail: Eva.Dulf@aut.utcluj.ro

C.-M. Ionescu
Dynamical Systems and Control Research Group, Ghent University, Ghent, Belgium

EEDT Core Lab on Decision and Control, Flanders Make Consortium, Ghent, Belgium

J. Awrejcewicz (ed.), *Perspectives in Dynamical Systems II: Mathematical and Numerical Approaches*, Springer Proceedings in Mathematics & Statistics 363,
https://doi.org/10.1007/978-3-030-77310-6_7

1 Introduction

Patient well-being and healthcare can be best described as bringing longer life and an improved quality of life for patients. The value of the treatment is determined by the amount of clinical benefit it can achieve balanced against its cost, beside its effects. Next to chronic pain and rehabilitation management, epilepsy, stroke and aneurysm are significant phenomena with long tails in the socio-economic impact on total costs for healthcare. A brain aneurysm is a bulge in an artery in the brain that has the potential to burst or rupture. A ruptured aneurysm can cause a type of stroke called a subarachnoid hemorrhage. An estimated 3% of United States of America population may have or develop a brain aneurysm each year, according to the Mayfield Clinic [1]. Not all aneurysms cause stroke, and vice-versa. However, if a person is at risk for a burst aneurysm, treatment is often required to prevent this potentially life-threatening occurrence. Medical specialists do not pre-detect an estimated 85% of aneurysms (only after they burst). A stroke may occur due to either the blood supply to the brain being blocked or a blood vessel in the brain rupturing. Two stroke types exists: hemorrhagic and ischemic. Hemorrhagic strokes are usually the result of one of two causes: an aneurysm or a collection of abnormal blood vessels in the brain that can rupture. Ischemic strokes are those that result from a blockage in an artery in the brain. When a blood clot breaks free from its place in an artery, it can lodge in a portion of the brain. This keeps blood from flowing freely to the brain. Without the oxygen and nutrients that the blood brings to the brain tissue, the tissue dies. The result can be impaired body functioning or death, while some of impairment is indirectly related to other dysfunctions such as multiple sclerosis, semi-paralysis, speech/mobility impairment, etc. In practice, the decision of whether to treat incidental intracranial saccular aneurysms is complicated by limitations in current knowledge of their natural history. A systematic review and pooled analysis of individual patient data from 8382 participants in six prospective cohort studies with subarachnoid haemorrhage as outcome was reported in [2]. Rupture occurred in 230 patients during 29,166 person-years of follow-up. The mean observed 1-year risk of aneurysm rupture was 14% (with 95% confidence interval 1.1–1.6) and the 5-year risk was 34% (with 95% confidence interval 2.9–4.0). Prediction factors were age, hypertension, history of subarachnoid haemorrhage, aneurysm size, aneurysm location, and geographical region. In study populations from North America and European countries other than Finland, the estimated 5-year absolute risk of aneurysm rupture ranged from 0.25% in individuals younger than 70 years without vascular risk factors with a small-sized (<7 mm) internal carotid artery aneurysm, to more than 15% in patients aged 70 years or older with hypertension, a history of subarachnoid haemorrhage, and a giant-sized (>20 mm) posterior circulation aneurysm. By comparison with populations from North America and European countries other than Finland, Finnish people had a 3.6-times increased risk of aneurysm rupture and Japanese people a 2.8-times increased risk. SAFE (Stroke Alliance For Europe) commissioned the Burden of Stroke study to show each EU country where it stands compared to others in terms of the stroke burden and how well it is meeting the need for acute and follow-up care, including examples of good practice (https://strokeeurope.eu/). The Burden of Stroke in Europe research shows in 2017 shocking disparities between and within countries along the entire stroke care pathway. Europe-wide comparisons of stroke and stroke care are vital to help each country prevent stroke and provide better care and support for everyone affected by it. To make accurate comparisons between different countries, populations and health systems, we need coordinated Europe-wide data collection. Therefore, European policy-makers, in particular the European Commission and the Joint Research Centre, should support and promote the use of a robust Europe-wide stroke register to assess quality of care along the whole stroke pathway. In Romania, the majority of population is covered through contributions to social insurance system; free at point of use for all [3]. For stroke epidemiology, on the population of 19,043,767 the incidence estimate (Global Burden of Disease—GBD 2015) is 61,552 strokes/year, 191 strokes per 100,000 inhabitants annually, the prevalence estimate (GBD 2015) is 252,774 strokes, 833 per 100,000 inhabitants and mortality (GBD 2015) is 54,272 deaths due to stroke/year, 156 deaths per 100,000 inhabitants annually, all numbers age- and sex-adjusted [4]. The information is based on registries from Targu Mures Registry (local, only hospitalized patients) [5] and the healthcare cost of stroke: total 163.1 million EUR, i.e. 8 EUR per capita [6]. In Belgium, for a total population of 11,007,020 we have an incidence estimate of (GBD 2015): 10,397 strokes/year, 50 strokes per 100,000 inhabitants annually, a prevalence estimate of (GBD 2015): 63,535 strokes, 348 per 100,000 inhabitants age- and sex-adjusted [4]. The case fatality of ischemic stroke is: 9.2 per 100 discharges, adults aged 45 or older, age- and sex-adjusted, and a mortality rate of 9501 deaths due to stroke/year, 387 deaths per 100,000 inhabitants annually age- and sex-adjusted [7]. The information is based on mandatory hospitalization data, Belgian Sentinel Network of General Practitioners, Institute of Health population surveys and the healthcare cost of stroke in Belgium: total 3937 million EUR, i.e. 35 EUR per capita [6]. Several major risk factors of aneurysm growth and rupture have been identified. There exist recommendations on diagnostic work up, monitoring and general management (blood pressure, blood glucose, temperature, thromboprophylaxis, anti-epileptic treatment, use of steroids). Apart from the above, leading risks for ischemic stroke are (1) hypertension and (2) surgery and anesthesia [8]. In today's EU vision and mission for health and wellbeing, an important role is played by decision support systems. It has been long acknowledged that medicine and engineering must go hand-in-hand for better results [9]. There is evidence to maintain the claim that decision support systems related to computer based (and implicitly mathematical patient model based) systems for titrating drugs during anesthesia have positive effect on reducing post-surgery secondary effects (time to recovery, post-surgery depression, etc.) [10]. One of the most debilitating complications in the perioperative period with serious clinical sequelae is cerebral ischemia [11]. In this paper, we propose a first hand solution and preliminary results for detecting 'out of the ordinary' events in the EEG signal which may indicate a prevalence for stroke in post-surgery anesthetized patients. To the best of the author's knowledge, this is the first work using fractional Fourier transform for stroke analysis. The motivation of the work was the good results of this powerful tool in other domains of signal processing [12].

The paper is organized as follows. The next section provides a summary of the method used to extract information from the signals followed by a section on results and discussion on further use of the tools proposed. A conclusion section summarizes the main outcome of this work and points to further steps.

2 Materials and Methods

It is possible to characterize a signal in time or frequency domain. However none of these two can cover the main features of the signal completely. Some characteristics are better shown in frequency domain while the other features may be determined more effectively in time domain [12]. In other word, we may not be able to capture some important information in time domain; while, they are clearly apparent in frequency domain and vice versa. Time-Frequency analysis is one of the approaches that gives a wider view towards the signal, because it has the advantages of both time- and frequency-analysis. Fourier Transform is a classic tool to analyze and process stationary signals, but it appears incapable for non-stationary or time varying signals. To estimate the spectral density of the EEG signal strength, the Fractional Fourier method is used in the present study, being a generalization of conventional Fourier method with an order parameter α. Mathematically, the αth order fractional Fourier transform is the αth power of the Fourier transform operator. The $\alpha =$ first order fractional transform is the ordinary Fourier transform. In this paper Discrete Fractional Fourier Transform (DFFT) with different fractional orders is proposed as feature extraction technique for EEG signals. When $\alpha = \pi/2$, it is obtained the Fourier transform, while for $\alpha = 0$, it can be obtained the signal itself. Any intermediate value of α produces a signal representation that can be considered as a rotated time–frequency representation of the signal. The fractional Fourier transform enables continuous movement between the time and frequency domains, allowing the simultaneous retention of both frequency and time domain information. Being a generalization of the ordinary Fourier transform, the fractional Fourier transform is more flexible in its applications and hence of potential interest to any area in which the Fourier transform is frequently implemented [13]. The used definition of the fractional Fourier transform is (FrFT) [14]:

$$f_p(u) = \int_{-\infty}^{+\infty} K_p\left(u, t\right) f(t)dt \tag{1}$$

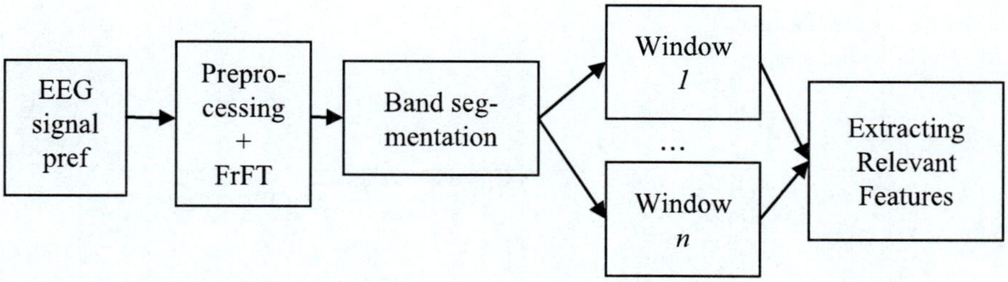

Fig. 1 The EEG signal processing workflow

where $K_p(u,t) = \begin{cases} A_\alpha \exp\left[j\pi\left(u^2 cot\alpha - 2utcsc\alpha + t^2\, cot\alpha\right)\right], \alpha \neq n\pi \\ \delta(u-t), \alpha = 2n\pi \\ \delta(u+t), \alpha = (2n+1)\pi \end{cases}$ is the kernel function of the fractional Fourier transform, $A_\alpha = \frac{\exp\left[-j\pi \frac{\mathrm{sgn}(sin\alpha)}{4} + \frac{j\alpha}{2}\right]}{|sin\alpha|^{1/2}}$, $\alpha = \frac{p\pi}{2}$, n is integer and δ represents the Dirac function.

The workflow is presented in Fig. 1.

The processing method functions are performed in Matlab® software.

The Bonn EEG database has been employed in this study, which is a well-known benchmark dataset for this problem [15].

3 The Results Obtained with the Proposed Tool

With the goal to provide a powerful tool for accurately localizing the source of stroke in brain activity by doctors, the authors combine the latest techniques for determining electrical activity in the brain by EEG signal processing.

The used EEG signal can be obtained on maximum 32 channels, as it is illustrated in Fig. 2. By selecting the corresponding button in the user interface, it can be selected one or more channel for interest, Fig. 3. The position of electrodes can be visualized in 2D or 3D plots, in order to obtain the best distribution by the medical staff, Fig. 4. The preprocessing stage of the signal consists on changing the data sampling rate (for example to reduce the sampling rate to save memory and disk storage) and filtering the data. The proper filter and the corresponding filter order and frequency specifications can be selected as it is presented in Fig. 5.

In order to estimate the spectral density of the signal strength, FrFT method is used, obtaining the plot from Fig. 6. The feature extraction result is marked with red squares for each channel, in order to be evaluated by the doctor for final diagnosis, Fig. 7.

Although these are preliminary first hand results, they suggest that the proposed methodology and tools may be suitable for the objective of stroke detection and later on prevention (by detecting pre-stroke events in the EEG signal). The

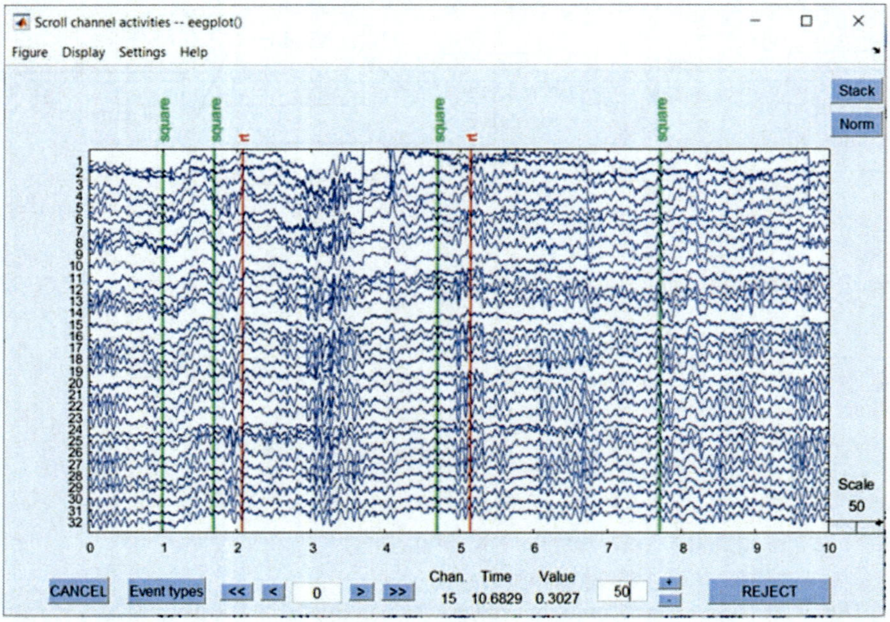

Fig. 2 The used EEG signal type

proposed model structure seems over-parameterized, but investigation into other events related to stroke visible in EEG signal may require the extra parameters - this study is ongoing.

4 Conclusions

This paper introduced emerging tools from fractional calculus and afferent models for EEG signal processing. It is provided a user-friendly tool for medical staff to establish a final diagnosis. The preliminary results indicate redundancy in model parameters, but further analysis of this study is ongoing. The next step consist on real patient data tests for a large number of acquired signal.

Acknowledgments This work was supported by the ÚNKP-19-4-OE-64 New National Excellence Program of the Ministry for Innovation and Technology and by a grant of the Romanian National Authority for Scientific Research and Innovation, project number PN-III-P2-2.1-PED-2019-0844, contract no.323PED.

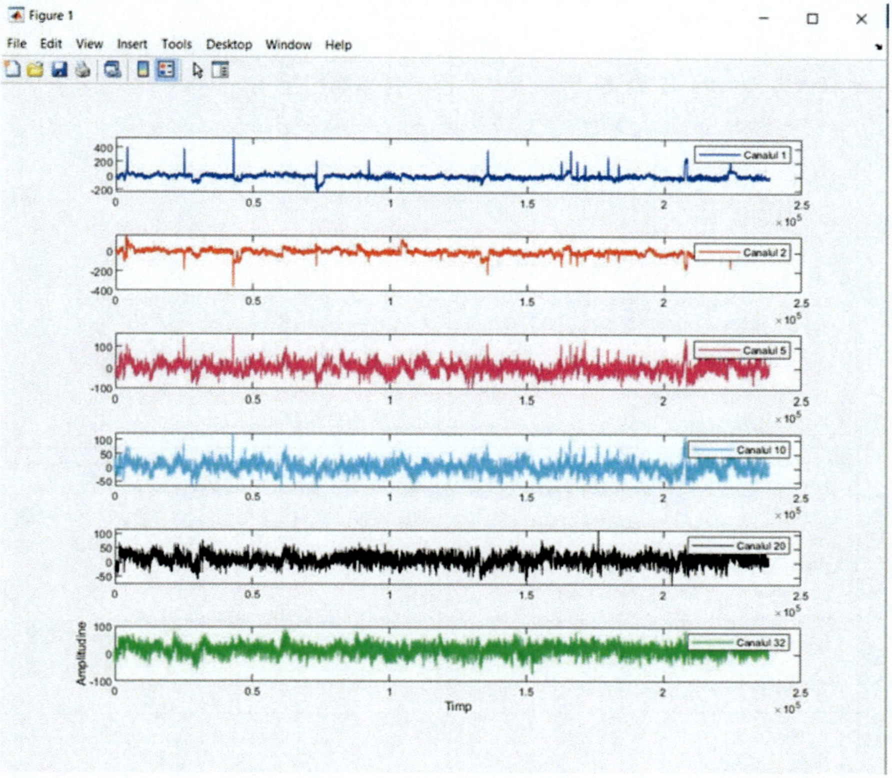

Fig. 3 Channel selection for analysis. In this case study are selected channels no. 1, 2, 5, 10, 20 and 32

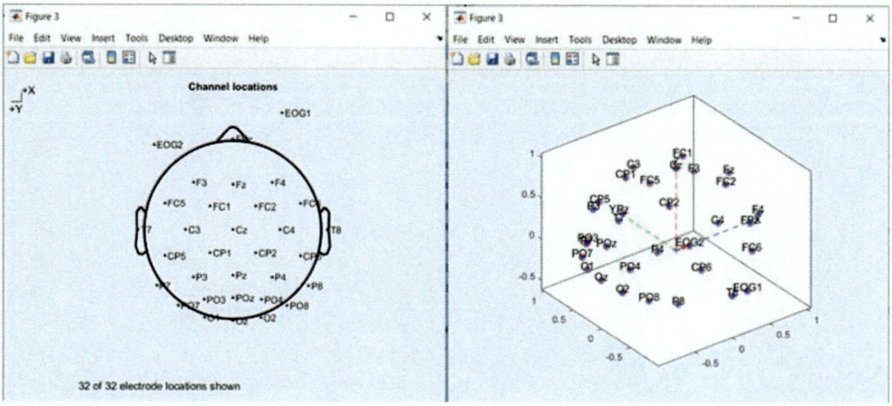

Fig. 4 2D or 3D plot of the electrodes

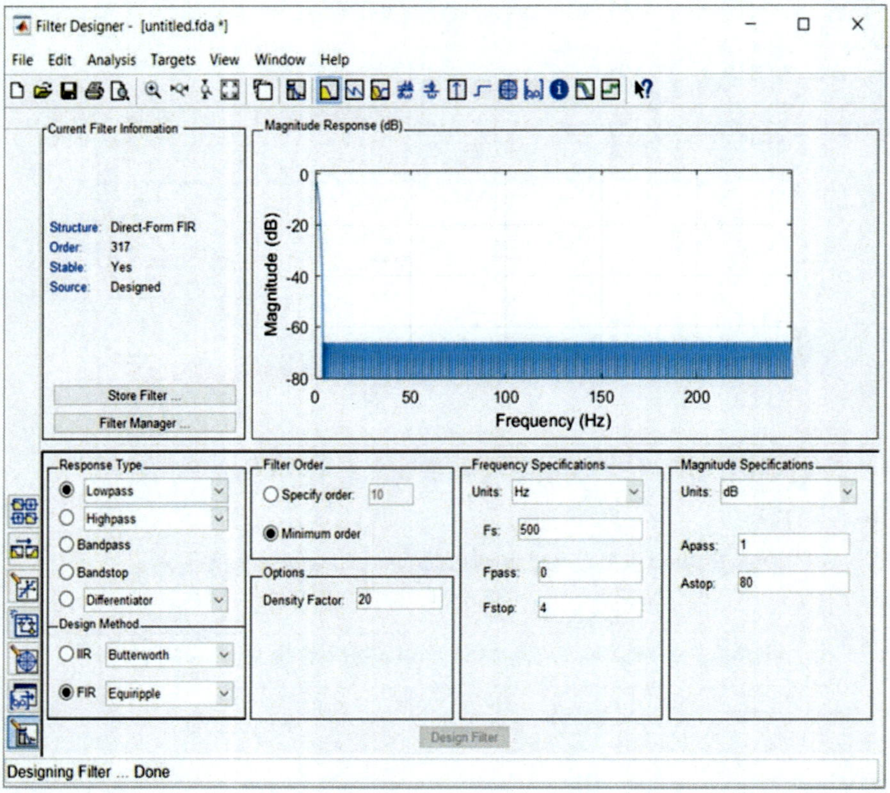

Fig. 5 User interface for filter design

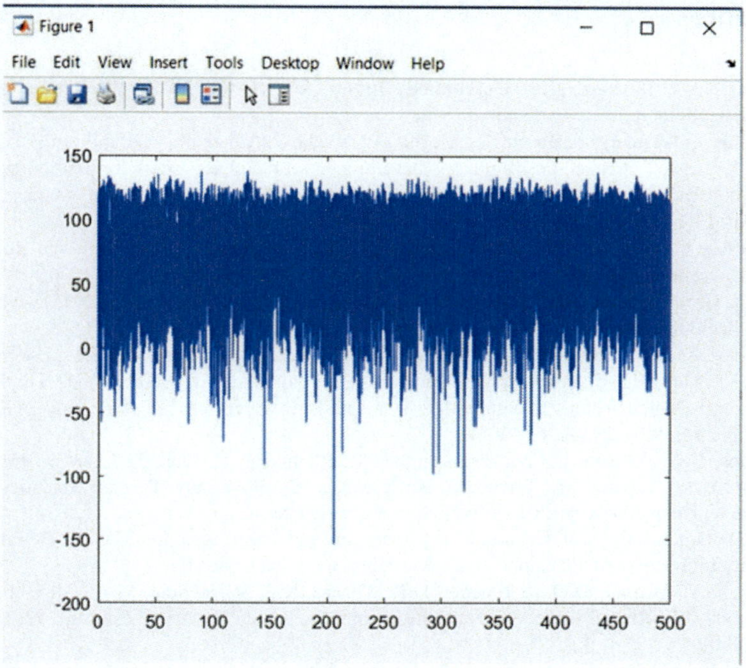

Fig. 6 Power spectra using fractional Fourier method

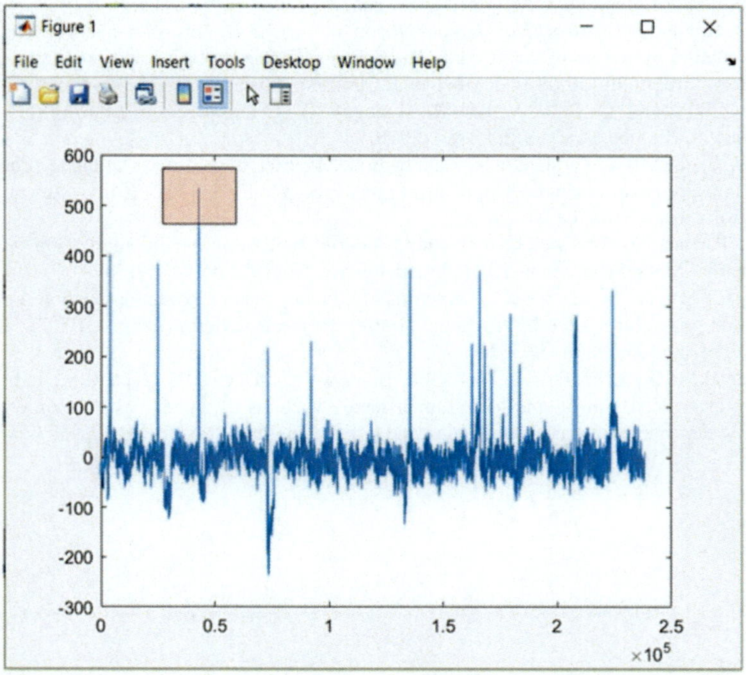

Fig. 7 Peak mark for channel 1

References

1. Ringer, A., Jimenez, L.: Unruptured Brain Aneurysm (2018). Available: https://www.mayfieldclinic.com/pe-aneurun.htm
2. Greving, J., Wermer, M., Brown, R., Morita, A., Juvela, S., Yanekura, M., Ishibashi, T., Torner, J., Nakayama, T., Rinkel, G., Algra, A.: Development of the phases score for prediction of risk of rupture of intracranial aneurysms: a pooled analysis of six prospective cohort studies. Lancet Neurol. **13**(1), 59–66 (2014)
3. Vladescu, C., Scintee, S.G., Olsavszky, V., Hernandez-Quevedo, C., Sagan, A.: Romania: health system review. Health Syst. Transit. **18**(4), 1–170 (2016)
4. GHDx. Global Health Data Exchange. Global Burden of Disease Study 2015 (2015). Available: ghdx.healthdata.org/gbd-2015
5. Szatmari, S., Pascu, I., Mihalka, L., Mulesa, S.V., Fekete, I., Fulesdi, B., Csiba, L., Zselyuk, G., Szasz, J., Gebefugi, J., Nicolescu, S., Vasiesiu, D., Smolanka, V.I., Bereczki, D.: The Mures-Uzhgorod-Debrecen study: a comparison of hospital stroke services in central-eastern Europe. Eur J Neurol. **9**(3), 293–296 (2002)
6. Wilkins, E.W., Wilson, L., Wickramasinghe, K., Bhatnagar, P., Leal, J., Luengo-Fernandez, R., Burns, R., Rayner, M., Townsend, N.: European Cardiovascular Disease Statistics 2017, European Heart Network (2017). Available: www.ehnheart.org
7. OECD. Organisation for Economic Co-operation and Development (OECD), Health Care Quality Indicators: Acute Care (2014). Available: https://stats.oecd.org/
8. Wong, G.Y., Warner, D.O., Schroeder, D.R., Offord, K.P., Warnerand, M.A., Maxson, P.M., Whisnant, J.P.: Risk of surgery and anesthesia for ischemic stroke. Anesthesiology. **92**(2), 425–432 (2000)
9. Magin, R., Vinagre, B., Podlubny, I.: Can cybernetics and fractional calculus be partners? Searching new ways to solve complex problems. IEEE Syst. Man Cybern. Mag. **4**(3), 23–28 (2018)
10. Neckebroek, M., Ionescu, C.M., van Amsterdam, K., De Smet, T., Debaets, P., Decruyenaere, J., De Keyser, R., Struys, M.: A comparison of propofol-to-bis post-operative intensive care sedation by means of target controlled infusion, Bayesian-based and predictive control methods. A feasibility study, under review. J. Clin. Monit. Pract. **33**, 675–686 (2019)
11. Zhou, Z.B., Meng, L., Gelb, A., Lee, R., Huang, W.Q.: Cerebral ischemia during surgery: an overview. J. Biomed. Res. **30**(2), 83–87 (2016)
12. Sejdic, E., Djurovic, I., Stankovic, L.: Fractional Fourier transform as a signal processing tool: an overview of recent developments. Signal Proc. **91**, 1351–1369 (2011). https://doi.org/10.1016/j.sigpro.2010.10.008
13. Zhao, T., Ran, Q.: The weighted fractional Fourier transform and its application in image encryption. Math. Probl. Eng. (2019). https://doi.org/10.1155/2019/4789194
14. Wang, T., Liu, N., Su, Z., Li, C.: A new time–frequency feature extraction method for action detection on artificial knee by fractional Fourier transform. Micromachines. **10**, 333 (2019). https://doi.org/10.3390/mi10050333
15. Andrzejak, R.G., Lehnertz, K., Mormann, F., Rieke, C., David, P., Elger, C.E.: Indications of nonlinear deterministic and finite-dimensional structures in time series of brain electrical activity: dependence on recording region and brain state. Phys. Rev. E. **64**, 061907 (2001)

Chaos, Bifurcations and Strange Attractors in Environmental Radioactivity Dynamics of Some Geosystems

Alexander V. Glushkov, Olga Yu. Khetselius, Sergiy M. Stepanenko, and Eugeny V. Ternovsky

Abstract The theoretical foundations and further application of an effective universal chaos-geometric approach to analysis and processing the data of radioactivity dynamics in environment are presented. The approach presented includes a group of advanced available methods or new ones (the correlation integral and fractal analysis methods, the average mutual information and false nearest neighbors algorithms, the Lyapunov's exponents and Kolmogorov entropy analysis, the surrogate data method, different algorithms of non-linear prediction models, spectral methods, etc.) to provide accurate numerical modeling and analysis of temporal dynamics of the atmospheric pollutants. The numerical results of analysis, modelling the radon concentration in the atmospheric environment are listed. The topological and dynamical invariants data for the ^{222}Rn concentration time series are computed with using the measurements data by the US Environmental Measurements Laboratory and Goddard Institute of Space Studies.

Keywords Dynamical system · Chaos · Environmental radioactivity · Radon

1 Introduction

One of the most urgent and important problems of theory of environmental systems and environmental protection is related to the correct description of the quantitative dynamics of environmental radioactivity (see, for example, [1–6]). Usually it is worth mentioning problems that are so urgent, such as long-term study of the evolutionary (fluctuation) dynamics of radionuclides in various environments as in spatial as in temporal aspects, elucidation of mechanisms of radionuclide transfer in hydro- and atmospheric systems, atmospheric transfer, transportation of radioactive substances with accounting for meteorological and other factors, etc.

A. V. Glushkov · O. Y. Khetselius · S. M. Stepanenko · E. V. Ternovsky (✉)
Odessa State Environmental University, Odessa, Ukraine

J. Awrejcewicz (ed.), *Perspectives in Dynamical Systems II: Mathematical and Numerical Approaches*, Springer Proceedings in Mathematics & Statistics 363,
https://doi.org/10.1007/978-3-030-77310-6_8

The key tasks of the study of the dynamics of atmospheric radionuclides include the study of the transport of radionuclides in various atmospheric conditions using the hierarchy of models of air circulation, modeling the spatial and temporal structure of the fields of concentrations of impurities in the atmosphere. In this case, possible scenarios for the propagation of impurities are based on the analysis of their behavior, depending on the variation of deterministic and random parameters of the models. Recall [1–4] that most models are currently used to estimate the state (as well as forecast) and dynamics of the environmental systems within deterministic models, or simplified ones, based on the simple statistical regressions. The success of these models, however, is limited by their inability to describe the nonlinear characteristics of the pollutant concentration behavior and the lack of understanding of the physical and chemical processes involved.

It is possible to distinguish multifactor, hydrodynamic (for example, the models of "torch", "molecular diffusion", "shallow water", "equation of hydrodynamics of the surface layer"), probabilistic-statistical approaches, as well as numerous numerical approaches [1–20]. Of particular importance are the torch models and standard classical diffusion models. The scattering of pollutants entering the natural environment is subject to the laws of classical or turbulent diffusion [1–3]. In whole one could remind about such widespread methods as MLDP0 (Modèle Lagrangien de Dispersion de Particules d'ordre 0), HYSPLIT (Hybrid Single-Particle Lagrangian Integrated Trajectory Model), NAME (Numerical Atmospheric-dispersion Modelling Environment), RATM (Regional Atmospheric Transport Model), FLEXPART (Lagrangian Particle Dispersion Model), model ECMWF (the European Center for Medium-Range Weather Forecasts) and others [1–8].

The following factors have a significant influence on the process of their dispersion of the radioactive substances in atmosphere: the thermodynamic state of atmosphere, the physical and chemical properties of harmful substances, the height and diameter of the emission source, the location of the sources and many others. But, in any case, classical-diffusion and standard probability-statistical models allow, strictly speaking, to obtain mainly only qualitative features of the process of distribution of substances, but the best quantitative level of description based on, for example, classical-diffusion models, as a rule, is achieved in the absence of vortex turbulence elements in the environment.

Adequate correct analysis, modeling and prediction of the propagation of radioactive material in natural environments should be based on a complex of both stochastic-diffusion models and apparatus of complex nonlinear dynamic systems theory and chaos theory.

New field of investigations of the environmental dynamical systems has been provided by a great progress in a development of a chaos and dynamical systems theory methods [14–25]. In our previous papers [4–8, 20–30] we have given a review of new methods and algorithms to analysis of different systems in the fields of Environmental and Earth sciences, quantum physics, electronics and photonics and used the nonlinear method of chaos theory and the recurrence spectra formalism to study stochastic futures and chaotic elements in dynamics of the environmental, chemical, biological and physical (namely, atomic, molecular, nuclear systems in an free state and an external electromagnetic field) systems. The non-trivial manifestations of a chaos phenomenon have been discovered.

In this paper we present the theoretical foundations and further application of an effective universal complex chaos-geometric and quantum-dynamic approach to analysis, processing, forecasting data of radioactivity dynamics in various environments and list some new numerical results of analysis, modelling the atmospheric radon ^{222}Rn concentration in the atmospheric environment. The complex chaos-geometric and quantum-dynamic approach includes a group of advanced available methods or new ones (the known correlation integral and fractal analysis methods, the mutual information and false nearest neighbors algorithms, the Lyapunov's exponents and Kolmogorov entropy analysis, different algorithms of non-linear prediction models, different spectral methods and algorithms, etc.) to provide accurate numerical modeling and analysis of temporal dynamics of the atmospheric pollutants, in particular, the atmospheric ^{222}Rn. The topological and dynamical invariants data for the radon ^{222}Rn concentration time series are computed with using the measurements data by the US Environmental Measurements Laboratory and the Goddard Institute of Space Studies [2, 3].

2 A Chaos-Geometric Approach to Analysis, Processing, Modelling and Forecasting the Environmental Radio-Activity Geosystem Dynamics

In our previous papers [2–8, 22–26] we have given a review of new methods and algorithms of the chaos-geometric (combined with quantum-dynamic) approach to analysis, processing, modeling and forecasting a temporal evolution of different dynamical systems.

As many blocks of the used approach have been developed earlier and need only to be reformulated regarding the problem studied in this paper, here we pay the attention at the most principal points and some new elements. The main stages of a chaos-geometric (combined with quantum-dynamic) approach to analysis, processing and forecasting data of the environmental radioactivity dynamics are as follows:

1. General qualitative analysis (in terms of ordinary differential equations or the Arnold analysis) of the radioactivity dynamics; the transport of radionuclides in various atmospheric conditions using the hierarchy of models of air circulation, modeling the spatial and temporal structure of the fields of concentrations of impurities in the atmosphere.
2. Application of the different chaos-geometric tests on the presence of chaotic (stochastic) elements, functions and modes in a system; the Gottwald-Melbourne test, etc.

3. Fractal and quantum geometry of a phase space (choice of time delay, determination of embedding dimension by methods of correlation dimension algorithm and false nearest neighbors algorithm).
4. Analysis and computing the dynamic and topological invariants of a chaotic system and nonlinear forecasting of a temporal (spatial) evolution of radioactive system dynamics.
5. The key points of the whole approach are reflected in the flowchart in Table 1.

The fundamental ideas of the combined chaos-geometric and quantum-dynamic (plus differential equations one and the Armold analysis) approach to modelling, processing and prediction of chaotic dynamics are ideologically reduced to reproduction (and reconstruction) of a phase space of the radioactive geosystems, prediction of the temporal evolution of the main parameters of a system. From the viewpoint of mathematical modelling it is a question of consideration of unambiguous representations of a kind:

$$\boldsymbol{F}_{i+1} = \mathbf{G}\left(F_i\right), \tag{1}$$

where $F \in \mathbf{R}^D$—is the state vector, D is the dimension, i—discrete time, $\mathbf{G}$ is the D-dimensional mapping. To implement the ideology of simulation of a compact geometric attractor and the use of chaos-cybernetic algorithm of predicted phase trajectories of the system to restore the phase space of the system, it is possible to use several concepts, first, the concept of average mutual information, and secondly, the concept of using the properties of the corresponding linear autocorrelation function (see Table 1). An alternative to the method of correlation dimension (integral) is the method of false nearest neighboring points. This approach allows to determine the dimensionality of the embedding for the reconstruction of the phase space, as well as to verify the results obtained by the method of correlation dimensionality. The technical details of realization these conceptions can be found in Refs. [4–10, 20–26].

The basic model for determining G can be, for example, a polynomial of order K:

$$\mathrm{G}\left(F_1, F_2, \ldots, F_D\right) = \sum_{l_1 = l_2 = \cdots = l_D = 0}^{K} b_{l_1 l_2 \ldots l_D} \prod_{i=1}^{D} t_i^{l_i} \prod_{i=1}^{D} l_j, \tag{2}$$

where the corresponding coefficients are determined so that the standard error of the approximation ε is minimal as follows:

$$\varepsilon^2 = \frac{1}{N_{tr} - D} \sum_{i=1}^{N_{tr}-D} \left[F_{i+D} - G\left(F_i, F_{i+1}, \ldots, F_{i+D-1}\right]^2 = \min \tag{3}$$

One of the principally important points of the whole approach to modeling and forecasting chaotic dynamics of the radioactive geosystems is computing the topo-

Table 1 Flowchart of the combined chaos-geometric and quantum-dynamic (plus differential equations one and the Armold analysis) approach to modelling, processing and prediction of chaotic dynamics of the radioactive geosystems (ecosystems)

I. General qualitative analysis (in terms of ordinary differential equations or the Arnold analysis) of the radioactivity dynamics in atmosphere; the atmospheric general dynamics differential equations analysis)

⇓

II. Application of the different chaos-geometric and quantum-dynamic tests on a presence of chaotic (stochastic) elements, functions and modes in the radioactive geosystem;

1. Gotwald-Melbourne test; Chirikov test; Naïve model tests;

⇓

2. Energy and spectral methods and algorithms: energy and power spectra, energy level statistics, random matrix analysis; characteristic distributions of the Wigner-Dyson type

⇓

III. Multifractal and quantum geometry of a phase space and dynamics of resonances

3. Computing the fractal parameters, multifractal spectra; wavelet analysis; cepstrum analysis

⇓

4. The Packard-Takens algorithm; the advanced autocorrelation function or average initial information algorithms; The Green's function method

⇓

5. Reconstruction of a phase space; Computing embedding dimension, correlation dimensions; using the methods of the correlation integral by Grassberger-Procaccia OR the false nearest neighbor points formalism

⇓

IV. Forecasting chaotic dynamics of the radioactive geosystems

6. Computing the invariants; The global Lyapunov's dimension analysis; Kolmogorov entropy analysis; The Kaplan-York dimension analysis; Method of nearest neighboring points

⇓

7. New methods and algorithms of nonlinear forecasting chaotic dynamics of the radioactive geosystems and ecosystems ("minmax" algorithms, methods of the stochastic propagators; Neural networks modelling algorithms with application of the polynomial or B-spline models, wavelets etc ...

logical and dynamical invariants. The latter include, in particular, local and global Lyapunov's dimensions or Lyapunov's exponents. It is worth to remind the classical definition of the Lyapunov's exponents through e logarithms of absolute values of eigen-values of linearized dynamics focused on the attractor, more precisely:

$$\lambda = \lim_{\substack{t \to \infty \\ d(0) \to 0}} \left(\frac{1}{t}\right) \log_2 \left[\frac{d(t)}{d(0)}\right]$$
$$d(t) = \left[\sum_{i=1}^{n} \delta F_i^2(t)\right]^{1/2} \tag{4}$$

Here, the norm determines the degree of divergence of two adjacent trajectories, that is, the master trajectory and the adjacent trajectory with initial conditions $S(0) + \delta S(0)(S = F)$. It is important to note that the negative dimensions indicate the local average compression rate and the positive ones indicate the expansion one. The most significant is the maximal Lyapunov's exponent, the positivity of which indicates the existence of a chaos in the system. In fact, if one manages to derive the whole spectrum of the Lyapunov's exponents, other invariants and parameters of the system, i.e. Kolmogorov entropy (K_{ent}) as well as an average predictability, the Kaplan-Yorke conjecture and the attractor's dimensions, can be determined and computed. It should be noted that there are several algorithms for computing a spectrum of the Lyapunov's exponents, among which the most common is the method based on the Jacobian mapping. All technical details can be found in Refs. [4–10, 20–26]. All calculations are performed with using "Geomath", "Superatom" and "Quantum Chaos", "ScanPoints" PC computational codes [31–45].

3 Analysis of Atmospheric Radon Concentration Time Series, the Topological and Dynamical Invariants and Conclusion

Here we list the results of applying a chaos-geometric (quantum-dynamic) approach to studying temporal dynamics (time series) of the atmospheric radon concentration fluctuations. The data of measurements of the radon concentrations at 5 USA sites were presented by the US Environmental Measurements Laboratory, Department of Energy and Goddard Institute for Space Studies (e.g., [2, 3] and references therein) and were actually integrated into a long-term measurement system. Meteorological data (wind characteristics, air temperature, etc.) were obtained simultaneously. Measurements were made during 1979–1981 at 15 heights (6–200 m). In [3], the radon-222 was actually used as a quantitative indicator (tracer; 3-D chemical tracer model) of convective transfer in the general circulation model (the so-called GISS general circulation model). As the input data we have used the time Series of the daily average radon concentrations at Chester (New Jersey) during 1978 (8192 points).

Below we present the data of numerical experiments on the restoration of the embedding dimension (d_E), using the method of correlation integral and the algorithm of false nearest neighboring points. In order to calculate the correlation dimension d_2 it one should calculate the correlation integrals C(r) for different

Table 2 Calculational data for the K chaotic index (K_{ch}) and different dynamical and topological invariants: time delay τ, correlation dimension (d_2), embedding space dimension (d_E), Lyapunov exponent (λ_i), Kolmogorov entropy (K_{ent}), Kaplan-York dimension (d_L), predictability limit (Pr_{max}) for the atmospheric radon time series (1978; Chester, New Jersey)

^{222}Rn	K_{ch}	τ	d_2	d_E
1978	0.91	16	6.03	7
λ_1	λ_2	K_{ent}	d_L	Pr_{max}
0.0194	0.0086	0.028	5.88	35

embedding dimensions. The correlation dimension of the attractor (d_A) is defined as the value of the correlation dimension, in which it does not change as the embedding dimension increases.

Table 2 summarizes all the results for the recovery of attractors, as well as the calculational data for the K chaotic index (K_{ch}) and different dynamical and topological invariants (time delay τ, correlation dimension (d_2), embedding space dimension (d_E), Lyapunov's exponent (λ_i), Kolmogorov entropy (K_{ent}), Kaplan-York dimension (d_L), etc. for the atmospheric radon time series) (1978; Chester, New Jersey).

In the case considered, the values of the chaos parameter K in all cases exceed 0.8, that is, the considered time series are subject to the influence of chaotic dynamics. The analysis of the dynamical and topological invariants shows that, for example, the resulting Kaplan-York dimension is very close to the correlation dimension and is smaller than the dimension of attachment, which confirms the correctness of the choice of the latter. This conclusions is fully analogous the conclusions [6].

The Lyapunov's exponents computational data demonstrate the chaotic elements in the corresponding time series. It is interesting that the randomness inherent in the studied series of radon concentrations for the Chester (New Jersey) site is much higher than in the analogous studying the radon dynamics in the Southern Finland [6, 10]. This resulted in a greater value of the Kolmogorov entropy and a correspondingly lower predictability limit.

To conclude, we have presented the theoretical foundations of an universal, complex chaos-geometric (plus quantum-dynamic) approach to analysis, processing, forecasting data of radioactivity dynamics. We have listed the numerical results of application of this approach to analysis and processing the radon ^{222}Rn concentration time series (1978; Chester, New Jersey) and obtained values of the topological and dynamical invariants. The data allows to reveal the deterministic chaos elements. The results presented show that the application of chaos theory methods to analysis and processing the time series of radon concentrations in the atmosphere is quite effective from both a theoretical and practical viewpoints.

Using the methods and algorithms [4–20], there is a possibility to solve a problem of recovery and forecasting temporal dynamics of the radioactive radon concentration fluctuations in both the short- and medium-term interval. It is clear that for the full implementation of this program, in addition to knowing the necessary fractal-chaotic properties, one will also need to use the formalism of conformal mapping of the corresponding modified series of the atmospheric pollutant concentrations within, say, a neural network approach (look in details [28, 42]).

It is of a great importance the development of a complex of the nonlinear chaotic-geometric models for the analysis, modeling of contamination, description of the transfer of radionuclides in the territories for which the relevant data on radioactive contamination is very scanty. The use of fractal sets, chaos and dynamical systems theories allows to analyze, predict and compute a temporal chaotic dynamics of arbitrary chaotic radioactive geosystems (ecosystems).

References

1. Chen, X., Paatero, J., Kerminen, V.-M., Riuttanen, L., et al.: Responses of the atmospheric concentration of radon-222 to the vertical mixing and spatial transportation. Boreal Environ. Res. (Helsinki). **21**, 299–318 (2016)
2. Jacob, D.J., Prather, M.J.: Radon-222 as a test of convective transport in a general circulation model. Tellus. **42b**, 118–134 (1990)
3. Pinault, J.-L., Baubron, J.-C.: Signal processing of diurnal and semidiurnal variations in radon and atmospheric pressure: a new tool for accurate in situ measurement of soil gas velocity, pressure gradient, and tortuosity. J. Geophys. Res. **102**(b8), 101–120 (1997)
4. Bunyakova, Y.Y., Glushkov, A.V.: Analysis and Forecast of the Impact of Anthropogenic Factors on Air Base in of an Industrial City. Ecology, Odessa (2010)
5. Glushkov, A.V.: Methods of a Chaos Theory. Astroprint, Odessa (2012)
6. Glushkov, A.V., Safranov, T.A., Khetselius, O.Y., Ignatenko, A.V., Buyadzhi, V.V., Svinarenko, A.A.: Analysis and forecast of the environmental radioactivity dynamics based on methods of chaos theory: general conceptions. Environ. Prob. **1**(2), 115–120 (2016)
7. Stepanenko, S.N., Shnaidman, V.N.: Comparison of the results of a generalization of experimental data with calculations from a three-parameter model of the atmospheric boundary layer. Fluid Mech. Soviet Res. **12**, 123–131 (1983)
8. Buyadzhi, V.V., Glushkov, A.V., Khetselius, O.Y., Ternovsky, V.B., Serga, I.N., Bykowszczenko, N.: An advanced analysis and modelling the air pollutant concentration temporal dynamics in atmosphere of the industrial cities: Odessa city. IOP Conf. Series Earth Environ. Sci. **92**, 012006 (2017)
9. Buyadzhi, V.V., Glushkov, A.V., Khetselius, O.Y., Bunyakova, Y.Y., Florko, T.A., Agayar, E.V., Solyanikova, E.P.: An effective chaos-geometric computational approach to analysis and prediction of evolutionary dynamics of the environmental systems: atmospheric pollution dynamics. J. Phys. Conf. Ser. **905**, 012036 (2017)
10. Glushkov, A.V., Bunyakova, Y.Y., Buyadzhi, V.V., Dubrovskaya, Y.V., Kuznetsova, A.A., Khetselius, O.Y.: New approach and microsystem technology of advanced non-linear analysis and modelling chaotic environmental radioactivity dynamics. Sensor Electr. Microsyst. Techn. **14**(3), 24–37 (2017)
11. Bunyakova, Y.Y., Ternovsky, V.B., Dubrovskaya, Y.V., Ignatenko, A.V., Svinarenko, A.A., Vitavetskaya, L.A.: Analysis of the beryllium-7 activity concentration dynamics in the atmospheric environment time series after the Fukushima Daiichi nuclear power plants emergency. Sensor Electr. Microsyst. Techn. **14**(4), 73–82 (2017)
12. Glushkov, A.V., Khetselius, O.Y., Bunyakova, Yu.Ya., Grushevsky, O.N., Solyanikova, E.P.: Studying and forecasting the atmospheric and hydroecological systems dynamics by using chaos theory methods. In: Awrejcewicz, J., Kazmierczak, M., Olejnik, P., Mrozowski, J. (eds.) Dynamical Systems Theory, T1, pp. 249–258. Lodz Univ. Press, Lodz (2013)

13. Schreiber, T.: Interdisciplinary application of nonlinear time series methods. Phys. Rep. **308**, 1–64 (1999)
14. Packard, N.H., Crutchfield, J.P., Farmer, J.D., Shaw, R.S.: Geometry from a time series. Phys. Rev. Lett. **45**, 712–716 (1980)
15. Kennel, M., Brown, R., Abarbanel, H.: Determining embedding dimension for phase-space reconstruction using a geometrical construction. Phys. Rev. A. **45**, 3403–3411 (1992)
16. Abarbanel, H.D.I., Brown, R., Sidorowich, J.J., Tsimring, L.S.: The analysis of observed chaotic data in physical systems. Rev. Mod. Phys. **65**, 1331–1392 (1993)
17. Fraser, A.M., Swinney, H.L.: Independent coordinates for strange attractors from mutual information. Phys. Rev. A. **33**, 1134–1140 (1986)
18. Grassberger, P., Procaccia, I.: Measuring the strangeness of strange attractors. Physica D. **9**, 189–208 (1983)
19. Gallager, R.G.: Information Theory and Reliable Communication. Wiley, New York (1986)
20. Glushkov, A.V., Khetselius, O.Y., Brusentseva, S.V., Zaichko, P.A., Ternovsky, V.B.: Studying interaction dynamics of chaotic systems within a non-linear prediction method: application to neurophysiology. In: Balicki, J. (ed.) Advances in Neural Networks, Fuzzy Systems and Artificial Intelligence Series: Recent Advances in Computer Engineering, vol. 21, pp. 69–75. WSEAS Pub., Gdansk (2014)
21. Glushkov, A.V., Svinarenko, A.A., Buyadzhi, V.V., Zaichko, P.A., Ternovsky, V.B.: Chaos-geometric attractor and quantum neural networks approach to simulation chaotic evolutionary dynamics during perception process. In: Balicki, J. (ed.) Advances in Neural Networks, Fuzzy Systems and Artificial Intelligence Series: Recent Advances in Computer Engineering, vol. 21, pp. 143–150. WSEAS Pub., Gdansk (2014)
22. Khetselius, O.Y.: Forecasting evolutionary dynamics of chaotic systems using advanced non-linear prediction method. In: Awrejcewicz, J., Kazmierczak, M., Olejnik, P., Mrozowski, J. (eds.) Dynamical Systems Applications, T2, pp. 145–152. Lodz Univ. Press, Lodz (2013)
23. Bunyakova, Yu.Ya., Khetselius, O.Yu.: Non-linear prediction statistical method in forecast of atmospheric pollutants. In: Proc. of 8th International Carbon Dioxide Conference, T2-098, Jena, Germany (2009)
24. Khetselius, O.Y., Glushkov, A.V., Bunyakova, Y.Y., Buyadzhi, V.V., Bondar, O.I., Vaschenko, V.N., Bykowszczenko, N.: New approach and microsystem technology to modelling dynamics of atmosphere ventilation of industrial city and elements of the "Green-City" construction technology. Sensor Electr. Microsyst. Techn. **14**(4), 37–46 (2017)
25. Glushkov, A.V., Khetselius, O.Y., Agayar, E.V., Buyadzhi, V.V., Romanova, A.V., Mansarliysky, V.F.: Modelling dynamics of atmosphere ventilation and industrial city's air pollution analysis: new approach. IOP Conf. Series Earth Environ. Sci. **92**(012014) (2017)
26. Glushkov, A.V., Khetselius, O.Y., Bunyakova, Y.Y., Prepelitsa, G.P., Solyanikova, E.P., Serga, E.N.: Non-linear prediction method in short-range forecast of atmospheric pollutants: low-dimensional chaos. In: Dynamical Systems—Theory and Applications, LIF111. Lodz Univ. Press, Lodz (2011)
27. Glushkov, A.V., Prepelitsa, G.P., Svinarenko, A.A., Zaichko, P.A.: Studying interaction dynamics of the non-linear vibrational systems within non-linear prediction method (application to quantum autogenerators). In: Awrejcewicz, J., Kazmierczak, M., Olejnik, P., Mrozowski, J. (eds.) Dynamical Systems Theory, T1, pp. 467–477. Lodz Univ. Press, Lodz (2013)
28. Glushkov, A.V., Svinarenko, A.A., Loboda, A.V.: Theory of Neural Networks on Basis of Photon Echo and Its Program Realization. TEC, Odessa (2003)
29. Glushkov, A.V., Kuzakon, V.M., Ternovsky, V.B., Buyadzhi, V.V.: Dynamics of laser systems with absorbing cell and backward-wave tubes with elements of a chaos. In: Awrejcewicz, J., Kazmierczak, M., Olejnik, P., Mrozowski, J. (eds.) Dynamical Systems Theory, T1, pp. 461–466. Lodz Univ. Press, Lodz (2013)
30. Khetselius, O.Y., Brusentseva, S.V., Tkach, T.B.: Studying interaction dynamics of chaotic systems within non-linear prediction method. In: Awrejcewicz, J., Kazmierczak, M., Olejnik, P., Mrozowski, J. (eds.) Application to Neurophysiology. Dynamical Systems Applications, T2, pp. 251–259. Lodz Univ. Press., Lodz (2013)

31. Khetselius, O.Y.: Hyperfine structure of radium. Photo-Dermatology. **14**, 83–85 (2005)
32. Glushkov, A.V., Buyadzhi, V.V., Kvasikova, A.S., Ignatenko, A.V., Kuznetsova, A.A., Prepelitsa, G.P., Ternovsky, V.B.: Nonlinear chaotic dynamics of quantum systems: molecules in an electromagnetic field and laser systems. In: Tadjer, A., Pavlov, R., Maruani, J., Brändas, E., Delgado-Barrio, G. (eds.) Quantum Systems in Physics, Chemistry, and Biology Series: Progress in Theoretical Chemistry and Physics (Book 30), pp. 71–84. Springer, Dordrecht (2016)
33. Glushkov, A.V., Khetselius, O.Y., Kuzakon, V.M., Prepelitsa, G.P., Solyanikova, E.P., Svinarenko, A.A.: Modeling of interaction of the non-linear vibrational systems on the basis of temporal series analyses (application to semiconductor quantum generators). In: Dynamical Systems—Theory and Applications. Lodz Univ. Press, Lodz (2011)
34. Svinarenko, A.A., Glushkov, A.V., Khetselius, O.Y., Ternovsky, V.B., Dubrovskaya, Y.V., Kuznetsova, A.A., Buyadzhi, V.V.: Theoretical spectroscopy of rare-earth elements: spectra and autoionization resonances. In: Orjuela, J.E.A. (ed.) Rare Earth Element, pp. 83–104. InTech, Rijeka (2017)
35. Ignatenko, A.V., Buyadzhi, A.A., Buyadzhi, V.V., Kuznetsova, A.A., Mashkantsev, A.A., Ternovsky, E.V.: Nonlinear chaotic dynamics of quantum systems: molecules in an electromagnetic field. In: Jenkins, S., Kirk, S.R., Maruani, J., Brändas, E. (eds.) Advances in Quantum Chemistry, vol. 78, pp. 149–170. Elsevier, Cambridge, MA (2019)
36. Svinarenko, A.A.: Study of spectra for lanthanides atoms with relativistic many-body perturbation theory: Rydberg resonances. J. Phys. Conf. Ser. **548**, 012039 (2014)
37. Florko, T.A., Ambrosov, S.V., Svinarenko, A.A., Tkach, T.B.: Collisional shift of the heavy atoms hyperfine lines in an atmosphere of the inert gas. J. Phys. Conf. Ser. **397**, 012037 (2012)
38. Svinarenko, A.A., Ignatenko, A.V., Ternovsky, V.B., Nikola, L.V., Seredenko, S.S., Tkach, T.B.: Advanced relativistic model potential approach to calculation of radiation transition parameters in spectra of multicharged ions. J. Phys. Conf. Ser. **548**, 012047 (2014)
39. Buyadzhi, V.V., Zaichko, P.A., Gurskaya, M.Y., Kuznetsova, A.A., Ponomarenko, E.L., Ternovsky, V.B.: Relativistic theory of excitation and ionization of Rydberg atomic systems in a Black-body radiation field. J. Phys. Conf. Ser. **810**, 012047 (2017)
40. Buyadzhi, V.V., Zaichko, P.A., Antoshkina, O.A., Kulakli, T.A., Prepelitsa, G.P., Ternovsky, V.B., Mansarliysky, V.F.: Computing of radiation parameters for atoms and multicharged ions within relativistic energy approach: advanced code. J. Phys. Conf. Ser. **905**, 012003 (2017)
41. Glushkov, A.V., Kuzakon, V.M., Khetselius, O.Y., Bunyakova, Y.Y., Zaichko, P.A.: Geometry of chaos: consistent combined approach to treating chaotic dynamics atmospheric pollutants and its forecasting. Proc. Int. Geometry Center. **6**(3), 6–13 (2013)
42. Buyadzhi, V.V., Glushkov, A.V., Mansarliysky, V.F., Ignatenko, A.V., Svinarenko, A.A.: Spectroscopy of atoms in a strong laser field: new method to sensing AC Stark effect, multiphoton resonances parameters and ionization cross-sections. Sensor Electr. Microsyst. Techn. **12**(4), 27–36 (2015)
43. Gubanova, E.R., Glushkov, A.V., Khetselius, O.Y., Bunyakova, Y.Y., Buyadzhi, V.V., Pavlenko, E.P.: New Methods in Analysis and Project Management of Environmental Activity: Electronic and Radioactive Waste. FOP, Kharkiv (2017)
44. Glushkov, A.V., Khetselius, O.Y., Svinarenko, A.A., Buyadzhi, V.V.: Spectroscopy of Autoionization States of Heavy Atoms and Multiply Charged Ions. TEC, Odessa (2015)
45. Glushkov, A.V., Khetselius, O.Y., Svinarenko, A.A., Buyadzhi, V.V.: Methods of Computational Mathematics and Mathematical Physics. P.1. TES, Odessa (2015)

Dynamics of Chains as a Tool to Study Thermomechanical Properties of Proteins

Piotr Weber

Abstract Polymer dynamics can be formulated on different levels of detail. One approach eliminates microscopic degrees of freedom and a polymer molecule is represented by a simplified structure—a chain. In the simplest case monomers of ideal chain have fixed length, and their orientation is independent of the orientations and positions of neighbouring monomers. This is reason that two monomers can co-exist at the same place. Ideal chain model doesn't describe correctly the local structure of polymer, but correctly describe the property on large-scale. In this scale chain can be treated as a thermodynamical system, which their dynamic have to fulfill laws of nonequilibrium thermodynamics. In a living systems there are a special polymers-proteins, that can operate under non-equilibrium conditions. During biochemical processes, they changes its states and are treated as free energy transducers. I will present a certain formalism of non-equilibrium thermodynamic when non-Markovian processes appear.

Keywords Mesoscopic nonequilibrium thermodynamics · Protein dynamics · Non-markovian processes

1 Introduction

Properties of proteins dynamics are crucial for understanding a biological processes at molecular level. They are fundamental constituent of all known organisms and play many diverse functions in organisms [1, 2]. All these specific functions are associated with their three-dimensional structure, that is maintain under physiological conditions. These structure, co-called native states, are determine by sequence of α-amino-acids, that was used by organism to build its proteins. Studies of

P. Weber (✉)
Gdańsk University of Technology, Gdańsk, Poland
e-mail: piotr.weber@pg.edu.pl

J. Awrejcewicz (ed.), *Perspectives in Dynamical Systems II: Mathematical and Numerical Approaches*, Springer Proceedings in Mathematics & Statistics 363,
https://doi.org/10.1007/978-3-030-77310-6_9

these structures, conditions of their change and stability are important to know the mechanism of the action of a protein in a cell [3].

Coarse-grained models are one possible way to model protein activity. In this approach some microscopic degrees of freedom are eliminated and a polymer molecule is represented by a simplified structure. There are many methods for building coarse-grained models that can be formulated at different levels of detail. Sometimes only consideration about a classes of atoms in a molecule can give us a valuable information about dynamics of biomolecule [4]. In other situations, a more detailed model must be used. One can assume that length of monomers of the polymeric molecule has fixed length that way instead a molecule one can descibe a chain. Its state can be described by parameters: θ_i—bond angles and γ_i dihedral angles [3], where $i = 1, 2, \ldots, N-1$ and N is a number of monomers. Additionally another parts of the protein, like a functional group can be also simplified to a beads, which have some volume and simplified shape. This simplification is the basis for some methods of protein simulation described in [3]. From the theoretical point of view, the probability density of the above mentioned angles can be considered. These parameters can changes its values due to thermal energy, which is a reason a transitions between polymer's conformational states and simultaneously changes in distribution. Dynamics of this transitions can be described by continuum diffusion-like model. In the simplest case the Fokker-Planck equation can be used, which results of assumption about Markovian character of transitions. However there are also evidence that non-Markovian processes appear and dynamics of a biomolecule can be describe by subdiffusive-like model [5].

Not all angles must always be taken into account. If we can distinguish Kuhn segments in a protein, for example α—helises, then the angles between these rigid segments significantly change. Other angles located inside these rigid structures change rapidly under the influence of thermal energy and have a certain average value. The protein system can be then described by reduced number of variables. The above mentioned time-dependent probability density is also described in this reduced set of parameters.

Biologically-functional state of the protein changes according to function which are realize in physiological phenomena. For example there is one biological state when a ion channel in the cell membrain is open and there is another when it is closed. However each biological state (chemical states as was named in [6]) is realize by many conformational substates. It has consequences in kinetic description of dynamics between biologically-functional states of protein.

In the simplest case one can consider transitions between two biologically-functional states of the one protein:

$$E_1 \leftrightarrow E_2, \tag{1}$$

where E_1 represents one biologically—functional state of given protein and E_2 another biologically—functional state of the same protein. Arrow between E_1 and E_2 means that they can transform from one state into another. From macroscopic point of view the solution of kinetic equations of that system, which describe their

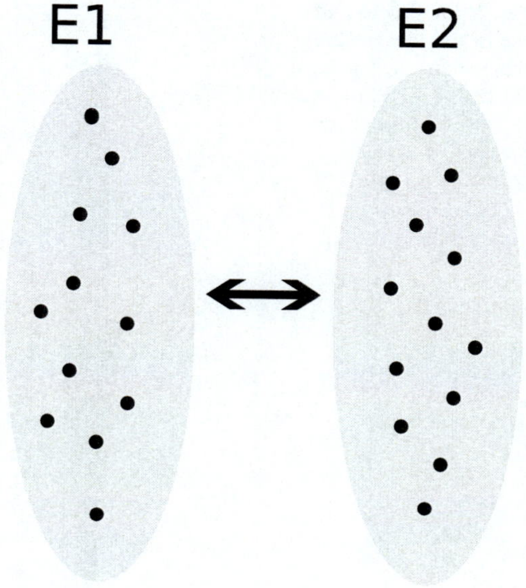

Fig. 1 Schematic representation of two biologically functional form of the protein $E1$ or $E2$. Dots represents substates, that realize the same biological function

dynamics as a time dependent concentrations of these two states, is non-exponential and depends on the initial values [6]. Relaxation function of the system, has a power law character. As explained at work [7] it is due to protein's self-similar energy landscape. In the Fig. 1 there is schematic representations of the idea. Each point in the E_1 or E_2 represents one conformational substate. One set of points represent one biologically—functional state of the protein.

One can see protein as a thermodynamic system operating in isothermal condition. Such treatment of the molecule is possible due to assumption, that the biomolecule is large enough to make a such thermodynamic description applicable. Isothermal conditions are assumed because most of biochemical reactions is much faster that changes of temperature in organism (for example during its circadian rhythm). Finite speed of change during biochemical transformations reason that assumption about quasi-static nature of this processes is very poor. Therfore transformtions between biologically-functional states has to fulfill laws of nonequilibrium thermodynamics. During this process the derivative of entropy in time consists with two components:

$$\frac{dS_{sys.}}{dt} = \frac{dS_e}{dt} + \frac{dS_i}{dt}, \tag{2}$$

where first term describes transfer velocity of entropy across the boundaries of the system and second describe velocity of entropy production within the system.

All processes in nature have to fullfils the second law of thermodynamics. Also chemical transformations, that constitued a methabolism of the organism, as a whole, occure in this way that the amount of chemical energy dissipated is positive.

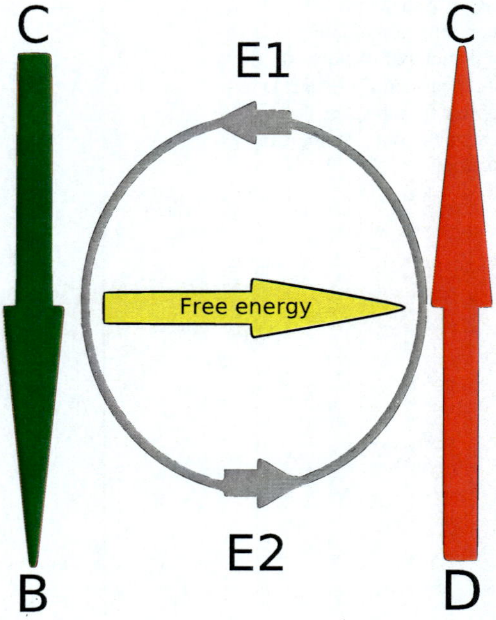

Fig. 2 Enzyme as a free energy transducer. Free energy donating chemical reaction $A \to B$. Reaction $D \to C$ is a free energy acceptor. It can also symbolize a work

However when one analyze a single reaction catalysing by enzyme one can see that some of them perform according to second law of theromdynamics, but other one are forced by them to proceed against the second law. Such situations can occur when two reactions are coupled by the same enzyme [8]. During this process first reaction transfers a part of its free energy recovered from dissipation to force the second reaction [9]. In the Fig. 2 there is schematic picture of this process. Reaction $A \to B$ (green arrow) is free energy donor, catalyse by biologically—functional state of ensyme E_1, and reaction $D \to C$ (red arrow) is a free energy aceptor, catalyze by biologically—functional state of enzyme E_2. Arrow (yellow) from reaction $A \to B$ to $D \to C$ represents a free energy transfer.

There are many examples of processes that take place in an organism according to the scenario presented above. As an example is the reaction of ATP hydrolysis which is coupled to transport of ions across membranes in the direction of increasing ion concentration. Such structures are called ion pump. Another example is the ATP hydrolysis that can result in a mechanical motion of single protein along microtubules or nucleic acid chains. Such proteins are called molecular motors. In this coupled reactions protein play a role of a free energy transducer [10]. Taking into account the physical meaning of changes of the free energy in isothermal conditions as a work, one can treat proteins as a chemochemical machines—an idea that is wildly presented in [6, 8, 9].

The results of the chemochemical machines operation can be measure on the micrometer spatial scale or sometimes macroscopically (for example muscle contraction). These results depend on its spatial organization in the cell. For example ionic pomps are embedded in the two—dimensional phospholipid membrain. They transport of a selected ions in one direction across the membrane which results in appearance a difference in the electrochemical potential between two side of the membrane. Another example is transport inside the cell. It is possible due to microtubules that are organized around centrosome (only in animal cell) [11] and molecular motors moving on them. Motors change the location of cell organelles and transports many substances in a small vesicles. Results of the transport are visible in microscope [12].

From perspective of the protein as a free energy transducer, catalyzing the dissipating reaction and then force other reaction or performing work are only a ways to changes its thermodynamical state. During this process protein changes its biologically—functioning state to another. If one come back to idea of a chain, it is equivalent to change one conformation, described by angles, to another. In this way, all processes in the cell and the thermomechanical properties of proteins can be considered through the use of common formalism.

To describe these changes only selected angles, important in the process, can be used. These n paramerters are noted by the $(\alpha_1, \ldots, \alpha_n)$. The one biological—functioning state of the protein is represented by specific configurations of angle ranges. Using these parameters time dependent density of probability can be define and then an unspecified currents in that space. Such currents are results of thermodynamic forces [13], which also have to be defined in that space. Using continuity equation with modified current and a mesoscopic nonequilibrium thermodynamics [14, 15] it is possible to obtain formula for derivative of entropy in time, which is similar to Eq. 2, but refers to more detailed description of the system.

2 Thermodynamical Considerations and Formalism

Dynamics of protein in the simplest case can be defined by a system of discrete master equation [6]:

$$\frac{d}{dt} P_i(t) = \sum_j \left[v_{ij} P_j(t) - v_{ji} P_i(t) \right], \tag{3}$$

where $P_i(t)$ represents a probability of a i state and the coefficient v_{ij} describe the transition probability per unit time from state j to state i.

Let's consider a sum of probabilities, which is connected with state A:

$$P_A = \sum_{i \in A} p_i \tag{4}$$

This quantity describe probability of a chemical state A. It is proportional to the molar concentrations $[A]$. Similar quantity P_B can be calculated for chemical state B and also it is proportional to the molar concentrations $[B]$.

Equation 3 can be written in a more compact way using appropriate matrices:

$$\frac{d}{dt}\hat{P}(t) = \hat{V}\hat{P}(t), \tag{5}$$

where $\hat{P}(t)$ is a vector that presents a probabilities of a states and matrix $\hat{V}$ obtains all coefficiens v_{ij}. Distinction of a chemical state, for example A, can be formally carried out by projection operator $\hat{A}$ which act on a vector $\hat{P}(t)$. This operation distinguishes a subset from a set of states according to formula $\hat{A}\hat{P}(t) = \hat{P}_A(t)$. The projector operator onto chemical state B can be seen as $\hat{B} = \hat{1} - \hat{A}$, where $\hat{1}$ is a identity operator. In this case we obtains a vector of probabilities for a subset representing chemical state B according to formula $\hat{B}\hat{P}(t) = \hat{P}_B(t)$.

One can use projection operator $\hat{A}$ and obtains equation of motion for $\hat{P}_A(t)$. To obtain this goal operator $\hat{A}$ on Eq. 3 and uses the Laplace transform technique. Equation of motion for $\hat{P}_A$ has a form [16]:

$$\frac{d}{dt}\hat{P}_A(t) = \hat{V}_A\hat{P}_A(t) + \int_0^t K(t-\tau)\hat{P}_A(\tau)d\tau + F(t), \tag{6}$$

where: $\hat{V}_A = \hat{A}\hat{V}$, $\hat{V}_B = \hat{B}\hat{V}$, $K(t) = \hat{V}_A \exp\left(\hat{V}_B t\right)\hat{V}_B$ and

$$F(t) = \hat{V}_A \exp\left(\hat{V}_B t\right)\hat{B}\hat{P}(0). \tag{7}$$

The last term in Eq. 6 represents influence of omitted states on evolution of state A. Probabilities connected with chemical state B appear only in Eq. 7 as a initial conditions $\hat{B}\hat{P}(0)$. It is possible to choose initial value in a such way that the last term vanish. Another possibilities is to treat $\hat{B}\hat{P}(0)$ as noise with average equal zero. That way $\hat{P}_A(t)$ satisfies its own equation. Now equation of evolution is non-Markovian. Second term on the right side of Eq. 6 there is term which represents a memory effect in the evolution.

A different view on dynamics of biomolecule, than this presented by Eq. 3, is to consider the biomolecule transformation as a progressively transforms. Biomolecule passing through successive molecular configurations. Then the continuous master equation should be consider as a starting point in simulations.

Let's introduce a vector of molecular parameters $\mathbf{X} = (x_1, \ldots, x_n)$. Time dependent probability density function of a molecular parameters $\mathbf{X}$ is denote by a symbol $\rho(\mathbf{X})$. For this density of probability multidimensional master equation has the following form:

$$\frac{\partial \rho(\mathbf{X}, t)}{\partial t} = \int_{\Gamma'} \left(W(\mathbf{X}|\mathbf{X}')\rho(\mathbf{X}', t) - W(\mathbf{X}'|\mathbf{X})\rho(\mathbf{X}, t)\right) d\mathbf{X}', \tag{8}$$

where Γ is a space of the parameters $\mathbf{X}$.

In this other approach the chemical states, A or B, can be distinguished as a subset in the Γ space. In this perspective here also we have a Markov process. In the simplest case is when there is one dimensional space, wchih represents reaction coordinate. Equation 8 can be write as a Fokker-Planck equation:

$$\frac{\partial \rho(x,t)}{\partial t} = -\frac{\partial}{\partial x}\left(A(x)\rho(x,t)\right) + \frac{\partial^2}{\partial x^2}\left(B(x)\rho(x,t)\right). \tag{9}$$

All processes in nature following rather Eq. 6 than Markov 3, therefore one can use generalization of Eq. 5 in the following form:

$$\frac{d}{dt}\hat{P}(s) =_0 D_t^{1-\alpha}\left[\hat{V}\hat{P}(t)\right]. \tag{10}$$

where:

$${}_0D_t^{1-\alpha} f(t) = \frac{1}{\Gamma(\alpha)}\left(\frac{\partial}{\partial t}\right)\int_0^t \frac{f(\tau)}{(t-\tau)^{1-\alpha}} d\tau \tag{11}$$

is the so-called Riemann-Liouville fractional derivative and $0 < \alpha < 1$ [17]. Memory effect is here incorporated by using this non-local operator. For continuos variables we can obtain continuos general master equation:

$$\frac{\partial \rho(x,t)}{\partial t} =_0 D_t^{1-\alpha}\left[\int_{-\infty}^{+\infty}\left(W(x|x')\rho(x',t) - W(x'|x)\rho(x,t)\right)dx'\right]. \tag{12}$$

This equation can be use to obtain Generalize Fokker-Planck equation [18, 19]:

$$\frac{\partial \rho(x,t)}{\partial t} =_0 D_t^{1-\alpha}\left[-\frac{\partial}{\partial x}\left(A(x)\rho(x,t)\right) + \frac{\partial^2}{\partial x^2}\left(B(x)\rho(x,t)\right)\right]. \tag{13}$$

In the approach treating the molecule as a chain simultaneously as a thermodynamic system, we can distinguish two types of parameters—angles. Some of them are equilibrated and have a certain average value. Another angles are not equilibrated and give the opportunity to define, mentioned in introduction, an important parameters $(\xi_1, \ldots, \xi_n)$ of the system during some process. Following [22] they measure differences between equilibrated angles and nonequlibrated angles:

$$\xi_i = \alpha_i - \alpha_i^0 \tag{14}$$

where:

$$\alpha_i^0 = \int_\Gamma \alpha_i P_{eq}(\mathbf{A})d\mathbf{A}, \tag{15}$$

represents values α_i in equlibrium state described by $P_{eq}(\mathbf{A})$. State of polymer is defined by time dependent density of probability $P(\boldsymbol{\xi}, t)$ in a space of molecular parameters $\boldsymbol{\xi} = (\xi_1, \ldots, \xi_n)$ necessary to determine its configuration. When the system is not at equilibrium, the distribution $P(\boldsymbol{\xi}, t)$ will change in time according to continuity equation:

$$\frac{\partial P(\boldsymbol{\xi}, t)}{\partial t} = -\nabla_{\xi} \cdot \mathbf{J}(\boldsymbol{\xi}, t), \tag{16}$$

where $\mathbf{J}(\boldsymbol{\xi}, t)$ is a current in a space of parameters $\boldsymbol{\xi}$ and ∇_{ξ} is a divergence defined in the same space.

Subdiffusion is describe by Eq. 13, which can be written in the form a modified continuity equation:

$$\frac{\partial P(\boldsymbol{\xi}, t)}{\partial t} = -{}_0D_t^{1-\alpha}\left[\nabla_{\xi} \cdot \mathbf{J}(\boldsymbol{\xi}, t)\right], \tag{17}$$

where ${}_0D_t^{1-\alpha}$ is Riemann-Liouvill derivative defined by Eq. 11.

Derivatives can be taken in any order, therefore subdiffusive Eq. 17 can be written as continuity equation with modification of the current:

$$\frac{\partial P(\boldsymbol{\xi}, t)}{\partial t} = -\nabla_{\xi} \cdot \tilde{\mathbf{J}}_i(\boldsymbol{\xi}, t), \tag{18}$$

where:

$$\tilde{\mathbf{J}}(\boldsymbol{\xi}, t) = {}_0D_t^{1-\alpha}\mathbf{J}(\boldsymbol{\xi}, t). \tag{19}$$

In nonequilibrium thermodynamics there is the Gibbs entropy postulate [13], which can be presented in a form [20]:

$$S = S_{eq} - k \int P(\boldsymbol{\xi}, t) \ln \frac{P(\boldsymbol{\xi}, t)}{P_{eq}(\boldsymbol{\xi})} d\boldsymbol{\xi}, \tag{20}$$

where $P_{eq}(\boldsymbol{\xi})$ is the equilibrium distribution, k—Boltzmann's constant and S_{eq} is the entropy of the equilibrium state. Differentiating expression 20 with respect to time one can obtains:

$$\frac{dS}{dt} = -k \int \frac{\partial P(\boldsymbol{\xi}, t)}{\partial t} \ln \frac{P(\boldsymbol{\xi}, t)}{P_{eq}(\boldsymbol{\xi})} d\boldsymbol{\xi}. \tag{21}$$

In Eq. 21 time derivative of $P(\boldsymbol{\xi}, t)$ can be replaced according to Eq. 17. Using properties of divergence of the n dimensional vector field, which is defined in arbitrary coordinate, one can obtain:

$$\frac{dS}{dt} = -k \int \nabla_{\xi} \cdot \mathbf{J}_S(\xi, t) d\xi + \sigma, \tag{22}$$

where:

$$\mathbf{J}_S(\xi, t) = \tilde{\mathbf{J}}(\xi, t) \ln \frac{P(\xi, t)}{P_{eq}(\xi)} \tag{23}$$

is a flux of entropy exchange with an environment and

$$\sigma = -k \int \tilde{\mathbf{J}}(\xi, t) \cdot \nabla_{\xi} \ln \frac{P(\xi, t)}{P_{eq}(\xi)} d\xi \tag{24}$$

presents an entropy production in a system. Gradient in 24 is a vector:

$$\nabla_{\xi} \ln \frac{P(\xi, t)}{P_{eq}(\xi)} = \left(\nabla_{\xi_1} \ln \frac{P(\xi, t)}{P_{eq}(\xi)}, \nabla_{\xi_2} \ln \frac{P(\xi, t)}{P_{eq}(\xi)}, \ldots, \nabla_{\xi_n} \ln \frac{P(\xi, t)}{P_{eq}(\xi)} \right). \tag{25}$$

This notation allow to write entropy production in the following form:

$$\sigma = -k \sum_{i=1}^{n} \int \tilde{\mathbf{J}}_i(\xi, t) \cdot \nabla_{\xi_i} \ln \frac{P(\xi, t)}{P_{eq}(\xi)} d\xi_i. \tag{26}$$

Equation 26 is a sum of products, each of which consists with flux vector component and thermodynamic force component:

$$\mathbf{A}_i = -k \nabla_{\xi_i} \ln \frac{P(\xi, t)}{P_{eq}(\xi)}, \tag{27}$$

which is associated with ξ_i coordinate. Equation 26 is extension a one-dimension case presented in [21]. One can say that thermodynamic forces are not affected by subdiffusive environment however the fluxes are affected.

3 Conclusions

Presented in this work formalism can be applied to many kind polymers, in particular to protein, where there are natural ways of parametrization of a molecules. It combines many other ideas presented in literature into one idea, where protein (polymer) is a nonequilibrium termodynamical system, free energy transducer and chain. Such protein operate in subdiffusive environment therfore this formalism can be treated as extension of mesoscopic nonequilibrium thermodynamics formalism.

References

1. Stryer, L.: Biochemistry. W.E. Freeman, New York (1995)
2. Keener, J., Sneyd, J.: Mathematical Physiology. Springer, New York (1998)
3. Liwo, A., Ołdziej, S., Kaźmierkiewicz, R., Groth, M., Czaplewski, C.: Design of a knowledge-based force filed for off-latice simulations of protein structure. Acta Biochem. Pol. **44**(3), 527–548 (1997)
4. Bełdowski, P., Weber, P., Dedinaite, A., Claesson, P., Gadomski, A.: Physical crosslinking of hyaluronic acid in the presence of phospholipids in an aqueous nano-environment. Soft Matt. **14**, 8997–9004 (2018)
5. Jung, G., Hanke, M., Schmid, F.: Generalized langevin dynamics: construction and numerical integration of non-Markovian particle based models. Soft Matt. **14**, 9368–9382 (2018)
6. Kurzyński, M., Chełminiak, P. Mean first-passage time in the stochastic theory of biochemical processes. Application to actomyosin molecular motor. J. Stat. Phys. **110**(112), 137–180 (2003)
7. Glöckle, W.G., Nonnenmacher, T.F.: A fractional calculus approach to self-similar protein dynamics. Biophys. J. 68(1), 46–53 (1995)
8. Hill, T.L.: Free Energy Transduction and Biochemical Cycle Kinetics. Springer, New York (1989)
9. Kurzyński, M.: Statistical physics of bilogical molecular machines. Acta Phys. Pol. B **36**(5), 1663–1675 (2005)
10. Kurzyński M.: The Thermodynamic Machinery of Life. Springer, Berlin (2006)
11. Solomon, E.P., Berg, L.R., Martin, D.W.: Biology. Belmont, Brooks/Cole, Thomson Learning, San Diego (2006)
12. Hancock W.O.: Bidirectional cargo transport: moving beyond tug-of-war. Nat. Rev. Mol. Cell Biol. **15**(9), 615–628 (2014)
13. De Groot, S.R., Mazur, P.: Non-equilibrium Thermodynamics. North-Holland Publishing, Amsterdam (1962)
14. Gadomski, A., Siódmiak, J., Santamaria-Holek, I., Rubi, J.M., Ausloos, M.: Kinetics of growth process controlled by mass-convective fluctuations and finite-size curvature effects. Acta Phys. Pol. B **36**, 1537–1559 (2005)
15. Perez-Madrid, A., Rubi, J.M., Mazur, P.: Brownian motion in the presence of a temperature gradient. Physica A **212**, 231–238 (1994)
16. Zwanzig, R.: Nonequilibrium Statistical Mechanics. Oxford University Press, Oxford (2001)
17. Podlubny, I.: Fractional Differential Equations. Academic, San Diego (1999)
18. Metzler, R., Klafter, J.: The random walk's guide to anomalous diffusion: a fractional dynamics approach. Phys. Rep. **339**, 1–77 (2000)
19. Metzler, R., Klafter, J.: The restaurant at the end of the random walk: recent developments in the description of anomalous transport by fractional dynamics. J. Phys. A Math. Gen. **37**, R161–R208 (2004)
20. Rubi, J.M., Perez-Madrid, A.: Mesoscopic non-equilibrium thermodynamics approach to the dynamics of polymers, Physica A **298**, 177–186 (2001)
21. Weber, P., Bełdowski P., Bier, M., Gadomski, A.: Entropy production associated with aggregation into granules in a subdiffusive environment. Entropy **20**(9), 1–5 (2018)
22. Mazur, P.: Fluctuations and non-equilibrium thermodynamics. Physica A **261**, 451–457 (1998)

Evaluation of the Crane's Actuators Strength Based on the Results Obtained from Dynamics Model

Andrzej Urbaś and Krzysztof Augustynek

Abstract The strength analysis of the crane's actuators is presented in the paper. The analysis is performed using the loads obtained from the dynamics analysis. The mathematical model of the flexible supported crane is formulated using the Lagrange equations of the second kind. The main structure of the crane is built of the five bodies forming an open-loop kinematic chain. The actuators form the closed-loop kinematics chains. The crane performs an assumed motion aimed at transferring a load in the form of lumped mass of various values. The mathematical model takes into account the jib's flexibility, which is discretized by means of the Rigid Finite Element Method (RFEM). The formalism of the joint coordinates and homogeneous transformation matrices are used to describe the crane's kinematics. The equations of motion are supplemented by the constraint equations formulated for the cut-joints. The Lagrange multipliers corresponding to reaction forces at the cut-joints, are used to the actuators' quasi-statics analysis. The Finite Element Method (FEM) is used to model the actuators' flexibility. The numerical calculations present the influence of the jib's flexibility and load's mass on the maximum stresses due to the deformations of the actuators at a given crane's working moment.

Keywords Crane · Dynamics analysis · Quasi-statics analysis

1 Introduction

Nowadays it can be observed a tendency to design transportation devices like cranes whose construction is light, can carry higher loads, can be efficiently controlled and satisfy safety requirements. Mathematical models for statics and dynamics allow us to evaluate deformations and stresses of the cranes' components forced by

A. Urbaś (✉) · K. Augustynek
Department of Mechanical Engineering Fundamentals, University of Bielsko-Biala, Bielsko-Biala, Poland
e-mail: aurbas@ath.bielsko.pl; kaugustynek@ath.bielsko.pl

J. Awrejcewicz (ed.), *Perspectives in Dynamical Systems II: Mathematical and Numerical Approaches*, Springer Proceedings in Mathematics & Statistics 363,
https://doi.org/10.1007/978-3-030-77310-6_10

operating conditions at the design stage. There are many papers devoted to model cranes' dynamics [1–12]. Analyzing literature, two approaches to modeling crane dynamics can be distinguished lumped mass and distributed mass models [12]. The first approach is the most widely used due to applied simplifications: the hoisting rope is treated as a massless cable and the load with the hook is modelled as a mass point. The flexibility of the carried structure can be approximated using the Finite Element Method [13] together with modal reduction methods [5]. Another method, developed mainly in Poland, is the Rigid Finite Element Method [8–11, 14]. In this method, a flexible link is replaced by means of a set of rigid elements interconnected by means of spring damping elements.

This paper is a continuation of the previous ones dedicated to develop models of the cranes with flexible supports, drive and links and also frictions in joints [8–11]. In first models, actuators were omitted and the carrier structure was modelled as the open-loop kinematics chains [8, 10]. In further works, the models were developed by introduction force actuators [9, 11]. As a result, cranes form the structure of the closed-loop kinematic chain. The force actuators were treated as rigid and their motion was enforced using assumed driving functions. The contribution of the paper is the proposed method of evaluating the elastic deformations of the actuators. In the proposed approach, deformations of the actuators in each integrations step are calculated by a solution of the quasi-statics problem formulated for replacement models of actuators. The replacement models are obtained by means of the Finite Element Method using planar beam elements with 3 dof in each node [13]. The joint coordinates together with homogeneous transformations matrices are used to describe the kinematics of the crane. The dynamics equations of motion are derived using the Lagrange equations of the second kind. In the numerical simulations, the stresses due to the deformations of the actuators during the motion of the crane are analyzed.

2 Dynamics Analysis of the Crane

Figure 1 presents the simplified model of the crane. The crane contains the main structure (m_c) mounted on the chassis (c). The chassis is supported by means of eight supports ($n_s = 8$).

The main structure contains four bodies $\left(n_b^{(m_c)} = 4\right)$ which form the open-loop kinematic chain. Additional the main structure consists the actuators (two closed-loop kinematics sub-chains $\left(s_{c_i}\right)\big|_{i=1,2}$). Each of the actuator contains two bodies $\left(n_b^{(s_{c_i})}\Big|_{i=1,2} = 2\right)$. The bodies of the crane are driven by means of torque $\left(\mathbf{t}_{dr}^{(m_c,1)}\right)$ and forces $\left(\mathbf{f}_{dr}^{(s_{c_1},2)}, \mathbf{f}_{dr}^{(s_{c_2},2)}, \mathbf{f}_{dr}^{(m_c,4)}\right)$.

The carried load (l) is treated as a mass point connected with the main structure by flexible rope (r). It is assumed that link (m_c, 2) can be treated as flexible.

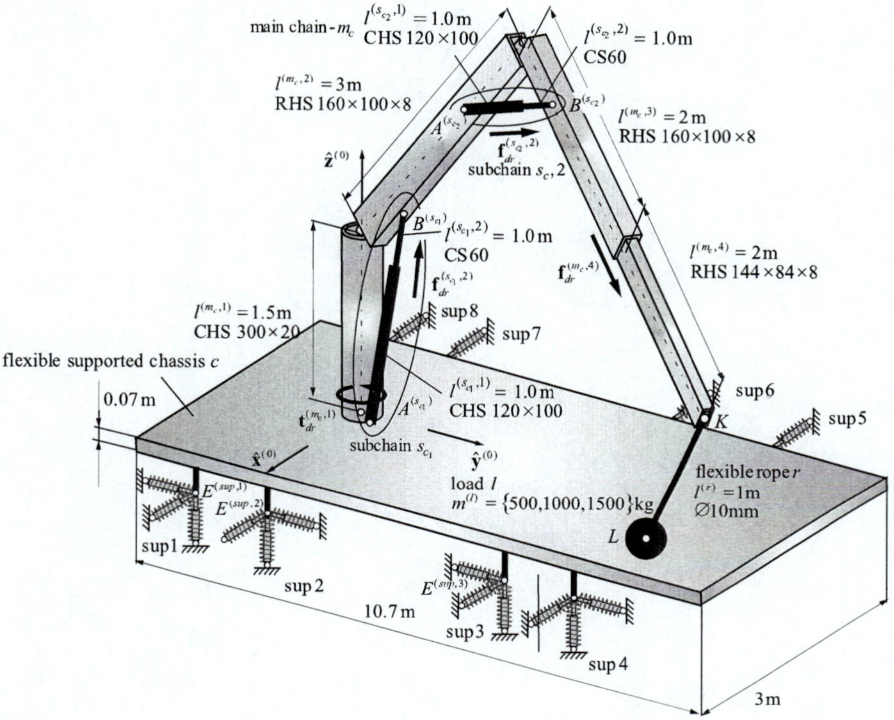

Fig. 1 Model of the crane

2.1 *Modelling the Flexibility of the Link: RFEM*

The Rigid Finite Element Method is used to discretise $(m_c, 2)$ link [14]. As a result, the flexible link is replaced by the system of rigid elements $\left(\text{rfe }(m_c, 2, r)\Big|_{r=1,\ldots,n_{rfe}^{(m_c,2)}}\right)$ interconnected by means of spring-damping elements $\left(\text{sde }(m_c, 2, r)\Big|_{r=1,\ldots,n_{sde}^{(m_c,2)}}\right)$—Fig. 2.

2.2 *Generalized Coordinates and Homogeneous Transformation Matrices*

The kinematics of the crane is described by the joint coordinates formalism and homogeneous transformation matrices [15]. Figure 3 shows model of the crane with assigned local frames and assumed joint coordinates.

The vector of the generalised (joint) coordinates can be written in the following form:

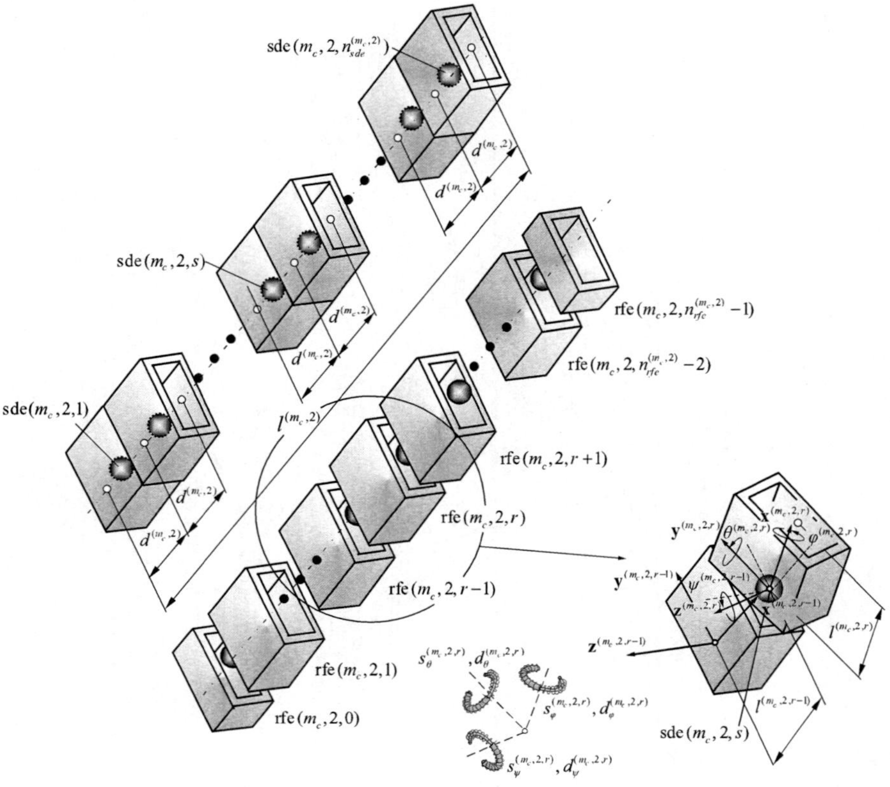

Fig. 2 Discretization link $(m_c, 2)$—RFEM

$$\mathbf{q} = \left(q_j\right)_{j=1,\ldots,n_{dof}} = \left[\mathbf{q}^{(c)^T} \ \overline{\mathbf{q}}^{(m_c)^T} \ \overline{\mathbf{q}}^{(s_{c_1})^T} \ \overline{\mathbf{q}}^{(s_{c_2})^T} \ \mathbf{q}^{(l)^T}\right]^T, \tag{1}$$

where:

- chassis c: $\mathbf{q}^{(c)} = \left(q^{(c)}\right)_{j=1,\ldots,n_{dof}^{(c)}} = \left[x^{(c)} \ y^{(c)} \ z^{(c)} \ \psi^{(c)} \ \theta^{(c)} \ \varphi^{(c)}\right]^T$,
- main chain m_c: $\mathbf{q}^{(m_c)} = \left(q^{(m_c)}\right)_{j=1,\ldots,n_{dof}^{(m_c)}} = \left[\mathbf{q}^{(c)^T} \ \overline{\mathbf{q}}^{(m_c)^T}\right]^T$,
  $\overline{\mathbf{q}}^{(m_c)} = \left[\tilde{\mathbf{q}}^{(m_c,1)^T} \ \tilde{\mathbf{q}}^{(m_c,2)^T} \ \tilde{\mathbf{q}}^{(m_c,3)^T} \ \tilde{\mathbf{q}}^{(m_c,4)^T}\right]^T$, $\tilde{\mathbf{q}}^{(m_c,1)} = \left[\psi^{(m_c,1)}\right]$,
  $\tilde{\mathbf{q}}^{(m_c,2)} = \left[\tilde{\mathbf{q}}^{(m_c,2,0)^T} \ldots \tilde{\mathbf{q}}^{(m_c,2,r)^T} \ldots \tilde{\mathbf{q}}^{\left(m_c,2,n_{rfe}^{(m_c,2)}\right)^T}\right]^T$, $\tilde{\mathbf{q}}^{(m_c,2,0)} = \left[\psi^{(m_c,2,0)}\right]$,
  $\tilde{\mathbf{q}}^{(m_c,2,r)}\Big|_{r=1,\ldots,n_{rfe}^{(m_c,2)}} = \left[\psi^{(m_c,2,r)} \ \theta^{(m_c,2,r)} \ \varphi^{(m_c,2,r)}\right]^T$, $\tilde{\mathbf{q}}^{(m_c,3)} = \left[\psi^{(m_c,3)}\right]$,
  $\tilde{\mathbf{q}}^{(m_c,4)} = \left[z^{(m_c,4)}\right]$

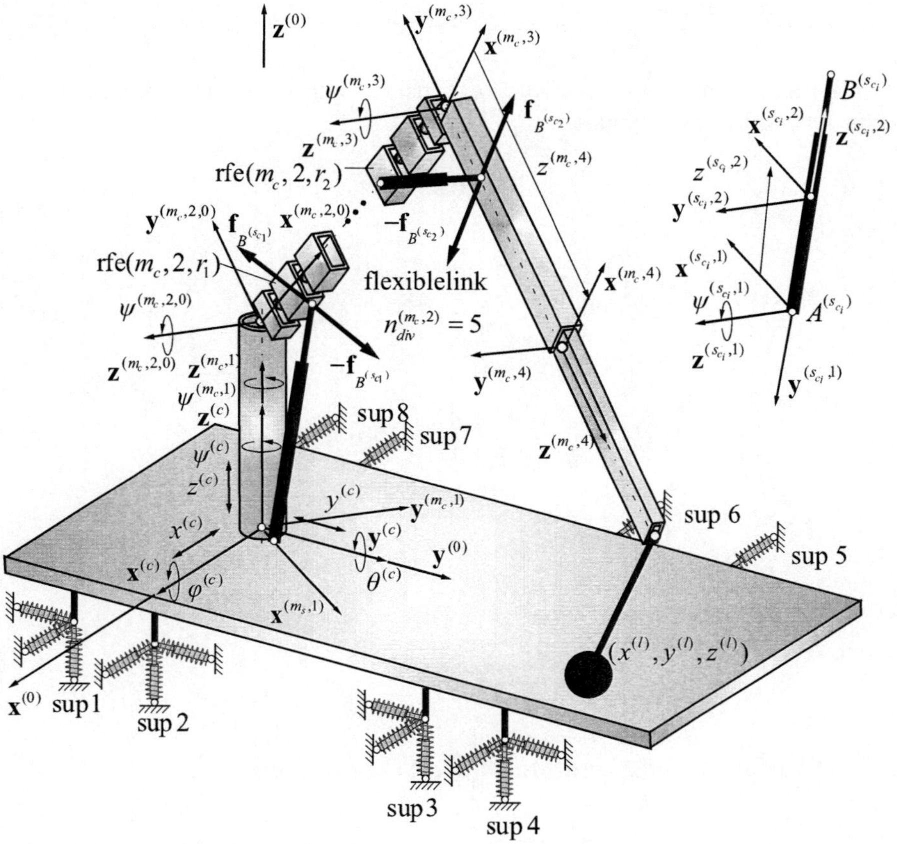

Fig. 3 Joint coordinates

– sub-chain (s_{c_1}): $\mathbf{q}^{(s_{c_1})} = \left(q_j^{(s_{c_1})}\right)_{j=1,\dots,n_{dof}^{(s_{c_1})}} = \left[\mathbf{q}^{(c)^T}\ \tilde{\mathbf{q}}^{(m_c,1)^T}\ \overline{\mathbf{q}}^{(s_{c_1})^T}\right]^T$,

$\overline{\mathbf{q}}^{(s_{c_1})} = \left(\overline{q}_j^{(s_{c_1})}\right)_{j=1,\dots\overline{n}_{dof}^{(s_{c_1})}} = \left[\tilde{\mathbf{q}}^{(s_{c_1},1)^T}\ \tilde{\mathbf{q}}^{(s_{c_1},2)^T}\right]^T$, $\tilde{\mathbf{q}}^{(s_{c_1}1)} = \left[\psi^{(s_{c_1},1)}\right]$,

$\tilde{\mathbf{q}}^{(s_{c_1},2)} = \left[z^{(s_{c_1},2)}\right]$

– sub-chain (s_{c_2}): $\mathbf{q}^{(s_{c_2})} = \left(q_j^{(s_{c_2})}\right)_{j=1,\dots n_{dof}^{(s_{c_2})}}$

$= \left[\mathbf{q}^{(c)^T}\ \tilde{\mathbf{q}}^{(m_c,1)^T}\ \tilde{\mathbf{q}}^{(m_c,2,0)^T}\ \dots\ \tilde{\mathbf{q}}^{(m_c,3,r_2)^T}\ \overline{\mathbf{q}}^{(s_{c_2})^T}\right]^T$,

$\overline{\mathbf{q}}^{(s_{c_2})} = \left(\overline{q}_j^{(s_{c_2})}\right)_{j=1,\dots\overline{n}_{dof}^{(s_{c_2})}} = \left[\tilde{\mathbf{q}}^{(s_{c_2},1)^T}\ \tilde{\mathbf{q}}^{(s_{c_2},2)^T}\right]^T$, $\tilde{\mathbf{q}}^{(s_c,2,1)} = \left[\psi^{(s_c,2,1)}\right]$,

$\tilde{\mathbf{q}}^{(s_{c_2},2)} = \left[z^{(s_{c_2},2)}\right]$

- load l: $\mathbf{q}^{(l)} = \left(q_j^{(l)}\right)_{j=1,\ldots,n_{dof}^{(l)}} = \left[\, x^{(l)} \; y^{(l)} \; z^{(l)} \,\right]^T$.

The transformation matrices from local frames to the global reference frame are defined by following formulas:

$$\mathbf{T}^{(c)} = \tilde{\mathbf{T}}^{(c)}, \tag{2a}$$

$$\mathbf{T}^{(m_c,b)}\Big|_{b=1,\ldots,n_b^{(m_c)}} = \mathbf{T}^{(m_c,b-1)}\tilde{\mathbf{T}}^{(m_c,b)}, \tag{2b}$$

$$\mathbf{T}^{(s_{c_1},b)}\Big|_{b=1,\ldots,\tilde{n}_b^{(s_{c_1})}} = \mathbf{T}^{(m_c,1)}\tilde{\mathbf{T}}^{(s_{c_1},b)}, \tag{2c}$$

$$\mathbf{T}^{(s_{c_2},b)}\Big|_{b=1,\ldots,\tilde{n}_b^{(s_{c_2})}} = \mathbf{T}^{(m_c,2,r_2)}\tilde{\mathbf{T}}^{(s_{c_2},b)}, \tag{2d}$$

where: $\mathbf{T}^{(m_c,0)} = \mathbf{T}^{(c)}$.

2.3 *The Lagrange Equation of the Second Kind*

The dynamics equations are derived using the formalism of Lagrange equations of the second kind [14]:

$$\frac{\mathrm{d}}{\mathrm{d}t}\frac{\partial E_k}{\partial \dot{\mathbf{q}}} - \frac{\partial E_k}{\partial \mathbf{q}} + \frac{\partial E_p}{\partial \mathbf{q}} + \frac{\partial R}{\partial \dot{\mathbf{q}}} = \mathbf{0}, \tag{3}$$

where: E_k is kinetic energy of the system, $E_p = E_{p,g} + E_{p,s}^{(\mathrm{sup})} + E_{p,s}^{(fl)} + E_{p,s}^{(dr)} + E_{p,s}^{(r)}$ is potential energy of gravity forces and spring deformations (supports, link, drives and rope), $R = R^{(\mathrm{sup})} + R^{(fl)} + R^{(dr)} + R^{(r)}$ is the Rayleigh function. Detailed formulas for determining components of Eq. (3) are presented in the following sections.

2.4 *Kinetic Energy and Potential Energy of Gravity Forces*

The total kinetic energy of the crane E_k is a sum of the kinetic energy of all subsystems and can be determined in the following way [14, 16]:

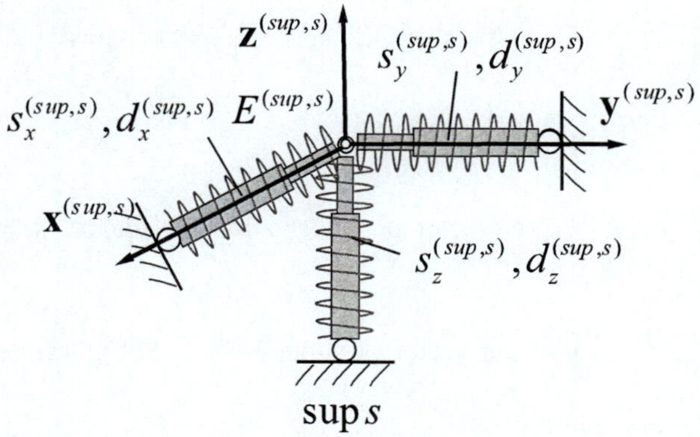

Fig. 4 Model of support

$$E_k = E_k^{(c)} + E_k^{(m_c)} + \sum_{i=1}^{n_{sc}} E_k^{(s_{c_i})} + E_k^{(l)}, \tag{4}$$

where: $E_k^{(\bullet)} = \frac{1}{2}\text{tr}\left\{\dot{\mathbf{T}}^{(\bullet)}\mathbf{H}^{(\bullet)}\dot{\mathbf{T}}^{(\bullet)^T}\right\}$, $\mathbf{H}^{(\bullet)}$ is pseudo-inertia matrix.

The potential energy of gravity forces can be calculated in a similar way:

$$E_{p,g} = E_{p,g}^{(c)} + E_{p,g}^{(m_c)} + \sum_{i=1}^{n_{sc}} E_{p,g}^{(s_{c_i})} + E_{p,g}^{(l)}, \tag{5}$$

where: $E_{p,g}^{(\bullet)} = m^{(\bullet)} g \mathbf{j}_3 \mathbf{T}^{(\bullet)} \mathbf{r}_{C^{(\bullet)}}^{(\bullet)}$, $E_{p,g}^{(l)} = m^{(l)} g$, $m^{(\bullet)}$ is mass of body, g is gravity acceleration, $\mathbf{j}_3 = \begin{bmatrix} 0 & 0 & 1 & 0 \end{bmatrix}$, $\mathbf{r}_{C^{(\bullet)}}^{(\bullet)}$ is position vector of centre of mass.

2.5 *Modelling of the Supports*

The supports $\left(\sup s|_{s=1,\dots,n_s}\right)$ are modelled by means of one directional spring-damping elements (Fig. 4).

Forces due to element deformations ($\mathbf{e}^{(sup,s)}$) are included in the equations of motion as the vector generalized forces in the following form:

$$\mathbf{s}^{(\text{sup})} = \left(s_i^{(\text{sup})}\right)_{i=1,\dots,n_{dof}} = \left(\frac{\partial E_{p,s}^{(\text{sup})}}{\partial \mathbf{q}} + \frac{\partial R^{(\text{sup})}}{\partial \dot{\mathbf{q}}}\right), \tag{6}$$

where: $E_{p,s}^{(\text{sup})} = \frac{1}{2}\sum_{s=1}^{n_{sup}} \left(\mathbf{e}^{(\text{sup},s)}\right)^T \mathbf{S}^{(\text{sup},s)} \mathbf{e}^{(\text{sup},s)}$, $R^{(\text{sup})} = \frac{1}{2}\sum_{s=1}^{n_{sup}} \left(\dot{\mathbf{e}}^{(\text{sup},s)}\right)^T \mathbf{D}^{(\text{sup},s)} \dot{\mathbf{e}}^{(\text{sup},s)}$,

$$s_i^{(\text{sup})} = \begin{cases} \sum\limits_{s=1}^{n_{sup}} \left(\left(\frac{\partial \mathbf{e}^{(\text{sup},s)}}{\partial q_i} \right)^T \mathbf{S}^{(\text{sup},s)} \mathbf{e}^{(\text{sup},s)} + \left(\frac{\partial \dot{\mathbf{e}}^{(\text{sup},s)}}{\partial \dot{q}_i} \right)^T \mathbf{D}^{(\text{sup},s)} \dot{\mathbf{e}}^{(\text{sup},s)} \right) & \text{if } q_i \in \mathbf{q}^{(c)}, \\ 0 & \text{otherwise,} \end{cases}$$

$\mathbf{e}^{(\text{sup},s)} = \mathbf{J}\mathbf{T}^{(m_c,1)} \mathbf{r}_{E(\text{sup},s)}^{(m_c,1)}$ is vector of deformation of support, $\mathbf{J} = \begin{bmatrix} \mathbf{j}_1 \\ \mathbf{j}_2 \\ \mathbf{j}_3 \end{bmatrix} = \begin{bmatrix} 1 & 0 & 0 & 0 \\ 0 & 1 & 0 & 0 \\ 0 & 0 & 1 & 0 \end{bmatrix}$, $\mathbf{r}_{E(\text{sup},s)}^{(m_c,1)}$ is position vector of sup s, $\mathbf{S}^{(sup,s)}$, $\mathbf{D}^{(sup,s)}$ are stiffness and damping matrices of sup s.

2.6 *Modelling of the Flexible Body*

Flexible deformation forces due to link flexibility are included in the equations of motion as the vector generalized forces in the following form:

$$\mathbf{s}^{(fl)} = \left(s_i^{(fl)} \right)_{i=1,\dots,n_{dof}} = \left(\frac{\partial E_{p,s}^{(fl)}}{\partial \mathbf{q}} + \frac{\partial R^{(fl)}}{\partial \dot{\mathbf{q}}} \right), \tag{7}$$

where: $E_{p,s}^{(fl)} = \frac{1}{2} \sum\limits_{s=1}^{n_{sde}^{(m_c,2)}} \left(\tilde{\mathbf{q}}^{(m_c,2,s)} \right)^T \mathbf{S}^{(m_c,2,s)} \tilde{\mathbf{q}}^{(m_c,2,s)}$, $R^{(fl)} = \frac{1}{2} \sum\limits_{s=1}^{n_{sde}^{(m_c,2)}} \left(\dot{\tilde{\mathbf{q}}}^{(m_c,2,s)} \right)^T \mathbf{D}^{(m_c,2,s)} \dot{\tilde{\mathbf{q}}}^{(m_c,2,s)}$,

$$s_i^{(fl)} = \begin{cases} \sum\limits_{s=1}^{n_{sde}^{(m_c,2)}} \left(\frac{\partial \tilde{\mathbf{q}}^{(m_c,2,s)}}{\partial q_i} \right)^T \mathbf{S}^{(m_c,2,s)} \tilde{\mathbf{q}}^{(m_c,2,s)} + \left(\frac{\partial \dot{\tilde{\mathbf{q}}}^{(m_c,2,s)}}{\partial \dot{q}_i} \right)^T \mathbf{D}^{(m_c,2,s)} \dot{\tilde{\mathbf{q}}}^{(m_c,2,s)} \\ \qquad \text{if } q_i \in \left\{ \tilde{\mathbf{q}}^{(m_c,2,1)}, \dots, \tilde{\mathbf{q}}^{\left(m_c,2,n_{rfe}^{(m_c,2)}-1\right)} \right\}, \\ 0 \qquad \text{otherwise,} \end{cases}$$

$\mathbf{S}^{(m_c,2,s)}$, $\mathbf{D}^{(m_c,2,s)}$ are stiffness and damping matrices of sde $(m_c, 2, s)$.

2.7 *Modelling of the Flexible Drives*

The selected links of the crane are driven by means of the flexible drives (Fig. 5).

The forces due to driving torque $\left(t_{dr}^{(m_c,1)} \right)$ and driving forces $\left(f_{dr}^{(\alpha)} \Big|_{\alpha \in \left\{ (m_c,4), \left(s_c,1,2 \right), \left(s_c,2,2 \right) \right\}} \right)$ can be presented as vector generalised forces

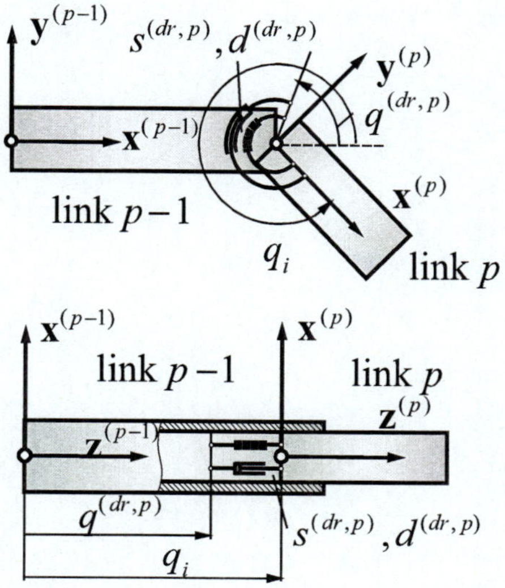

Fig. 5 Model of drive

vector in the following form:

$$\mathbf{s}^{(dr)} = \left(s_i^{(dr)}\right)_{i=1,\ldots,n_{dof}} = \left(\frac{\partial E_{p,s}^{(dr)}}{\partial \mathbf{q}} + \frac{\partial R^{(dr)}}{\partial \dot{\mathbf{q}}}\right), \tag{8}$$

where:

$$E_{p,s}^{(dr)} = \frac{1}{2} s^{(dr,1)} \left(\psi^{(dr,1)} - \psi^{(m_c,1)}\right)^2 + \frac{1}{2} s^{(dr,4)} \left(z^{(dr,4)} - z^{(m_c,4)}\right)^2 + \frac{1}{2} \sum_{i \in \{1,2\}} s^{(dr,2)} \left(z^{(dr,2)} - z^{(s_{c_i},2)}\right)^2$$

$$R^{(dr)} = \frac{1}{2} d^{(dr,1)} \left(\dot{\psi}^{(dr,1)} - \dot{\psi}^{(m_c,1)}\right)^2 + \frac{1}{2} d^{(dr,4)} \left(\dot{z}^{(dr,4)} - \dot{z}^{(m_c,4)}\right)^2 + \frac{1}{2} \sum_{i \in \{1,2\}} d^{(dr,2)} \left(\dot{z}^{(dr,2)} - \dot{z}^{(s_{c_i},2)}\right)^2$$

$$s_i^{(dr)} = \begin{cases} -\left(s^{(dr,\bullet)}\left(q^{(dr,\bullet)} - q_i\right) + d^{(dr,\bullet)}\left(\dot{q}^{(dr,\bullet)} - \dot{q}_i\right)\right) & \text{if } q_i \in \left\{\psi^{(m_c,1)}, z^{(m_c,4)}, z^{(s_{c_1},2)}, z^{(s_{c_2},2)}\right\}, \\ 0 & \text{otherwise.} \end{cases}$$

2.8 *Modelling of the Load and Rope*

It is assumed that the load is suspended on the flexible rope (Fig. 6). Forces due to rope deformation ($e^{(r)}$) are included in the equations of motion as the vector generalized forces as follows:

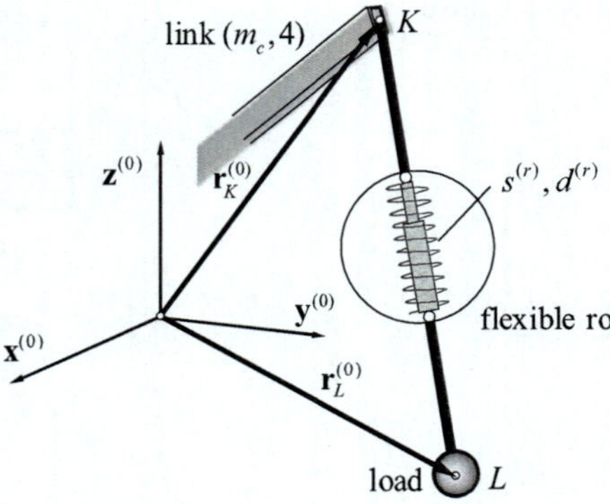

Fig. 6 Model of rope

$$\mathbf{s}^{(r)} = \left(s_i^{(r)}\right)_{i=1,\dots,n_{dof}} = \left(\frac{\partial E_{p,s}^{(r)}}{\partial \mathbf{q}} + \frac{\partial R^{(r)}}{\partial \dot{\mathbf{q}}}\right), \tag{9}$$

where: $E_{p,s}^{(r)} = \frac{1}{2}\delta^{(r)} s^{(r)} \left(e^{(r)}\right)^2$, $R^{(r)} = \frac{1}{2}\delta^{(r)} d^{(r)} \left(\dot{e}^{(r)}\right)^2$,

$$s_i^{(r)} = \begin{cases} \delta^{(r)} \left(s^{(r)} \frac{e^{(r)}}{l^{(r)}} \mathbf{r}_{KL}^T + d^{(r)} \dot{\mathbf{r}}_{KL}^T\right) \mathbf{J}\mathbf{T}_i^{(m_c,4)} \mathbf{r}_K^{(m_c,4)} & \text{for } q_i \in \mathbf{q}^{(m_c)}, \\ -\,\delta^{(r)} \left(s^{(r)} \frac{e^{(r)}}{l^{(r)}} \mathbf{r}_{KL}^T + d^{(r)} \dot{\mathbf{r}}_{KL}^T\right) \mathbf{j}_i, & \text{for } q_i \in \mathbf{q}^{(l)}, \end{cases}$$

$$e^{(r)} = l^{(r)} - l_0^{(r)},\ \delta^{(r)} = \begin{cases} 1, e^{(r)} > 0 \\ 0, e^{(r)} \le 0 \end{cases},\ l^{(r)} = \sqrt{\mathbf{r}_{KL}^T \mathbf{r}_{KL}}, \mathbf{r}_{KL} = \mathbf{J}\left(\mathbf{r}_K^{(0)} - \mathbf{r}_L^{(0)}\right).$$

2.9 Equations of Motion

The dynamics equations of motion with the constraint equations can be written in the following general form [8–11, 14]:

$$\begin{bmatrix} \mathbf{M}(\mathbf{q}) & -\mathbf{C}(\mathbf{q}, \dot{\mathbf{q}})^T \\ \mathbf{C}(\mathbf{q}, \dot{\mathbf{q}}) & \mathbf{0} \end{bmatrix} \begin{bmatrix} \ddot{\mathbf{q}} \\ \mathbf{f}_j \end{bmatrix} = \begin{bmatrix} -\mathbf{e}(\mathbf{q}, \dot{\mathbf{q}}) - \mathbf{s}(\mathbf{q}, \dot{\mathbf{q}}) \\ \mathbf{c}(\mathbf{q}, \dot{\mathbf{q}}) \end{bmatrix}, \tag{10}$$

where: $\mathbf{M}(\mathbf{q})$ is mass matrix, $\mathbf{C}(\mathbf{q}, \dot{\mathbf{q}})$ is constraint matrix, $\mathbf{f}_j$ is the vector of the reaction forces in the cut-joints, $\mathbf{e}(\mathbf{q}, \dot{\mathbf{q}})$ is the vector of the Coriolis, gyroscopic and centrifugal forces, $\mathbf{s}(\mathbf{q}, \dot{\mathbf{q}}) = \mathbf{s}^{(\text{sup})} + \mathbf{s}^{(fl)} + \mathbf{s}^{(dr)} + \mathbf{s}^{(r)}$ is the vector of the spring and damping forces formulated for the supports, link, drives and rope, $\mathbf{c}(\mathbf{q}, \dot{\mathbf{q}})$ is the vector of the right sides of constraint equations.

The dynamics of the crane forms the set of differential-algebraic equations with index 1. The dynamics equations of motion and state equations, are integrated using the Runge-Kutta method of the fourth order with a constant step size.

3 Quasi-Statics Analysis of the Actuators: FEM

The proposed model of the actuator is presented in Fig. 7. The Finite Element Method is used to model its flexibility [13]. It is assumed that actuator s_{c_i} is forced by reaction force acting in cut-joint ($\mathbf{f}_{B^{(s_{c_i})}}$) and described in α system and gravity forces due to the weight of the actuator. It is assumed that the flexibility of seals is omitted.

The cylinder body $(s_{c_i}, 1)$, piston rod $(s_{c_i}, 2)$ and common part of cylinder and rod (s_{c_i}, c) are discretized by means of a beam element with 3-dof in the node. The number of the deformable finite elements (dfe) is equal to $n_{dfe}^{(s_{c_i})} = n_{dfe}^{(s_{c_i},1)} + n_{dfe}^{(s_{c_i},2)} + n_{dfe}^{(s_{c_i},c)}$.

The nodal displacement vector of dfe $(s_{c_i}, 1, k)$ is defined as follows:

$$\mathbf{u}^{(s_{c_i},1,k)} = \left[x_i^{(s_{c_i},1,k)} \; y_i^{(s_{c_i},1,k)} \; \psi_i^{(s_{c_i},1,k)} \; x_j^{(s_{c_i},1,k)} \; y_j^{(s_{c_i},1,k)} \; \psi_j^{(s_{c_i},1,k)} \right]^T. \tag{11}$$

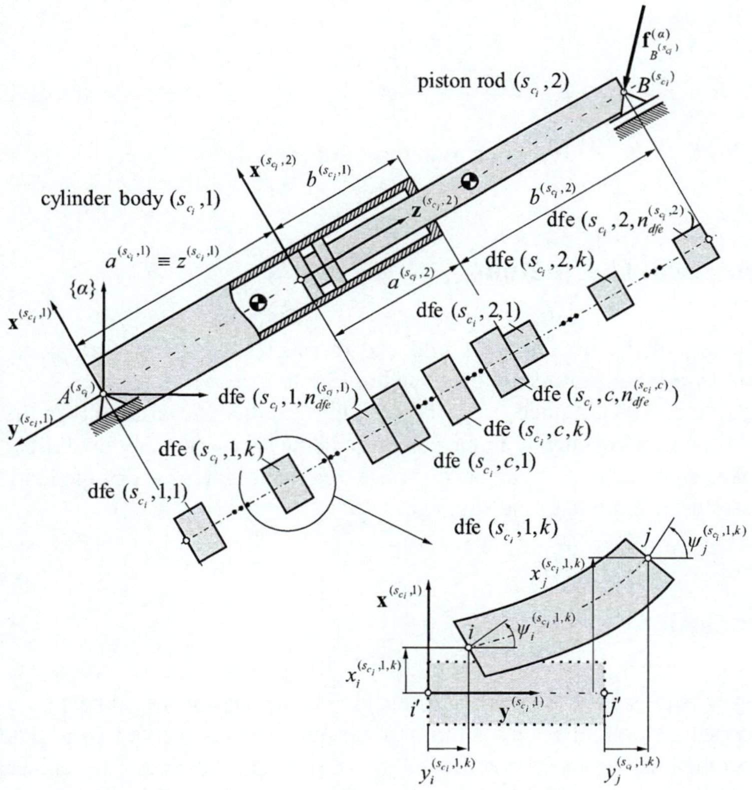

Fig. 7 Discretization of the actuator—FEM

The statics equations of the actuator based on the FEM are expressed in the following form:

$$\mathbf{K}^{(s_{c_i})}\mathbf{u}^{(s_{c_i})} = \mathbf{f}_{ext}^{(s_{c_i})}, \tag{12}$$

where: $\mathbf{K}^{(s_{c_i})}$ is global stiffness matrix, $\mathbf{u}^{(s_{c_i})}$ is nodal displacement vector, $\mathbf{f}_{ext}^{(s_{c_i})}$ is vector of external forces.

The vector of internal forces occurring in dfe $(s_{c_i}, 1, k)$ is calculated as follows:

$$\mathbf{f}_{int}^{(s_{c_i},1,k)} = \mathbf{K}^{(s_{c_i},1,k)}\mathbf{u}^{(s_{c_i},1,k)}, \tag{13}$$

where: $\mathbf{f}_{int}^{(s_{c_i},1,k)} = \left[f_{n,i}^{(s_{c_i},1,k)} \ f_{t,i}^{(s_{c_i},1,k)} \ m_{g,i}^{(s_{c_i},1,k)} \ f_{n,j}^{(s_{c_i},1,k)} \ f_{t,j}^{(s_{c_i},1,k)} \ m_{g,j}^{(s_{c_i},1,k)} \right]^T$, $\mathbf{K}^{(s_{c_i},1,k)}$ is stiffness matrix of dfe $(s_{c_i}, 1, k)$.

The nodal stress vector of dfe $(s_{c_i}, 1, k)$ is defined as follows:

$$\boldsymbol{\sigma}^{(s_{c_i},1,k)} = \left[\sigma_i^{(s_{c_i},1,k)} \ \sigma_j^{(s_{c_i},1,k)} \right]^T, \tag{14}$$

where: $\sigma_\alpha^{(s_{c_i},1,k)}\Big|_{\alpha\in\{i,j\}} = \frac{f_{n,\alpha}^{(s_{c_i},1,k)}}{A^{(s_{c_i},1,k)}} + \frac{m_{g,\alpha}^{(s_{c_i},1,k)}}{W^{(s_{c_i},1,k)}}$, $A^{(s_{c_i},1,k)}$ is cross-sectional area of dfe $(s_{c_i}, 1, k)$, $W^{(s_{c_i},1,k)}$ is section modulus of dfe $(s_{c_i}, 1, k)$.

4 Numerical Calculations

The influence of the flexibility of link and mass of load on the time courses of maximum stress in actuators are presented in Fig. 8.

Analyzing results obtained, it can be seen the significant oscillations in the time courses of the stress occurring in the actuators in the case when the flexibility of the link is taken into account. It can be also observed that higher mass of the load leads to increase the amplitude of the stresses.

5 Conclusions

In the paper is presented the method of evaluating stresses of the crane's actuators. The proposed method relies on a solution of the quasi-statics problem formulated for the replacement models of the actuators in each integration step. The replacement models are based on the planar finite beam elements which are loaded by reaction forces obtained from the dynamics model of the crane. Due to the changing

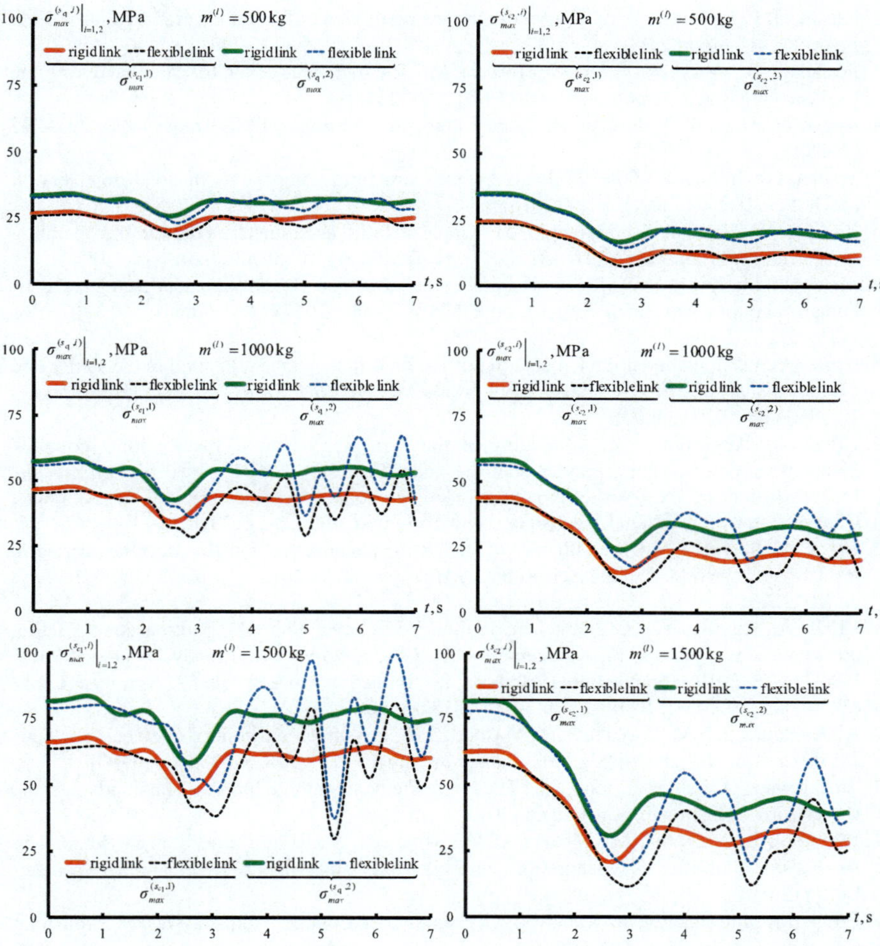

Fig. 8 Time courses of stress in actuators

configuration of the actuators during the simulation, the presented method requires discretization of the actuators in each integration step. In numerical simulations, the influence of the mass of the load on stresses in the actuators were analyzed. In the future, the dynamics model of the crane will be extended to analyze the interaction between flexible actuators and other components of the crane.

References

1. Ghigliazza, R.M., Holmes, P.: On the dynamics of crane, or spherical pendula with moving supports. Int. J. Non-Linear Mech. **37**, 1211–1221 (2002)

2. Kłosiński, J.: Swing-free stop control of the slewing motion of the mobile crane. Control. Eng. Pract. **13**, 451–460 (2005)
3. Posiadała, B., Warys, P., Cekus, D., Tomala, M.: The dynamics of the forest crane during the load carrying. Int. J. Struct. Stab. Dyn. **13**(7), 1–9 (2013)
4. Rupar, D., Hladnik, J., Jerman, B.: Loader crane inertial forces. FME Trans. **44**(3), 291–297 (2016)
5. Trąbka, A.: Influence of flexibilities of cranes structural components on load trajectory. J. Mech. Sci. Technol. **30**(1), 1–14 (2016)
6. Kacalak, W., Budniak, Z., Majewski, M.: Crane stability assessment method in the operating cycle. Transp. Probl. **12**(4), 141–151 (2017). https://doi.org/10.20858/tp.2017.12.4.14
7. Majewski, M., Kacalak, W.: Innovative intelligent interaction systems of loader cranes and their human operators. Adv. Intell. Syst. Comput. **573**, 474–485 (2017). https://doi.org/10.1007/978-3-319-57261-1_47
8. Urbaś, A.: Computational implementation of the rigid finite element method in the statics and dynamics analysis of forest cranes. Appl. Math. Modell. **46**, 750–762 (2017). https://doi.org/10.1016/j.apm.2016.08.006
9. Urbaś, A., Augustynek, K.: Modelling of the dynamics of grab cranes with a complex kinematic structure, taking into account the links' flexibility and advanced friction models. In: Proceedings of 13th World Congress on Computational Mechanics and 2nd Pan American Congress on Computational Mechanics, New York, USA, Jul 22–27 (2018)
10. Urbaś, A., Szczotka, M.: The influence of the friction phenomenon on a forest crane operator's level of discomfort. Maintenance Reliab. **21**(2), 197–210 (2019). https://doi.org/10.17531/ein.2019.2.3
11. Urbaś, A., Augustynek, K.: Mathematical model of a crane with taking into account friction phenomena in actuators. In: Kecskeméthy, A., Geu, F.F. (eds.) Multibody Dynamics 2019. ECCOMAS 2019 Computational Methods in Applied Sciences, vol. 53. Springer, Cham (2020). https://doi.org/10.1007/978-3-030-23132-3_36
12. Abdel-Rahman, E.M., Nayfeh, A.H., Masoud, Z.N.: Dynamics and control of cranes: a review. J. Vib. Control. **9**(7), 863–908 (2003). https://doi.org/10.1177/1077546303009007007
13. Zienkiewicz, O., Taylor, R., Zhu, J.Z.: The Finite Element Method: Its Basis and Fundamentals. Butterworth-Heinemann, Amsterdam (2013)
14. Wittbrodt, E., Szczotka, M., Maczyński, A., Wojciech, S.: Rigid Finite Element Method in Analysis of Dynamics of Offshore Structures Ocean Engineering & Oceanography. Springer, Berlin (2013)
15. Craig, J.J.: Introduction to Robotics. Mechanics and Control. Addison-Wesley Publishing Company, Inc., Reading, MA (1989)
16. Jurevič, E.I. (ed.): Dynamics of Robot Control. Nauka, Moscow (1984) (in Russian)

Nonlinear Dynamics of Atomic and Molecular Systems in an Electromagnetic Field: Deterministic Chaos and Strange Attractors

Alexander V. Glushkov, Anna V. Ignatenko, Anna A. Kuznetsova, Elena V. Bakunina, Oleg V. Dykyi, Alexandra O. Makarova, and Eugeny V. Ternovsky

Abstract We present a new mathematical approach to studying deterministic chaos and strange attractors in dynamics of nonlinear processes in atomic and molecular systems in an electromagnetic field. To treat chaotic dynamics of systems it is constructed effective scheme that includes new quantum-dynamic models (based on the finite-difference solution of the Schrödinger equation, optimized operator perturbation theory and realistic model potential for quantum systems) and advanced nonlinear analysis and a chaos theory methods such as power spectrum analysis, the correlation integral algorithm, the fractal method, the Lyapunov's exponents and Kolmogorov entropy analysis, etc. Availability of multiple resonances with super little widths in spectrum of an atom in external magnetic field is treated and provided by interference phenomena and fluctuations. Dynamics of resonances in spectrum of diatomic molecule in the infrared electromagnetic field is considered and the topological and dynamical invariants are recalculated.

Keywords Nonlinear dynamics · Atomic systems · Electromagnetic field · Chaos · Resonances

---

A. V. Glushkov (✉) · A. V. Ignatenko · A. O. Makarova · E. V. Ternovsky
Odessa State Environmental University, Odessa, Ukraine

A. A. Kuznetsova
National University "Odessa Maritime Academy", Odessa, Ukraine

E. V. Bakunina · O. V. Dykyi
National University "Odessa Law Academy", Odessa, Ukraine

J. Awrejcewicz (ed.), *Perspectives in Dynamical Systems II: Mathematical and Numerical Approaches*, Springer Proceedings in Mathematics & Statistics 363,
https://doi.org/10.1007/978-3-030-77310-6_11

1 Introduction

In the last decade, the theory of dynamical systems and the theory of chaos are characterized by significant progress both in the development of new concepts, as well as new methods and new applications. Modern theory of nonlinear dynamical systems has established the main mechanisms of instability and scenarios of transition to chaos in many nonlinear classical systems and devices with many applications in various sciences, including mechanics, chemistry, biology, physics and others.

In real, especially quantum systems, chaotic dynamics take much more complex, partially or not completely understood forms. According to modern concepts, the theory of quantum chaos actually studies quantum-mechanical systems that are chaotic in the classical limit.

Traditionally, quantum chaos refers to the set of effects observed in quantum-physical systems relating to purely nonlinear effects, which are manifested in quantum systems described by equations of the Schrödinger type or density matrix. It is well known that quantum mechanics, which has existed for over 60 years, allows us to describe both systems that integrate in the classical limit (such as a hydrogen atom) and classically unintegrated systems (such as a helium atom).

The well-known principle of correspondence indicates that quantum mechanics in the quasiclassical domain at scales of the system comparable to the de Broglie wavelength continuously goes to the classical one. On the other hand, in quantum mechanics, the concept of trajectory, at least in pragmatic or Copenhagen interpretation, loses its usual meaning (it reappears only in the quasiclassical domain).

There is an opinion that in the interpretation of the phenomenon of quantum chaos it is more correct to speak about such characteristic manifestations as the intersection of energy levels in multivariable space, elements of stochasticity in the spectra of particularly highly excited states of atomic and molecular systems, phenomena of clustering states, interference, fluctuations, and merging of resonances, etc. From the other side, it was considered the most natural study of the phenomenon of chaos on the basis of methods of classical mechanics and qualitative theory of differential equations, within which it is natural to operate concepts such as bifurcation, instability, boundary cycle, strange attractor, etc. [1–18].

Their application to quantum systems is also quite acceptable, moreover, often the scenario of emergence of a chaotic dynamics in quantum models is (not always) similar to the classical one. For example, the stochastic motion of an electron in an atom in the external fields is naturally interpreted in the language of consideration of a certain type of resonance interactions of modes corresponding to the motion in these fields, and its manifestation area narrows as an interaction increases.

The qualitative picture of the process of emerging chaotic dynamics in quantum systems in general is reduced to the following scenario: an external, for example, a magnetic field leads to the appearance of primary nonlinear resonances, a strong interaction between which leads to the appearance of secondary resonances and the emergence of stochastic oscillations, right down to the formation of the Arnold's web. When the external field strength is above a certain critical value, the various stochastic layers merge, resulting in global stochasticity in the system.

An analysis of the chaotic phenomena in quantum systems was carried out not only based on the methods of classical mechanics (in fact, within the framework of the Newtonian dynamics), but also on the basis of semiclassical or semi-quantum methods, in particular, the method of quantum trajectories (quantization of classical mechanics), and path integrals by Feynman-Higgs, the Gutswiller's theory of "periodic orbits", the Delos closed orbit method, complex coordinate method, a random matrix theory, diagonalization methods and some others (e.g. [1–8]).

New field of investigations of chaotic effects in theory of quantum systems has been provided by a great progress in a development of a chaos and dynamical systems theory methods [6–22].

In previous our papers [7, 8, 17–19] we have presented a few new computational quantum algorithms to study stochastic futures and chaotic elements in dynamics of atomic and molecular systems in an external electromagnetic fields. The known mathematical tools such as power spectrum analysis, correlation integral and fractal algorithms, the Lyapunov's exponents analysis and others have been applied to numerical analysis of chaotic features in dynamics of the quantum systems.

In this paper we present an effective mathematical approach to studying deterministic chaos and strange attractors in dynamics of nonlinear processes in atomic and molecular systems in an electromagnetic field.

To treat a chaotic dynamics of systems it is proposed the theoretical scheme that includes new quantum-dynamic models (based on the finite-difference solution of the Schrödinger equation, optimized operator perturbation theory and realistic model potential method) and advanced nonlinear analysis and a chaos theory methods such as power spectrum analysis, the correlation integral algorithm, the Lyapunov's exponents and Kolmogorov entropy analysis, etc. The approach is applied to study of chaotic phenomena in some atomic and diatomic systems in an external electromagnetic (magnetic) field.

2 A Quantum-Geometric Approach to Analysis, Processing, Modelling Chaotic Dynamics of the Quantum Systems in Electromagnetic Field

In Refs. [7, 8, 17–19] we have given a review of new methods and algorithms of the chaos-geometric approach to analysis, processing, modeling and forecasting a chaotic dynamics for different classical and quantum systems.

So, here we pay attention only at main elements. The total scheme for studying chaos-dynamical phenomena in quantum systems (in particular, atomic systems in magnetic, crossed electric and magnetic fields, Rydberg atoms in a electromagnetic field, molecular systems in a infrared electromagnetic field, etc.) and computing the topological and dynamical invariants in application to quantum systems include the following:

1. Quantum-dynamical computing of quantum systems: Schrödinger (Dirac) equation for quantum system in an external field (numerical solving, the finite differences, model potential, operator perturbation theory, etc. methods); Preliminary analysis and processing dynamical variable series of physical system;
2. Preliminary study and assessment of the presence of chaos: the Gottwald-Melbourne test; Fourier decompositions, irregular nature of change—chaos; Spectral analysis, Energy spectra statistics, the Wigner distribution, the spectrum of power, "Spectral rigidity";
3. The multi-fractal geometry: computation time delay τ using autocorrelation function or mutual information; Determining embedding dimension by the method of correlation dimension or algorithm of false nearest neighbouring points; Calculation of multi-fractal spectra; wavelet analysis;
4. Computing global Lyapynov's exponents, Kaplan-York dimension, Kolmogorov entropy, average predictability measure; Methods of nonlinear prediction (classical and quantum neural network algorithms, the algorithm optimized trajectories, stochastic propagators, memory functions, etc.)

The key idea in the study of the spectra of chaotic systems and, in particular, quantum systems, is provided by the fact that a definition of quantum chaos is interpreted primarily as a property of a group of states of the spectra of the system. It is the interpretation of one of the mechanisms of quantum chaos through the induction of resonances in the spectrum of the system, their strong interaction with subsequent overlapping, the emergence of stochastic layers and further transition to a global stochasticity in the system.

It has led to the most common criterion for chaos in spectral research (especially from the point of view of the mechanism of overlapping and merging of resonances), i.e. the criterion of a chaos by Chirikov.

In this scheme, the overlap of nonlinear resonances is defined as the ratio of the sum of the half-widths of the resonances to the distances between them

$$K = [(\Gamma_1/2) + (\Gamma_1/2)] / \mid E_2 - E_1 \mid, \tag{1}$$

where Γ_i and E_i are, respectively, the width and energy of the "i"th resonance. It is usually assumed that at sufficiently large values ($K \geq 4$) the phenomenon of chaos is realized in the system.

Among the spectral characteristics, which are usually calculated when studying the elements of chaos in the spectra of systems, one should include:

1. The relative value of the interlevel distances S_n, which is standardly defined as:

$$S_n = (E_n - E_{n-1})\, \rho\, (E_n)\,, \tag{2}$$

   where $\rho(E)$ is a density of levels.

2. Function P (S) of the distribution of the relative value of the interlevel distances S_n; if the position of the levels in the spectrum is not chaotic, then P(S), as a rule, has the form of a Poisson distribution $P_p(S) = \exp(-S)$; if there is chaos in the system, then the Wigner-Dyson distribution is realized (in general, the Brody distribution).
3. Characteristics of the degree of ordering of levels in the spectrum at large (in comparison with the interlevel distance) ε the spectral stiffness $\Delta_3(L)$, which is defined as follows:

$$\Delta_3(x, L) = \frac{1}{L}\min_{A,B} \int_{x}^{x+L} (n(\varepsilon) - A\varepsilon - B)^2 d\varepsilon \tag{3}$$

It should be borne in mind that for a sequence of levels ε_n, normalized to a unit density ($\varepsilon_n = \varepsilon_{n-1} + S_n$), a step function n($\varepsilon$), equal to the number of levels with $\varepsilon_n \leq \varepsilon$ is used.

By construction, n(ε), has the form of a ladder with a single average slope. The value of $\langle \Delta_3(x, L) \rangle$, averaged over the values of x from the region in which the nature of the fluctuations of the spectrum can be considered constant, depends only on L and is denoted by $\Delta_3(L)$. The function $\Delta_3(L)$ describes the ordering of the spectrum over large areas: the slower the growth of $\Delta_3(L)$ with increasing L, the less probable in the spectrum are close clusters of levels and gaps with reduced level density.

4. Correlation coefficients C(n) values of energy intervals, divided by a fixed number of levels, determined in the usual way:

$$C(n) = \frac{\sum_i (S_{i-n} - 1)(S_i - 1)}{\left[\sum_i (S_{i+n} - 1)^2 \sum_i (S_i - 1)^2\right]^{\frac{1}{2}}}. \tag{4}$$

Finally, another of the most common characteristics of nonlinear (chaotic) dynamics is the so-called power spectrum, which is determined in a standard way. More detailed characteristics of spectral methods are given, for example, in [7–19].

In Table 1 we present the main blocks of the combined quantum-dynamical and chaos-geometric approach to nonlinear analysis, modelling and prediction of chaotic dynamics of quantum system in an electromagnetic field.

In Refs. [7, 8] the total approach has been used for studying the chaotic features in spectrum of the hydrogen atom in a magnetic field, diatomic molecules interacting with a linearly polarized electromagnetic field. It is shown that the chaotic features are realized in the nonlinear dynamics of diatomic molecules in a linearly polarized electromagnetic field that is in a reasonable agreement with the classical modelling data by Berman, Kolovskii, Zaslavsky, Zganh et al. [1–5]. The detailed description of the every stage in the scheme (Table 1) is earlier presented in Refs. [7, 8, 17–

Table 1 Combined quantum-dynamical and quantum-geometric approach to nonlinear analysis, modelling and prediction of chaotic dynamics of quantum (atomic, molecular and nuclear) system in an electromagnetic field

I. Quantum-dynamical computing of quantum systems: 1. Schrödinger (Dirac) equation for quantum system in an external field (numerical solving, the finite differences , model potential , operator perturbation theory etc methods) 2. Preliminary analysis and processing dynamical variable series of physical system
⇓
II. Study and assessment of the presence of chaos: 3. Fourier decompositions, irregular nature of change – chaos; 4. Spectral analysis, Energy spectra statistics, the Wigner distribution, the spectrum of power, "Spectral rigidity";
⇓
III. The multi-fractal geometry of the phase space. : 5. Autocorrelation function and mutual information; 5. Method of correlation dimension 6. Wavelet analysis;
⇓
IV.Prediction model: 6. Computing global Lyapynov indicators; Kaplan-York dimension, 7. Methods of nonlinear prediction.

30]. All calculations are performed with using "Geomath", "Superatom", "Quantum Chaos", "ScanPoints" computational codes [7, 9, 17–19, 24, 31–43].

3 Chaotic Dynamics of Atoms and Molecules in Electromagnetic Field: Numerical Solution of the Schrödinger Equation and Power Spectrum Analysis

In this subsection we present the results of modeling the hydrogen and rubidium spectra in an external magnetic (crossed electric and magnetic) filed. In Refs. [7, 8, 18, 43] it has been developed an effective nonperturbative quantum and chaos-dynamic approach to modeling the chaotic dynamics of atomic systems in homogeneous magnetic field, which is based on the operator optimized perturbation theory and finite-difference solution of the Schrödinger equation for an atom in the field (in a cylindrical coordinate system $z||B$; $\Psi \sim e^{iM\varphi}$). The cited equation can be written as follows:

$$\left[\frac{\partial^2}{\partial\rho^2}+\frac{1}{\rho}\frac{\partial}{\partial\rho}+\frac{\partial^2}{\partial z^2}-\frac{M^2}{\rho^2}-4\gamma^2\rho^2+\frac{4}{r}+V_c(r)+\left(\frac{E}{R_y}-\gamma M\right)\right]\Psi(\rho,z)=0 \quad (5)$$

where $\gamma = B/B_o$, $B_o = 2.3505 \times 10^5$ T, $V_c(r)$—potential electron self-consistent field, including the Hartree potential plus the Kohn-Sham exchange-correlation potential (other notations are standard).

The quantitative modeling of regular and chaotic dynamics, computation power and spectral parameters for the atoms of hydrogen, neon in a uniform magnetic field ($\gamma = 0.01$–$10{,}000$) showed that the system generated quantum chaos, which is manifested in a very complex and irregular dependences of state energies upon the magnetic field amplitude, the presence of the level intersections (as example, for the Ne quasi-intersections in dependence of the energy states $|0_N>$ and $|2p_0>$ upon the magnetic field amplitude at $\gamma = 158.7$, $|2p_0>$ and $|1s^2>$ states at $\gamma = 40.2$), in a photoionization cross sections, power spectra, etc.

We have calculated and carried our analysis of the photoionization spectrum, power spectrum, the energies and widths of resonances, the distribution of resonances in the hydrogen atom in the special magnetic field with the strength 5.96 T (the energy interval 20–80 cm^{-1}).

According to our data, the density of states in the middle of each channel (Landau resonances) is 33 cm^{-1} for the average resonance width—0.004 cm^{-1}, which is consistent with experimental data Kleppner et al.: 0.004–0.006 cm^{-1} (e.g. [7, 8, 18, 43]).

Further we present the results of modelling the chaotic dynamics of atomic systems in the crossed electric F_I and magnetic γ fields, based on the numerical solution of the Schrödinger equation:

$$H = 1/2\left(p_\rho^2 + l_z^2/\rho^2\right) + \gamma l_z/2 + (1/8)\,\gamma^2\rho^2 + (1/2)\,p_z^2 + F_1 z\cdot\sin(\omega t) + V(r) \quad (6)$$

the operator perturbation theory and density functional method [18, 43].

Here we use the following denotations: $\tilde{f} = F_1\gamma^{-4/3}$, $\varepsilon = E^{ion}\gamma^{-2/3}$, where E^{ion} is an ionization energy of a free atom. We have carried out modelling a chaotic dynamics for the Rydberg Li, Rb (n ~ 100, m = 0) atoms in a static magnetic ($B = 4.5$ T) and oscillating electric field with frequency $\omega = 102$Мгц ($\varepsilon = -0.03$, $\varpi = 0.32$, γ-1/3 in the range 35–50; $f = 0.000$–0.070).

Figure 1 shows the power spectrum of Rb: (a) in a magnetic field ($f = 0$, the electric field is absent); (b) in a static magnetic field and oscillating electric field $f = 0.0035$ (our data).

The scenario of transition to chaos in the system includes the induction of nonlinear resonances by a magnetic field, their strong interaction and further merging with the appearance of global chaos when critical field strength is exceeded.

Further we shortly present the advanced data for the modeling the temporal dynamics (polarization parameter) of the diatomic molecule PbO in the resonant

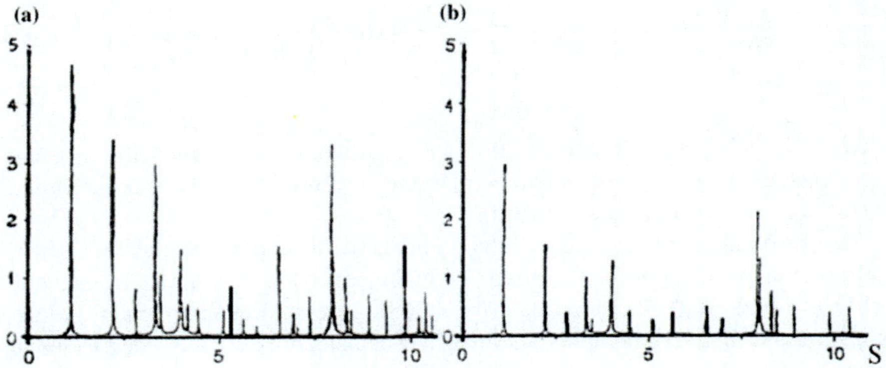

Fig. 1 The power spectrum of Rb: (**a**) in a magnetic field ($f = 0$, the electric field is absent); (**b**) in a static magnetic field and oscillating electric field $f = 0.0035$ (our data)

Table 2 The correlation dimension D_c, Lyapunov's exponents (L_i, $i = 1, 2$), Kaplan-York attractor dimension (D_L), Kolmogorov entropy (KE)

D_c	L_1	L_2	D_L	KE
2.83	0.153	0.0185	2.58	0.172

electromagnetic field. This molecule in the linearly polarized field has been studied, for example, in Refs. [7, 8]. All information about the key characteristics of electromagnetic field as well as spectral molecule parameters is listed in the cited Refs. The new element here is using more efficient approach to solving the Schrödinger equation with the realistic density functional theory potential curve of diatomic molecule $U(x)$ [7, 8]. We numerically studied the corresponding temporal dependence of a polarization (which is normalized to the intensity of the field interaction with the molecule) on the basis of the quantum-geometric approach to analysis of a chaotic dynamics of the molecule interacting with a resonant linearly polarized field.

The concrete step is an analysis of the corresponding time series with the $n = 7.6 \times 10^3$ and $\Delta t = 5 \times 10^{-14}$ s. In Table 2 we list the computed values of the correlation dimension D_c, the Kaplan-York attractor dimension (D_L), the Lyapunov's exponents (L_i, $i = 1$–3), the Kolmogorov entropy (KE). In conclusion of this subsection let us underline, that difference between the presented data and data of Ref. [8] on the topological and dynamical invariants can be explained by processing different polarization time series.

4 Conclusions

We presented the fundamentals of a computational approach to studying deterministic chaos and strange attractors in dynamics of nonlinear processes in atomic and molecular systems in an electromagnetic field. To treat chaotic dynamics of systems it is constructed effective scheme that includes new quantum-dynamic models (based on the finite-difference solution of the Schrödinger equation, optimized operator perturbation theory and realistic model potential for quantum systems) and advanced analysis methods of dynamical systems and chaos theory. As illustration we presented some numerical results for atoms of hydrogen and rubidium in a magnetic and crossed magnetic and oscillating electric field and diatomic molecule of PbO in a resonant resonant linearly polarized field. The data presented the numerical values for a set of dynamical and topological invariants are listed.

References

1. Berman, G.P., Bulgakov, E.N., Holm, D.D.: Nonlinear resonance and dynamical chaos in a diatomic molecule driven by a resonant IR field. Phys. Rev. A. **52**, 3074–3080 (1995)
2. Zhang, C., Katsouleas, T., Joshi, C.: Harmonic frequency generation and chaos in laser driven molecular vibrations. In: Proc. of Shortwavelength Physics with Intense Laser Pulses, San-Diego, CA, 29–31 Mar 1993
3. Zhai, L.-J., Zheng, Y.-J., Ding, S.-L.: Dynamics of vibrational chaos and entanglement in triatomic molecules: Lie algebraic model. Chin. Phys. B. **21**(7), 070503 (2012)
4. Lombardi, M., Matzkin, A.: Dynamical entanglement and chaos: the case of Rydberg molecules. Phys. Rev. A. **73**, 062335 (2006)
5. Arango, C.A., Kennerly, W.W., Ezra, G.S.: Classical and quantum mechanics of diatomic molecules in tilted fields. J. Chem. Phys. **122**, 184303 (2015)
6. Glushkov, A.V., Prepelitsa, G.P., Svinarenko, A.A., Zaichko, P.A.: Studying interaction dynamics of the non-linear vibrational systems within non-linear prediction method (application to quantum autogenerators). In: Awrejcewicz, J., Kazmierczak, M., Olejnik, P., Mrozowski, J. (eds.) Dynamical Systems Theory, T1, pp. 467–477. Lodz Univ. Press, Lodz (2013)
7. Glushkov, A.V., Buyadzhi, V.V., Kvasikova, A.S., Ignatenko, A.V., Kuznetsova, A.A., Prepelitsa, G.P., Ternovsky, V.B.: Nonlinear chaotic dynamics of quantum systems: molecules in an electromagnetic field and laser systems. In: Tadjer, A., Pavlov, R., Maruani, J., Brändas, E., Delgado-Barrio, G. (eds.) Quantum Systems in Physics, Chemistry, and Biology Series: Progress in Theoretical Chemistry and Physics, Vol 30, pp. 169–180, Springer, Cham (2017). https://doi.org/10.1007/978-3-319-50255-7_10
8. Ignatenko, A.V., Buyadzhi, A.A., Buyadzhi, V.V., Kuznetsova, A.A., Mashkantsev, A.A., Ternovsky, E.V.: Nonlinear chaotic dynamics of quantum systems: molecules in an electromagnetic field. In: Jenkins, S., Kirk, S.R., Maruani, J., Brändas, E. (eds.) Advances in Quantum Chemistry, vol. 78, pp. 149–170. Elsevier, Cambridge, MA (2019)
9. Gottwald, G., Melbourne, I.: Testing for chaos in deterministic systems with noise. Physica D. **212**, 100–110 (2005)
10. Packard, N.H., Crutchfield, J.P., Farmer, J.D., Shaw, R.S.: Geometry from a time series. Phys. Rev. Lett. **45**, 712–716 (1980)
11. Kennel, M.B., Brown, R., Abarbanel, H.: Determining embedding dimension for phase-space reconstruction using a geometrical construction. Phys. Rev. A. **45**, 3403–3411 (1992)
12. Abarbanel, H.D.I., Brown, R., Sidorowich, J.J., Tsimring, L.S.: The analysis of observed chaotic data in physical systems. Rev. Mod. Phys. **65**, 1331–1392 (1993)
13. Schreiber, T.: Interdisciplinary application of nonlinear time series methods. Phys. Rep. **308**, 1–64 (1999)
14. Fraser, A.M., Swinney, H.L.: Independent coordinates for strange attractors from mutual information. Phys. Rev. A. **33**, 1134–1140 (1986)

15. Grassberger, P., Procaccia, I.: Measuring the strangeness of strange attractors. Physica D. **9**, 189–208 (1983)
16. Gallager, R.G.: Information Theory and Reliable Communication. Wiley, New York (1986)
17. Glushkov, A.V.: Methods of a Chaos Theory. Astroprint, Odessa (2012)
18. Glushkov, A.V.: Relativistic Quantum Theory. Quantum Mechanics of Atomic Systems. Astroprint, Odessa (2008)
19. Glushkov, A.V., Svinarenko, A.A., Loboda, A.V.: Theory of Neural Networks on Basis of Photon Echo and Its Program Realization. TEC, Odessa (2003)
20. Khetselius, O.Y.: Forecasting evolutionary dynamics of chaotic systems using advanced non-linear prediction method. In: Awrejcewicz, J., Kazmierczak, M., Olejnik, P., Mrozowski, J. (eds.) Dynamical Systems Applications, T2, pp. 145–152. Lodz Univ. Press, Lodz (2013)
21. Bunyakova, Y.Y., Ternovsky, V.B., Dubrovskaya, Y.V., Ignatenko, A.V., Svinarenko, A.A., Vitavetskaya, L.A.: Analysis of the beryllium-7 activity concentration dynamics in the atmospheric environment time series after the Fukushima Daiichi nuclear power plants emergency. Sensor Electr. Microsyst. Technol. **14**(4), 73–82 (2017)
22. Khetselius, O.Y., Brusentseva, S.V., Tkach, T.B.: Studying interaction dynamics of chaotic systems within non-linear prediction method. In: Awrejcewicz, J., Kazmierczak, M., Olejnik, P., Mrozowski, J. (eds.) Application to Neurophysiology Dynamical Systems Applications, T2, pp. 251–259. Lodz Univ. Press, Lodz (2013)
23. Svinarenko, A.A., Ignatenko, A.V., Ternovsky, V., Nikola, L., Seredenko, S.S., Tkach, T.B.: Advanced relativistic model potential approach to calculation of radiation transition parameters in spectra of multicharged ions. J. Phys. Conf. Ser. **548**, 012047 (2014)
24. Florko, T.A., Ambrosov, S.V., Svinarenko, A.A., Tkach, T.B.: Collisional shift of the heavy atoms hyperfine lines in an atmosphere of the inert gas. J. Phys. Conf. Ser. **397**, 012037 (2012)
25. Buyadzhi, V.V., Zaichko, P.A., Gurskaya, M.Y., Kuznetsova, A.A., Ponomarenko, E.L., Ternovsky, V.B.: Relativistic theory of excitation and ionization of Rydberg atomic systems in a Black-body radiation field. J. Phys. Conf. Ser. **810**, 012047 (2017)
26. Svinarenko, A.A., Khetselius, O.Y., Buyadzhi, V.V., Florko, T.A., Zaichko, P.A., Ponomarenko, E.L.: Spectroscopy of Rydberg atoms in a black-body radiation field: Relativistic theory of excitation and ionization. J. Phys. C Ser. **548**, 012048 (2014)
27. Khetselius, O.Y.: Hyperfine Structure of Atomic Spectra. Astroprint, Odessa (2008)
28. Khetselius, O.Y.: Quantum Structure of Electroweak Interaction in Heavy Finite Fermi-Systems. Astroprint, Odessa (2011)
29. Khetselius, O.Y.: Spectroscopy of cooperative electron-gamma-nuclear processes in heavy atoms: NEET effect. J. Phys. Conf. Ser. **397**, 012012 (2012)
30. Khetselius, O.Y.: Atomic parity non-conservation effect in heavy atoms and observing P and PT violation using NMR shift in a laser beam: To precise theory. J. Phys. Conf. Ser. **194**, 022009 (2009)
31. Khetselius, O.Y.: Relativistic perturbation theory calculation of the hyperfine structure parameters for some heavy-element isotopes. Int. J. Quant. Chem. **109**, 3330–3335 (2009)
32. Khetselius, O.Y.: Relativistic calculation of the hyperfine structure parameters for heavy elements and laser detection of the heavy isotopes. Phys. Scr. **135**, 014023 (2009)
33. Khetselius, O.Y.: Relativistic calculating the spectral lines hyperfine structure parameters for heavy ions. AIP Conf. Proc. **1058**, 363–365 (2008)
34. Khetselius, O.Y.: Relativistic hyperfine structure spectral lines and atomic parity non-conservation effect in heavy atomic systems within QED theory. AIP Conf. Proc. **1290**(1), 29–33 (2010)
35. Svinarenko, A.A.: Study of spectra for lanthanides atoms with relativistic many-body perturbation theory: Rydberg resonances. J. Phys. Conf. Ser. **548**, 012039 (2014)
36. Buyadzhi, V.V., Kuznetsova, A.A., Buyadzhi, A.A., Ternovsky, E.V., Tkach, T.B.: Advanced quantum approach in radiative and collisional spectroscopy of multicharged ions in plasmas. In: Jenkins, S., Kirk, S.R., Maruani, J., Brändas, E. (eds.) Advances in Quantum Chemistry, vol. 78, pp. 171–191. Elsevier, Cambridge, MA (2019)

37. Dubrovskaya, Y.V., Khetselius, O.Y., Vitavetskaya, L.A., Ternovsky, V.B., Serga, I.N.: Quantum chemistry and spectroscopy of pionic atomic systems with accounting for relativistic, radiative, and strong interaction effects. In: Jenkins, S., Kirk, S.R., Maruani, J., Brändas, E. (eds.) Advances in Quantum Chemistry, vol. 78, pp. 193–222. Elsevier, Cambridge, MA (2019)
38. Khetselius, O.Y.: Optimized relativistic many-body perturbation theory calculation of wavelengths and oscillator strengths for li-like multicharged ions. In: Jenkins, S., Kirk, S.R., Maruani, J., Brändas, E. (eds.) Advances in Quantum Chemistry, vol. 78, pp. 223–251. Elsevier, Cambridge, MA (2019)
39. Glushkov, A.V., Efimov, V.A., Gopchenko, E.D., Dan'kov, S.V., Polishchyuk, V.N., Goloshchak, O.P.: Calculation of spectroscopic characteristics 4 of alkali-metal dimers on the basis of a model perturbation theory. Opt. Spectr. **84**(5), 670–678 (1998)
40. Glushkov, A.V., Malinovskaya, S.V., Loboda, A.V., Shpinareva, I.M., Prepelitsa, G.P.: Consistent quantum approach to new laser-electron-nuclear effects in diatomic molecules. J. Phys. Conf. Ser. **35**, 420–424 (2006)
41. Glushkov, A.V.: Relativistic multiconfiguration time-dependent self-consistent-field theory for molecules. Sov. Phys. J. **34**(10), 871–876 (1991)
42. Glushkov, A.V.: Atom in an Electromagnetic Field. KNT, Kiev (2005)
43. Huber, K.P., Herzberg, G.: Molecular Spectra and Molecular Structure. IV. Constants of Diatomic Molecules. Van Nostrand Reinhold Co., New York (1979)

Deterministic Chaos, Bifurcations and Strange Attractors in Nonlinear Dynamics of Relativistic Backward-Wave Tube

Alexander V. Glushkov, Andrey V. Tsudik, Valentin B. Ternovsky, Dmytro V. Astaykin, Andrii V. Bondarenko, Dmytro V. Danylenko, and Vasily V. Buyadzhi

Abstract We present the results of modelling, analysis, forecasting the dynamics of relativistic backward-wave tube (RBWT) with accounting for relativistic effects ($\gamma_0 = 1.5$–6.0), dissipation factor (factor D) and an effect of presence of the space charge. There are computed the temporal dependences of the normalized field amplitudes (power) in a wide range of variation of the controlling parameters (electric length of an interaction space N, bifurcation parameter proportional to current I, the Pirse parameter J and relativistic factor γ_0), which are characteristic for distributed relativistic electron-waved self-vibrational systems. The computed temporal dependence of a field amplitude (power) F_{max} in a physically reasonable agreement with theoretical estimates and the using the pulsed accelerator "Saturn" experimental data by Ginzburg et al. (IAP, Nizhny Novgorod). The nonlinear analysis technique (including a multi-fractal approach, the methods of correlation integral, false nearest neighbours, surrogate data, the Lyapunov's exponent's algorithm and others) is applied to analysis of numerical parameters of the RBWT chaotic dynamics. There are computed the dynamic and topological invariants of the RBWT dynamics in auto-modulation (AUM)/chaotic regimes. The bifurcation diagrams with definition of the dynamics self-modulation/chaotic areas in planes "J-γ_0", "D-J" are constructed.

Keywords Relativistic backward-wave tube · Dynamics · Chaos · Attractors

A. V. Glushkov · A. V. Tsudik · V. V. Buyadzhi
Odessa State Environmental University, Odessa, Ukraine

V. B. Ternovsky (✉) · D. V. Astaykin · A. V. Bondarenko · D. V. Danylenko
National University "Odessa Maritime Academy", Odessa, Ukraine

J. Awrejcewicz (ed.), *Perspectives in Dynamical Systems II: Mathematical and Numerical Approaches*, Springer Proceedings in Mathematics & Statistics 363,
https://doi.org/10.1007/978-3-030-77310-6_12

1 Introduction

At present, one of the most relevant and very complex areas of physics of elements, systems and devices of electronics is certainly a study of regular and chaotic dynamics of nonlinear processes in different classes of devices of so-called relativistic high-frequency or even ultrahigh-frequency microwave electronics [1–10]. Among the main problems of relativistic microwave electronics, of course, there is, first of all, the quantitative study of the mechanisms of energy conversion accelerated to relativistic velocities of high-intensity electron flows into powerful coherent electromagnetic radiation and their use in various devices for further application in science and technology, for example, for the purposes of nanosecond location, special radio applications, accelerators with ultra-fast particle energy rate. The generators of chaotic oscillations of the microwave range are of great practical interest in the problems of plasma heating in controlled fusion installations and other applications. However, their application in practice faces a number of problems, in particular, the need to increase the stability and efficiency of generation, increase the energy in the microwave pulse, maintain high coherence of radiation for high values of lasing power, the possibility of wide adjustment of lasing frequency and so on [1–12].

A fundamentally new direction in the study of the dynamics of complex processes in microwave generators, and in particular in one of the most known classes, namely, backward-wave tubes (BWT) was the application of methods of the dynamical systems and chaos theories. To date, non-relativistic BWT have been experimentally and theoretically well studied.

The authors [1–4] presented results of studying dynamics of a non-relativistic BWT, including computing the phase portraits, modeling the fundamental electromagnetic characteristics and their chaotic properties. Investigation of the quantitative role of such effects as spatial charge and relativistic effects, energy losses etc. has been studied within a nonlinear theory for the O-type BWT [3].

The peculiarity of relativistic BWT (RBWT) is that the interaction of the microwave field with the electron beam is carried out through a synchronous wave harmonic, which propagates towards the electron flow. In contrast to non-relativistic BWT, the study of nonlinear dynamics of complex processes, different modes of operation in the RBWT is characterized by a significantly lower level of understanding.

A consistent dynamic theory of relativistic BWT (RBWT) is developed in Ref. [7]. One of the most important details of these works is the serious attempts to study the conditions for the emergence of chaotic and super chaotic regimes in the RBWT dynamics.

The discrepancy between the calculated and experimental values of the RBWT dynamic parameters is due, in particular, to the use of simplified numerical models that do not adequately describe its relativistic dynamics, not a sufficiently correct approximation of the capacitive charge, etc. The description of such a new phenomenon in the dynamics of nonlinear processes in the RBWT as a relativistic chaos necessitates the construction of more quantitatively adequate dynamic models of RBWT functioning with elements of chaos. An effective opportunity to reduce the threshold for the transition to a regime of relativistic chaos can be provided through the use of a chain of two related RBWT etc.

In this paper we present the results of the modelling, analysis, forecasting the dynamics of relativistic backward-wave tube (RBWT) with accounting for key relativistic effects (the corresponding relativistic $\gamma_0 = 1.5$–6.0), dissipation factor (factor D) and an effect of presence of the space charge. There are computed the temporal dependences of the normalized field amplitudes (power) in a wide range of variation of the controlling parameters (electric length of an interaction space N, bifurcation parameter proportional to current *I*, the Pirse parameter *J* and relativistic factor γ_0), which are characteristic for distributed relativistic electron-waved self-vibrational systems. The computed temporal dependence of a field amplitude (power) F_{max} in a physically reasonable agreement with theoretical estimates and the using the pulsed accelerator "Saturn" experimental data by Ginzburg et al. (IAP, Nizhny Novgorod).

2 Nonlinear Dynamics of Relativistic Back-Ward Tube and Chaos-Geometric Approach to Analysis, Modelling and Forecasting Time Series of the Tube Power

2.1 Non-Stationary Dynamics of Relativistic Back-Ward Tube

Let us consider the RBWT nonlinear dynamics, using the standard non-stationary theory [2–8]. Despite the papers [3, 4], we will consider the RBWT model with direct accounting for relativistic, dissipation and effects [7–9].

As the dynamical system, RBWT has three governing parameters:

1. a standard relativistic factor, which is defined as follows:

$$\gamma_0 = \left(1 - \beta_0^2\right)^{-1.2} \tag{1}$$

   where $\beta_0 = v_0/c$, v_0 is the initial velocity of the electrons;
2. a parameter N, which defines an electric length of the interaction space;
3. a parameter, which is proportional to the current of the electron beam:

$$L = 2\pi CN/\gamma_0 \tag{2}$$

The Piers parameter C in Eq. (2) is as follows:

$$C = \sqrt[3]{I_0 K_0/(4U)}, \tag{3}$$

where I_0 is a constant component of the beam current, U is an accelerating voltage, and K_0 is a communication resistance of the deceleration system.

The equation in the usual dimensionless form for a phase $\theta(\zeta, \tau, \theta_0)$ of relativistic electron (that flew into the space of interaction with the phase θ_0 and has a coordinate ζ at time τ) and a dimensionless complex amplitude (of the high-frequency field $E(x,t) =$ Re $[\varepsilon(x,t)\exp(i\omega_0 t - i\beta_0 x)]$) $F(\zeta,\tau) = \tilde{E}/\left(2\beta_0 U C^2\right)$ are as follows:

$$\partial^2\theta/\partial\zeta^2 = -L^2\gamma_0^3\left[\left(1+\frac{1}{2\pi N}\partial\theta/\partial\zeta\right)^2-\beta_0^2\right]^{3/2} \mathrm{Re}\left[F\exp(i\theta)\right],$$

$$\partial F/\partial\tau - \partial F/\partial\zeta = -L\tilde{I}, \tag{4}$$

$$\tilde{I} = -\frac{1}{\pi}\int\limits_0^{2\pi} e^{-i\theta}d\theta_0$$

with the boundary and initial conditions:

$$\theta|_{\zeta=0} = \theta_0,$$
$$\partial\theta/\partial\zeta|_{\xi=0} = 0,$$

$$F|_{\zeta=1} = 0, \tag{5}$$

$$F|_{\tau=0} = F^0(\zeta),$$

and $\theta_0 \in [0; 2\pi]$ is the initial phase.

Based on the system (4) and (5), we formulate a generalized model of system dynamics, which takes into account the effect of dissipation, in particular, the impact of energy losses in the deceleration system of the RBWT. Remaining the equation of motion of relativistic electrons and the boundary conditions the same as in model (4) and (5), one could take into account the energy loss during wave propagation in the deceleration system by including an additional term in the left part of the excitation equation (4). Namely, the term $d{\cdot}F$. Here d is the normalized dissipation parameter, which is expressed through the level of losses in decibels in the "cold" system D and the normalized length L:

$$d = 0.115D/L. \tag{6}$$

It should be noted that, from the qualitative viewpoint, a formal consideration of the dissipation effect in the form of energy losses during wave propagation in the deceleration system in the BWT at realistic values of the dissipation parameter should be reduced to shifting the bifurcation values of L to increase. Indeed, according to the Refs. [5–7], in the absence of the required energy losses, the instability of the equilibrium state $F \equiv 0$ in an infinitely long system "electron flow—reverse wave" has a absolute character. The localized perturbation in the system evolves as it happens its distribution in space in both directions with increasing amplitude.

With an increase of the dissipation parameter there is the moment when the absolute instability changes to convective one. The increase in perturbation during propagation occurs only in the direction of beam motion.

It should also be added that the feedback modulation mechanism in BWT, which takes into account the propagation of wave perturbations toward the electron beam from the region where the regrouping takes place, is actually affected by dissipation to a greater extent than the mechanism responsible for self-excitation at the fundamental operating frequency. Indeed there is an additional important effect, namely, the wave reflections at the ends of the decelerating system of the RBWT. Obviously, from the viewpoint of the influence on the dynamics of processes it should be taken into account too.

In order to find the corresponding solutions of the systems of differential equations of type (4) and (5) and to make further numerical modeling dynamics of nonlinear processes in the RBWT. It is used the PC complex of programs based on the use of finite-difference schemes such as "predictor-corrector" and the Thomson's run method to solve the corresponding system of linear algebraic equations (e.g. [7–16]).

2.2 *Chaos-Geometric Approach to Analysis, Modelling and Forecasting Time Series of the Tube Power*

In this subsection we briefly describe a fundamental chaos-geometric approach to analysis, modeling and forecasting a temporal dynamics of the fundamental characteristic parameters of the system studied. In a series of the papers [11–14] the authors have given a review of new methods and algorithms of this approach, which includes a multi-fractal approach, the methods of correlation integral, false nearest neighbours, surrogate data, the Lyapunov's indicators's algorithm and others. Below it will be applied to analysis and processing numerical parameters of the RBWT chaotic dynamics.

As many blocks of the used approach have been described in detail earlier (e.g., [11–14]), here we are limited to presenting the key moments. The detailed flowchart of the combined chaos-geometric and dynamical differential equations approach to nonlinear analysis, processing and prediction of the chaotic dynamics is presented in Refs. [11–15].

The main stages of a chaos-dynamical approach in application to analysis and modeling the RBWT temporal dynamics the following stages [8, 11–14]:

1. General qualitative analysis of the RBWT temporal dynamics, including qualitative analysis of the non-stationary differential equations solutions; analytical and numerical solution of systems of dynamic equations describing the corresponding systems and devices of relativistic microwave electronics with obtaining the corresponding the time series of basic characteristics of systems.
2. A set of the procedures to check a presence of chaotic (stochastic) features and modes, including the known Gottwald-Melbourne test etc. It allows to present a preliminary study and conclusion about the presence of chaos in the system. Besides, it is useful to calculate the energy and power spectra too.
3. Geometry of a phase space and multi-fractal geometry; reconstruction of a phase space for the dynamical system (choice of time delay, determination of embedding dimension by methods of correlation integral and the and false nearest neighbors algorithm).
4. Prediction of the RMBT nonlinear dynamics; calculation of the dynamical and topological invariants of the system, including computing the Lyapunov's exponents, Kolmogorov energy, the Kaplan-York dimension and the predictability limits on the basis of the advanced algorithms; Determining the number of nearest neighboring points NN for the best forecast results (analysis of qualitative indicators) etc.
5. Prediction of temporal (spatial) fluctuations of the fundamental dynamical parameters of the system studied; such new methods and algorithms of nonlinear prediction as methods of predicted trajectories, stochastic propagators with blocks of the polynominal and other approximations etc.

In fact the latter point includes a high level chaos-geometric (cybernetic) study of the characteristics of chaos in the dynamics of systems and devices and construction of the first models for predicting relativistic chaos. The detailed description of every point is presented in Refs. [11–20]. All calculations are performed with using "Geomath", "Superatom" and "Quantum Chaos", "ScanPoints" computational codes [6–14, 20–34].

3 Solution of the Relativistic Back-Ward Tube Dynamical Equations System and Results of the Chaos-Geometric Analysis and Modelling the Power Time Series

As input data, we used the following parameters: the energy of electrons—150 keV, starting current of 7 A composed impedance connection 0.5 Ω, length of interaction space—0.623 m, the average radius waveguides—1.38 sm period corrugating—

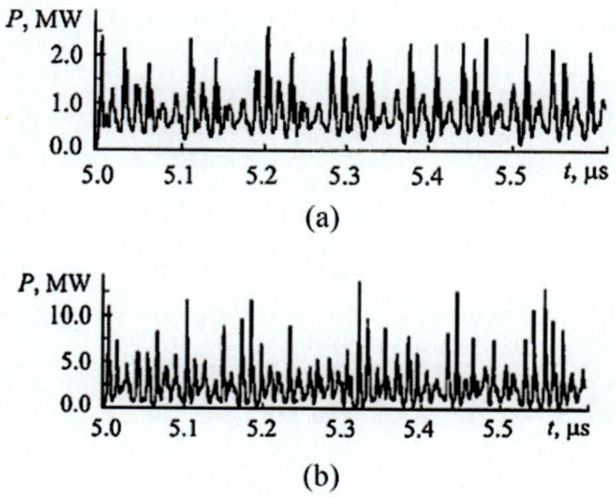

Fig. 1 Results for temporal dependence of power at injection currents: (**a**) 55 A, (**b**) 120 A

1.73 sm radius of the electron beam—0.67 sm. The dynamic model (1)–(4) has been implemented in two ways: (a) consideration of the effects of space charge; (b) accounting for and the effect of slowing the loss of energy in the system (at the ends of reflection and some other factors discussed more etc.). The bifurcation parameter is:

$$J = eI \mid Z \mid / \left(2\beta_0^2 mc^2\right).$$

Here Z—resistance connection, I—beam current; the parameter space charge is: $Q = Ieg/(mc\omega^2 b)$, where the transverse wave number $g = \omega/(c\beta_0\gamma_0)$, coefficient of reduction space charge $f_k = 0.55$ and space charge density is as follows:

$$q_k = (1/\pi)\int_0^{2\pi} e^{-k\theta} d\theta_0. \tag{7}$$

In Fig. 1 we list the calculated theoretical temporal dependence of the RMBT power at the injection currents: (a) 55 A, (b) 90 A, (c) 120 A.

At current 7 A it is set stationary mode that with increasing value of current strength transited to the periodic automodulation (I = 30 A, on our data, the period of T_a = 7.3 ns; experimental value [3, 4]: 8 ns), and then when I = 55 A it is realized the chaotic auto-modulation mode (Fig. 1a). By increasing the amount of current to 75 A there is the quasi-periodical auto-modulation (period 13.8 ns) and, finally, when the current value is more than 100 A it is realized essentially chaotic regime. Note that reset of the quasi-periodic auto-modulation mode can be explained by an effect of space charge. The similar theoretical estimates (however without the dissipation effect) and experiment results data have been obtained by Ginsburg et

Table 1 Correlation dimension d_2, embedding dimension, determined on the basis of false nearest neighbours algorithm (d) with percentage of false neighbours (%) calculated for different values of lag τ

Chaos (I)			Hyperchaos (II)		
τ	d_2	(d_N)	τ	d_2	(d_N)
60	3.6	5 (5.5)	67	7.2	10 (12)
6	3.1	4 (1.1)	10	6.3	8 (2.1)
8	3.1	4 (1.1)	12	6.3	8 (2.1)

Table 2 The Lyapunov exponents (λ_i), the dimension of the Kaplan-York attractor, the Kolmogorov entropy $K_{entr.}$ (our data)

Chaos	λ_1	λ_2	λ_3	λ_4	$K_{entr.}$
(I)	0.261	0.0001	−0.0004	−0.528	0.26
(II)	0.514	0.228	0.0000	−0.0002	0.74

al. [3, 4]. Let us note that all results are in a physically reasonable agreement with each other.

Further let us present the results of the chaos-geometric analysis and processing the data for the RMBT power time series. In Table 1 we list our data on the correlation dimension d_2, embedding dimension, determined on the basis of false nearest neighbours algorithm (d_N) with percentage of false neighbours (%), calculated for different values of lag τ (I—chaos, Fig. 1a; II—hyperchaos, Fig. 1b).

In Table 2 we list our calculational data on the Lyapunov's exponents (λ_i), the dimension of the Kaplan-York attractor, the Kolmogorov entropy ($K_{entr.}$). As analysis shows, there are the positive and negative Lyapunov's exponents values. The resulting Kaplan York dimension in both cases are very similar to the correlation dimension (calculated by the algorithm by Grassberger-Procaccia [19]).

Further, in Fig. 2 we present the firstly obtained original (continuous line) and predicted (dotted line) dependences of power in the chaotic regime (I): (a) without energy loss effect, (b) taking into account the effect of loss. We have used the prediction model [11, 12] of the Schreiber type [17] with the polynominal approximation. The details can be found in Refs. [11–14]. The presented mechanism of changing different modes in the RBWT dynamics due an increasing a current value and the bifurcation parameter J corresponds to certain value relativistic factor, namely $\gamma_0 = 1.3$. More important is the analysis of the RBWT nonlinear dynamics in the plane "relativistic factor—bifurcation parameter." Our analysis shows that there are the quantitative limits of auto-modulation in the plane of parameters: bifurcation parameter J-relativistic factor γ_0.

A characteristic feature of the chart is the presence of so-called effect of "beak", which is based on relativistic factor goes far deeper automodulation area. Firstly this effect was predicted in Refs. [3, 4]. In essentially relativistic limit (see Fig. 3) the frequency of auto-modulation falls by about half. Obviously, that all of the above characteristics is much more complicated compared to the dynamics of non-

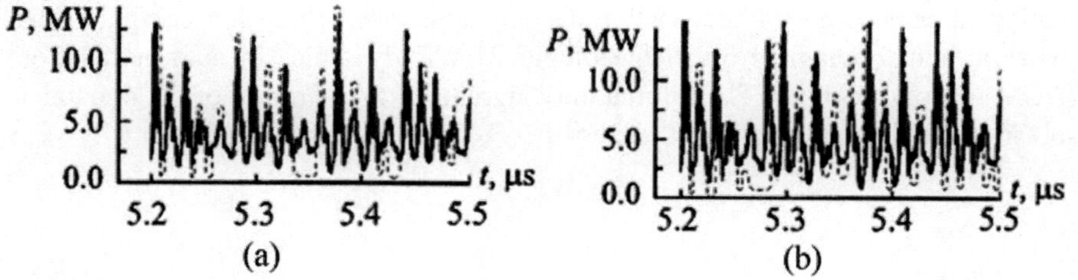

Fig. 2 Original (continuous line) and predicted (dotted line) dependences of power in the chaos mode (I): (**a**) without energy loss effect, (**b**) taking into account the effect of loss

Fig. 3 The dependence of the frequency of auto-modulation upon relativistic factor

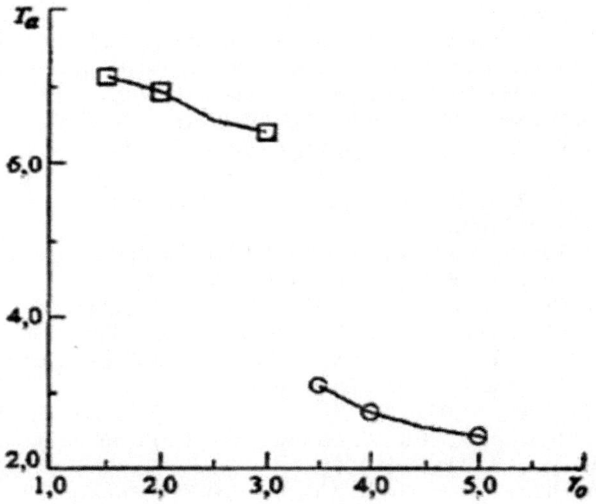

relativistic back-ward tube. So, it will be more correct to say about relativistic chaos phenomenon in the RBWT dynamics.

4 Conclusions

In this work we have performed quantitative modelling, analysis, forecasting the RWBT dynamics with accounting relativistic ($\gamma_0 = 1.5$–6.0) and dissipation effects and the effects of presence of a space charge, reflection of waves at the end of deceleration system etc. The computed temporal dependence of a field amplitude (power) Fmax in a physically reasonable agreement with theoretical estimates and the using the pulsed accelerator "Saturn" experimental data by Ginzburg et al. (IAP, Nizhny Novgorod). The nonlinear analysis technique (including a multi-fractal approach, the methods of correlation integral, false nearest neighbours, surrogate data, the Lyapunov's exponent's algorithm and others) is applied to analysis of numerical parameters of the RBWT chaotic dynamics. There are computed the dynamic and topological invariants of the RBWT dynamics in auto-modulation (AUM)/chaotic regimes. The bifurcation diagrams with definition of the dynamics self-modulation/chaotic areas in planes "J-γ_0", "D-J" are constructed.

References

1. Zheng, X., Tanaka, K., Minami, K., Granatstein, V.: Experimental study of a high-power backward-wave oscillator operating far from upper cutoff. J. Phys. Soc. Jpn. **67**(4), 1466–1472 (1998)
2. Trubetskov, D.I., Anfinogentov, V.G., Ryskin, N.M., Titov, V.N., Khramov, A.E.: Complex dynamics of electron microwave devices (nonlinear non-stationary theory from nonlinear dynamics). Radioengineering. **63**, 61–68 (1999)
3. Ginsburg, H.S., Kuznetsov, S.P., Fedoseyev, T.N., et al.: Theory of transients in relativistic BWO. Izv. Vuzov. Ser. Radiophys. **21**, 1037–1052 (1978)
4. Ginzburg, N.S., Zaitsev, N.A., Ilyakov, E., Kulagin, V.I., Novozhilov, Y., Rosenthal, P., Sergeev, V.: Chaotic generation in backward wave tube of the megawatt power level. J. Techn. Phys. **71**, 73–80 (2001)
5. Kuznetsov, S.P., Trubetskov, D.I.: Chaos and hyper-chaos in the backward wave tube. Izv. Vuzov. Ser. Radiophys. **XLVII**, 383–399 (2004)
6. Glushkov, A.V., Kuzakon, V.M., Ternovsky, V.B., Buyadzhi, V.V.: Dynamics of laser systems with absorbing cell and backward-wave tubes with elements of a chaos. In: Awrejcewicz, J., Kazmierczak, M., Olejnik, P., Mrozowski, J. (eds.) Dynamical Systems Theory, T1, pp. 461–466. Lodz Univ. Press, Lodz (2013)
7. Lepikh, Y.I., Glushkov, A.V., Ternovsky, V.B., Brusentseva, S.V., Duborez, A.V.: Nonlinear dynamics of relativistic backward-wave tube in self-modulation and chaotic regime. Photoelectronics. **24**, 77–87 (2015)
8. Glushkov, A.V., Tsudik, A.V., Novak, D.A., Dubrovsky, O.B.: Chaotic dynamics of relativistic backward-wave tube with accounting for space charge field and dissipation effects: new effects. Photoelectronics. **27**, 44–51 (2018)
9. Glushkov, A.V., Malinovskaya, S.V., Loboda, A.V., Shpinareva, I.M., Gurnitskaya, E.P., Korchevsky, D.A.: Diagnostics of the collisionally pumped plasma and search of the optimal plasma parameters of X-ray lasing: calculation of electron-collision strengths and rate coefficients for Ne-like plasma. J. Phys. Conf. Ser. **11**, 188–198 (2005)
10. Glushkov, A.V., Malinovskaya, S.V., Prepelitsa, G.P., Ignatenko, V.M.: Manifestation of the new laser-electron nuclear spectral effects in the thermalized plasma: QED theory of co-operative laser-electron-nuclear processes. J. Phys. Conf. Ser. **11**, 199–206 (2005)
11. Glushkov, A.V.: Methods of a Chaos Theory. Astroprint, Odessa (2012)
12. Glushkov, A.V., Svinarenko, A.A., Loboda, A.V.: Theory of Neural Networks on Basis of Photon Echo and Its Program Realization. TEC, Odessa (2003)
13. Khetselius, O.Y.: Forecasting evolutionary dynamics of chaotic systems using advanced non-linear prediction method. In: Awrejcewicz, J., Kazmierczak, M., Olejnik, P., Mrozowski, J. (eds.) Dynamical Systems Applications, T2, pp. 145–152. Lodz Univ. Press, Lodz (2013)
14. Ignatenko, A.V., Buyadzhi, A.A., Buyadzhi, V.V., Kuznetsova, A.A., Mashkantsev, A.A., Ternovsky, E.V.: Nonlinear chaotic dynamics of quantum systems: molecules in an electromagnetic field. In: Jenkins, S., Kirk, S.R., Maruani, J., Brändas, E. (eds.) Advances in Quantum Chemistry, vol. 78, pp. 149–170. Elsevier, Cambridge, MA (2019)
15. Packard, N.H., Crutchfield, J.P., Farmer, J.D., Shaw, R.S.: Geometry from a time series. Phys. Rev. Lett. **45**, 712–716 (1980)

16. Kennel, M., Brown, R., Abarbanel, H.: Determining embedding dimension for phase-space reconstruction using a geometrical construction. Phys. Rev. A. **45**, 3403–3411 (1992)
17. Schreiber, T.: Interdisciplinary application of nonlinear time series methods. Phys. Rep. **308**, 1–64 (1999)
18. Fraser, A., Swinney, H.: Independent coordinates for strange attractors from mutual information. Phys. Rev. A. **33**, 1134–1140 (1986)
19. Grassberger, P., Procaccia, I.: Measuring the strangeness of strange attractors. Physica D. **9**, 189–208 (1983)
20. Stepanenko, S.N., Shnaidman, V.N.: Comparison of the results of a generalization of experimental data with calculations from a three-parameter model of the atmospheric boundary layer. Fluid Mech. Soviet Res. **12**, 123–131 (1983)
21. Glushkov, A.V.: Relativistic Quantum Theory. Quantum Mechanics of Atomic Systems. Astroprint, Odessa (2008)
22. Khetselius, O.Y.: Spectroscopy of cooperative electron-gamma-nuclear processes in heavy atoms: NEET effect. J. Phys. Conf. Ser. **397**, 012012 (2012)
23. Khetselius, O.Y.: Atomic parity non-conservation effect in heavy atoms and observing P and PT violation using NMR shift in a laser beam: to precise theory. J. Phys. Conf. Ser. **194**, 022009 (2009)
24. Khetselius, O.Y.: Relativistic perturbation theory calculation of the hyperfine structure parameters for some heavy-element isotopes. Int. J. Quant. Chem. **109**, 3330–3335 (2009)
25. Khetselius, O.Y.: Relativistic calculation of the hyperfine structure parameters for heavy elements and laser detection of the heavy isotopes. Phys. Scr. **135**, 014023 (2009)
26. Khetselius, O.Y.: Relativistic calculating the spectral lines hyperfine structure parameters for heavy ions. AIP Conf. Proc. **1058**, 363–365 (2008)
27. Khetselius, O.Y.: Relativistic hyperfine structure spectral lines and atomic parity non-conservation effect in heavy atomic systems within QED theory. AIP Conf. Proc. **1290**(1), 29–33 (2010)
28. Svinarenko, A.A.: Study of spectra for lanthanides atoms with relativistic many-body perturbation theory: Rydberg resonances. J. Phys. Conf. Ser. **548**, 012039 (2014)
29. Florko, T., Ambrosov, S., Svinarenko, A., Tkach, T.: Collisional shift of the heavy atoms hyperfine lines in an atmosphere of the inert gas. J. Phys. Conf. Ser. **397**, 012037 (2012)
30. Buyadzhi, V.V., Zaichko, P.A., Gurskaya, M.Y., Kuznetsova, A.A., Ponomarenko, E.L., Ternovsky, V.B.: Relativistic theory of excitation and ionization of Rydberg atomic systems in a Black-body radiation field. J. Phys. Conf. Ser. **810**, 012047 (2017)
31. Svinarenko, A.A., Khetselius, O.Y., Buyadzhi, V.V., Florko, T.A., Zaichko, P.A., Ponomarenko, E.L.: Spectroscopy of Rydberg atoms in a Black-body radiation field: relativistic theory of excitation and ionization. J. Phys. Conf. Ser. **548**, 012048 (2014)
32. Buyadzhi, V.V., Zaichko, P.A., Antoshkina, O.A., Kulakli, T.A., Prepelitsa, G.P., Ternovsky, V.B., Mansarliysky, V.F.: Computing of radiation parameters for atoms and multicharged ions within relativistic energy approach: advanced Code. J. Phys. Conf. Ser. **905**, 012003 (2017)
33. Buyadzhi, V.V., Kuznetsova, A.A., Buyadzhi, A.A., Ternovsky, E.V., Tkach, T.B.: Advanced quantum approach in radiative and collisional spectroscopy of multicharged ions in plasmas. In: Jenkins, S., Kirk, S.R., Maruani, J., Brändas, E. (eds.) Advances in Quantum Chemistry, vol. 78, pp. 171–191. Elsevier, Cambridge, MA (2019)
34. Khetselius, O.Y.: Optimized relativistic many-body perturbation theory calculation of wavelengths and oscillator strengths for Li-like multicharged ions. In: Jenkins, S., Kirk, S.R., Maruani, J., Brändas, E. (eds.) Advances in Quantum Chemistry, vol. 78, pp. 223–251. Elsevier, Cambridge, MA (2019)

Detection of Chaotic Behavior in Dynamical Systems Using a Method of Deformable Active Contours

Alexander Ruchkin and Constantin Ruchkin

Abstract In this article we investigated the problem of detecting the chaotic behavior of the dynamical system with a Hamiltonian structure using pattern recognition methods. We carried out a numerical constructed of the phase space structure of these dynamical system which represented on 2D Poincare' sections of a special points cloud in chaotic cases. This chaotic regions are characterized by many various bad formalizing analytically forms. We are classified these forms on 2d sections in simply and multiply connected, inside and outside located. We adapted a deformable active contours method for closed curves to automatic detecting these chaotic regions of the dynamical system.

Keywords The dynamical systems · The Hamiltonian systems · The regular and chaotic behavior system · The recognition image methods · The active contour method

1 Introduction

Currently existing methods of the KAM theory of research of dynamical systems make it possible to unambiguously characterized of a dynamical system and indicate its behavior under given initial conditions: regular, quasi-regular, or chaotic. However, in practice it is quite difficult to conduct a full analytical study of dynamic systems since many systems have a large dimension and many parameters. As a result, the solutions obtained are very local and rude. In most cases, when dynamic systems are studied by numerical methods, there are also errors and cumulative calculation errors. It is rather difficult to investigated regular global cases by numerical methods.

A. Ruchkin · C. Ruchkin (✉)
National Technical University of Ukraine "Igor Sikorsky Kyiv Polytechnic Institute", IASA, Kyiv, Ukraine

J. Awrejcewicz (ed.), *Perspectives in Dynamical Systems II: Mathematical and Numerical Approaches*, Springer Proceedings in Mathematics & Statistics 363,
https://doi.org/10.1007/978-3-030-77310-6_13

Now for the analysis of nonlinear dynamical systems are increasingly used new computer-based methods: statistical methods of ergodic theory, statistical forecasting methods, numerical methods of research of high accuracy and performance, computer cognitive research methods, methods of computational and artificial intelligence. New approaches give better results if they combine analytical, numerical and algorithmic idea simultaneously.

Using pattern recognition methods in conjunction with numerical integration methods will reduce the effect of accumulation errors over time and give a more accurate result. The numerical researches of phase space, which consist of set of not intersected phase trajectories, have convenient to make by means of Poincare's sections. Poincare's section, which are constructed in the phase space, have dimensionality on unit is less than dimensionality of researched dynamical system. The exceptional interest the dynamical systems of the third and fourth order is represented. The result of these researches can be displayed graphically on the computer monitor. Poincare's sections images on a plane or in space accordingly. If points of a phase flow form on a curve it is possible to speak about the regular behavior of Hamilton systems. The cloud of points appearing in section of Poincare of a phase flow will testify to approach of a chaotic behavior of system.

In this paper, the problem of constructing an intelligent computer system for the automatic investigation of dynamic systems has continued [1, 2]. A mathematical and algorithmic apparatus has developed to highlight chaotic areas. The detection of the boundary of chaotic regions has allowed the detection of the boundary of regular regions. Therefore, the task of identifying chaotic regions is also important and necessary in the study of dynamical systems.

2 Actual Related Researches

Currently, many researchers pay much attention to the investigation of the regular and chaotic behavior of dynamical systems of various origin using new combined methods—analytical and numerical methods. Of particular interest are methods and models such as: models of an object, a pattern, a recognition and recognition training; the 0-1 test for chaos; method of smaller alignment index (SALI); method of swarm optimization localization periodic orbit; active contour method detection chaotic regions.

Next we consider them in more detail.

2.1 Pattern Recognition Method of Dynamical System

For the first time mathematical models of pattern recognition for dynamical system Neumark Yu. has discussed in the article [3]. Further the dynamic systems are studied by methods of pattern recognition scientists: Neumark Yu. I., Teklina L.G., Kotelnikov I.V in [3–6].

In the works [4, 5] authors propose used to models of an object, a pattern, a recognition and recognition training. The possibilities of using pattern recognition methods to study mathematical models with a large number of parameters are discussed. The principal point is to study models by constructing phase and parametric portraits. This allows one to solve problems of predicting the states of the object described by the mathematical model in hand and controlling the object and analyzing and studying problems that follow from the particular content of the model. Examples of three known mathematical models are given to illustrate this problem.

In [6] the problem of constructing the structure of the phase space of the computer methods of pattern recognition. Practical construction of a computer for a specific phase portrait of a dynamical system is based on a certain way calculated array of end segments of the phase trajectories in the forward and backward time. To build a computer structure of the phase space authors had to solve next problems: the definition of the form of the phase trajectory (state of equilibrium, limit cycle, chaotic motion); finding the limit of stable subsets (attractors) and select areas of attraction. To reach these goals, they used the methods of data mining and pattern recognition. So if the phase trajectory seen as a time series (data set), the task of analyzing the form of the phase trajectory is reduced to the construction of training set for this set of data. Cluster analysis of data from the training set and the construction of local decision rules are suitable for finding attractors. The problem of recognition of attractive region for sustainable building possible subsets solved by dividing the decision rules.

2.2 *The 0-1 Test for Chaos*

Another research method is called the 0-1 test for chaos. The development of this method for detecting chaotic cases is carried out Gottwald G.A., Melbourne I in series [7–15] and Zachilas L., Psarianos I. in [16].

In work [7] authors review theoretical and practical aspects of the 0-1 test for deterministic dynamical systems. The test is designed to distinguish between regular, i.e. periodic or quasi-periodic, dynamics and chaotic dynamics. It works directly with the time series and does not require any phase space reconstruction. This makes the test suitable for the analysis of discrete maps, ordinary differential equations, delay differential equations, partial differential equations and real world time series. To illustrate the range of applicability authors apply the test to examples of discrete dynamics such as the logistic map, Pomeau-Manneville intermittency maps with both summable and nonsummable autocorrelation functions, and the Hamiltonian standard map exhibiting weak chaos. In article authors also consider examples of continuous time dynamics such as the Lorenz-96 system and a driven and damped nonlinear Schrodinger equation. Finally, they show the applicability of the 0-1 test for time series contaminated with noise as found in real world applications.

In work [16] authors perform the stability analysis and they study the chaotic behavior of dynamical systems, which depict the 3-particle Toda lattice truncations through the lens of the 0-1 test, proposed by Gottwald and Melbourne. Authors prove that the new test applies successfully and with good accuracy in most of the cases. Authors perform some comparisons of the well-known maximum Lyapunov characteristic number method with the 0-1 method, and they claim that 0-1 test can be subsidiary to the LCN method. The 0-1 test is a very efficient method for studying highly chaotic Hamiltonian systems of the kind and is particularly useful in characterizing the transition from regularity to chaos.

2.3 Method of Smaller Alignment Index (SALI)

The development of method of Smaller Alignment Index (SALI) for detecting chaotic cases is carried out by a group of researchers: Ch. Skokos, Ch. Antonopoulos, T.C. Bountis, M.N. Vrahatis, N. Kyriakopoulos, V. Koukouloyannis, P. Kevrekidis in [17–21].

In article [17] they used the Smaller Alignment Index (SALI) to distinguish rapidly and with certainty between ordered and chaotic motion in Hamiltonian flows. This distinction is based on the different behavior of the SALI for the two cases: the index fluctuates around non–zero values for ordered orbits, while it tends rapidly to zero for chaotic orbits. They present a detailed study of SALI's behavior for chaotic orbits and show that in this case the SALI exponentially converges to zero. Exploiting the advantages of the SALI method, in work authors demonstrate how one can rapidly identify even tiny regions of order or chaos in the phase space of Hamiltonian systems of 2 and 3 degrees of freedom.

In [18] authors investigate a system of vortex dynamics in an atomic Bose-Einstein condensate (BEC), consisting of three vortices, two of which have the same charge. These vortices are modeled as a system of point particles which possesses a Hamiltonian structure. This tripole system constitutes a prototypical model of vortices in BECs exhibiting chaos. By using the angular momentum integral of motion they reduce the study of the system to the investigation of a two degree of freedom Hamiltonian model and acquire quantitative results about its chaotic behavior. This investigation tool is the construction of scan maps by using the Smaller Alignment Index (SALI) as a chaos indicator. Applying this approach to a large number of initial conditions they manage to accurately and eminently measure the extent of chaos in the model and its dependence on physically important parameters like the energy and the angular momentum of the system.

2.4 Method of Swarm Optimization for Locating Periodic Orbits

Method of swarm optimization for dynamical system used of C. Skokos, K. Parsopoulos, P. Patsis, M.Vrahatis, Antonopoulos Ch., Bountis T. in [18–20].

In [18] authors propose particle swarm optimization (PSO) as an alternative method for locating periodic orbits in a three-dimensional (3D) model of barred galaxies. Authors develop an appropriate scheme that transforms the problem of finding periodic orbits into the problem of detecting global minimizers of a function, which is defined on the Poincare surface section of the Hamiltonian system. The method succeeded in tracing the initial conditions of periodic orbits in cases where Newton iterative techniques had difficulties.

In work [19] the detection of periodic orbits bears significance for the study of nonlinear mappings, since they can reveal crucial information on their dynamics. Recently, population–based stochastic optimization algorithms were introduced to address problems where traditional gradient–based approaches failed. The efficiency of these approaches in a applications, triggered further research towards the development of more efficient variants. This work presents the principal concepts of applying concurrent stochastic population–based approaches for the detection of periodic orbits, and also reports new results attained by the application of Memetic Algorithms on well–known chaotic maps for periodic orbits with high period.

In [20] authors are proposed a new approach for the identification of the resonances appearing in symplectic maps. In the proposed methodology, they make use of Evolutionary Algorithms which are population based search strategies used for global optimization. Authors have applied the proposed methodology to the 2-dimensional (2D) Hénon map and obtained promising results which can be generalized to symplectic maps of higher (2m) dimensions. As is well-known, such maps are representative of Hamiltonian systems and occur in many physical applications.

2.5 Method of Clustering of the Phase Trajectory

The paper [21] describes an approach to quantitative analysis of multivariate dynamic system in phase space by means cluster analysis. The system is used as mathematical model for various living systems. The model is used in various applications. One of the related problems is to represent a phase trajectory as a sequence of clusters to classify the system's state.

The algorithm for partitioning a phase trajectory into clusters is presented. Input data for the algorithm is a data matrix which is corresponds to a set of sequential samples of the given phase trajectory. Optional parameters are dimension of the space in which the clusters lie, and phase trajectory noise variance. The algorithm results in a tree-like graph. The graph nodes contain given phase trajectory clusters and might be used for system's state classification.

Phase trajectory of a dynamic system with Lorenz attractor is considered as a test problem to demonstrate the approach. The initial phase trajectory lies in 3D-space. It was projected into N-dimensional space and distorted with non-correlated additive Gaussian noise. The given phase trajectory was partitioned into clusters using the described algorithm. The clusters make a tree T. The root of the tree corresponds to the phase trajectory that lies in r-dimensional space Rr. The next level of the tree consists of cluster nodes that lie in (r-l)-dimensional space, etc., up to the last level that corresponds to one dimensional cluster nodes. The algorithm was examined with various test trajectories.

The above studies show the importance and relevance of developing new methods for investigation the regular and chaotic behavior of dynamical systems. They allow obtaining new results for existing and new dynamical systems. One of such methods is a computer analysis of chaotic region of the Poincare section for dynamic systems using active contour methods, which is proposed in this article and will be considered in the next section. This method is a new technique used in the study of dynamical systems, and is of great interest for constructing a general system for the automatic detection of regular and chaotic behavior of dynamical systems [1, 2].

3 Active Contour Method for Dynamical System

3.1 Active Contour Method and There Modification

So, in this chapter we investigation images section Poincare for detection closed chaotic clouds. One of the general task in image processing to split the image into several parts like objects: foreground and background set of pixels. This procedure called segmentation. There is two way to define the segments of the image. The first if we locate the surface pixels of the object areas. The second way to do that if we define the boundaries of the area of interested. Wide spectrum of the different segmentation techniques known from the simple threes holding—during region growing and edge detection methods -until different kind of machine learning techniques, like clustering methods or neural networks segmentation. The active contour models belong to the class of the boundaries methods and the edge detection methods.

The active contour method considers various modifications, their features and applications. The active contour model algorithm (snake) introduced by Michael Kass, Andrew Witkin and Demetri Terzopoulos in 1988 [22]. A snake is an energy minimizing technique guided by external constraint and influenced by image forces. The algorithm deforms a contour to lock onto features of interest on an image. Usually the features are edges, lines or boundaries. It is working in 2D, 3D or in even higher dimensions. Its 3D version is often called as deformable models or active surfaces. But the original algorithm, introduced by Kass, suffers from some problems. Now the snake method has of two kind modifications (geometric active contours (geodesic active contours) and active model method).

Geometric active contours are designed to highlight the segments of an object based on the fact that there is a clear boundary between the object and its segment. In this method, the circuit tends to maximize smoothing and in the absence of clear boundaries—tends to the point, to solve the problem of closing the path to the point—added the force of the contour. This modification is designed to solve the problem of highlighting complex contours that could be divided into explicit individual segments. Since the force that breaks the circuit is set arbitrarily, in most algorithms, it becomes unclear exactly how accurately it will affect the overall functionality of the contour. To solve this problem, geodesic active contours have been developed in which the force of the contour is calculated on the basis of force, which determines the path of the path to the boundaries of the object. This modification solves the problem of interaction of forces that affect the contour, which allows you to enter the coefficients of forces that affect the contour and balance them.

The active shape method [23] is a modification of the classic active contour method. Its peculiarity is that it uses the original shape of the contour and the approximation goes to it. The main difference is in the parameters of internal energy—if the active contours internal energy tends to approach the shape of the circle, in active models—to the form, minimally different from the original. In the active model algorithm, at every step of the algorithm, a valid deviation from a given path is set. The advantage of this modification is that it improves the accuracy of calculating the contour of an object, but at the same time reduces the scope of the already configured algorithm. The reason for the emerging flaw is that because of the existing approximation pattern, the variability of the resulting contours decreases. But if you want to detect the boundaries of pre-known objects, or boilerplate objects—this modification will have a big advantage over the classic algorithm. This modification solves the problem of highlighting the outline of an object by active contours, in case the form of the initial approximation with acceptable deviations is known, but the exact position of the initial approach is not known.

3.2 Development Mathematical Model of Active Contour Model

The basis of the active contours methods (snakes) is that the contour even before the contouring of the desired object has a kind of initial form and, due to various conditions affecting it, changes its shape (contouring object)—deformed. In this model, the task of finding the boundaries of an object is formulated as a change in the positions of the contour points to the new ones, in which the functionality of E—“energy” reaches a minimum. The behavior of the active contour and its properties are completely determined by its functionality E (energy). The energy of the contour depends on its shape, the size of the contour and its position in the image. It is recorded as the sums of two functions: "internal" energy—$Eint$ and "external" energies—$Eext$ (1).

$$E\left(v_i\right) = aEint\left(v_i\right) + bEext\left(v_i\right) \tag{1}$$

where a and b—weights coefficients;
$Eint$—the inner energy of the points;
$Eext$—the external energy of the points.

The inner energy is the energy of the breaking of the contour. This option is responsible for regulating the shape of the contour. The inner energy minimizes the breaking of the contour. External energy is responsible for the inconsistency of the contour of the image. The outer contour seeks to minimize the difference between the contour and boundary of the object that is contoured, and the smaller the difference, the less the value of the external energy. As described above, the basic concept of this contouring method is to find a new position for each point of the contour, by identifying a position with a minimum amount of two energies that affect the contour: external and internal, among the matrix of energies For every point. The outer one is responsible for changing the shape of the contour, and the inner one is responsible for reducing the brokenness of the contour line. To get a new position for the point—calculated energy matrix (2). After calculation, they are added elementally and among them a position with a minimum value is searched.

$$Ematrix = \begin{pmatrix} E[0][0](v_i) & \dots & E[0][N](v_i) \\ \dots & E[j][k](v_i) & \dots \\ E[N][0](v_i) & \dots & E[N][N](v_i) \end{pmatrix} \tag{2}$$

To describe the classic algorithm, let the outline be labeled as V(2).

$$V = \{v_1, \dots, v_N\} \tag{3}$$

where $v_1, \dots v_N$—these are the dots of this contour;
N—the number of these points.

Each point is a pixel of an image, with coordinates x and y, so $v_i = (x_i, y_i)$, and $i = \{1, \dots, N\}$ Then, for each point of the circuit, the function of finding a mini mum of energy will be decided, for the amount of energies of those forces acting on the contour. Energy, in this case, is called a function that converts pixels of the input image and other parameters into a field of forces and is considered by formula (1) and consists of calculations of internal energy and external. Internal energy—energy is responsible for preserving the shape of the contour. It consists of two parts: continuity energy and energy expansion, and is defined by formula (4).

$$Eint\left(v_i\right) = c \cdot Econ\left(v_i\right) + d \cdot Ebal\left(v_i\right) \tag{4}$$

where c and d—weight coefficients;
$Econ$—the energy of continuity;
$Ebal$—expansion energy.

Continuity energy (1) is responsible for maintaining a sequence of dots in the contour (no intersections in the contour line). If the circuit is unclosed—this energy directs it to the deformation in a straight line, for cases with closed circuits—the direction goes into a circle. For each point of the circuit, this energy is calculated relative to the deviation of the point position relative to the adjacent points (5)–(7).

$$Econ_{jk}(v_i) = \frac{1}{l(V)} \cdot \left\| p_{jk}(v_i) - \gamma (v_{i+1} + v_{i-1}) \right\|^2 \quad (5)$$

$$\gamma = \frac{1}{2 \cdot cos\left(\frac{2\pi}{n}\right)} \quad (6)$$

$$l(V) = \frac{1}{n} \cdot \sum_{i=1}^{n} \| v_{i+1} - v_i \|^2 \quad (7)$$

where $p_{jk}(v_i)$—position of the point accordingly of the matrix element;
γ—smoothing factor;
$l(V)$—overall smoothness of the contour.

Expansion energy provides extension or compression of the contour, provides a general change in the area of the contour cover. The expansion energy matrix, for the contour points, is calculated relative to the deviation from the normal vector to the point (8).

$$Ebal_{jk}(v_i) = n_i \cdot \left(v_i - p_{jk}(v_i)\right) \quad (8)$$

where n_i—vector normal to point v_i.

External energy is the energy responsible for the movement of the contour points to the brightness of the image. It consists of two energies: the energy of the image and the energy of the gradient, and is determined by the formula (9).

$$Eext(v_i) = m \cdot Emag(v_i) + g \cdot Egrad(v_i) \quad (9)$$

where m and g—weight coefficients;
$Emag$– the image energy;
$Egrad$ is the energy of the gradient.

Image energy is responsible for the movement of contour points to positions with high image intensity values, and is defined by formula (10).

$$Emag_{jk} = I\left(p_{jk}(v_i)\right) \quad (10)$$

From the energy of the gradient, depends on the movement of points to places with the maximum difference in brightness of the image, as reflected in the formula (11).

$$Egrad_{jk} = \nabla \left[G_\sigma \cdot I \left(p_{jk} \left(v_i \right) \right) \right] \quad (11)$$

After calculating the total energy, the point takes a position in which there is a minimum of energy among the array of $E_{jk}(v_i)$ (12).

$$v_i^{'} = E_{jk} \left(v_i \right) \rightarrow min. \quad (12)$$

These are the basic principles of calculating the energies of the contour. In a large number of modifications, additional forces are added that affect the contour, and the influence of forces on the change of the contour changes. Although this method is not designed to work with video image, but if you solve the problem with computing power and reduce it to the required level—this method will be extremely effective in solving the problem. The reason why this method is appropriate is that the method partially performs the condition of adaptability (any information about the outline is needed only in the first steps of the method—the initial approximation). So, if you solve the problem with computing power and implement adaptive execution of the initial approximation, this method will be able to effectively solve the problem. Due to the interest in this method, analysis of its various modifications were carried out to obtain more information about the possibilities of the method.

As mentioned earlier, this paper will use the previously developed algorithm described in previous paragraph as the basis. By this method, the result of calculating the energies acting on the circuit will be vectors of action of forces and distances to the minimum energies of these forces. In this case, the action vector and distance can be expressed through the coordinates of the position of the point to which this force is trying to move the starting point of the path. Since the vector of force is the opposite of energy, and the task is to find a position with the value of the minimum energy amount of forces acting on the circuit, then for this method, the solution to the problem of energy minimization will be to find a point relative to which the amount the energies needed to shift it to the calculated positions of the active forces will be minimal. So the positions in which the contour point moves relative to the energies acting on it, have the appearance of the points of the image, the new position of the point of the contour can be calculated as a point, the sum of races (Figs. 1 and 2).

3.3 Adaptive Method of Active Contour for Dynamical System

In this method, classical energy calculations were replaced by calculations of energies along vectors and search for positions on each energy separately, instead of

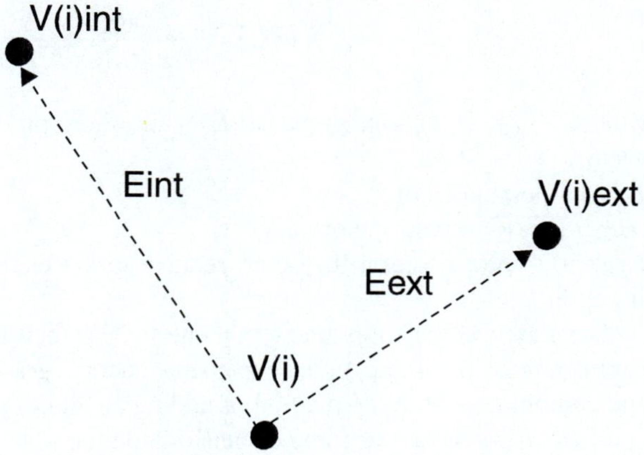

Fig. 1 Calculating the action of forces influencing the contour

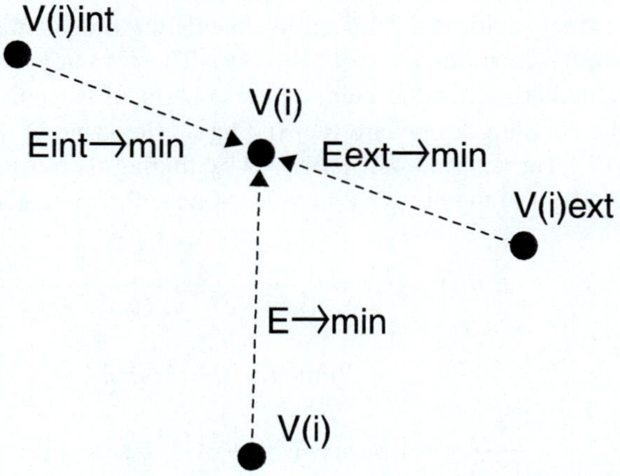

Fig. 2 Calculating a new position for the contour point

calculating the total matrix of energies. The classic internal energy calculation (4) has been replaced by a position calculation for a minimum of energy (13)–(14) over the vector (16), which is directed towards the middle between the adjacent points, for the contour point of the $\overline{vect_i}$ (17) and depends on the parameters of the image. All resulting vectors are normalized under formulas.

$$Eint_j\,(v_i) = \frac{1}{\left(\overline{vect_i} \cdot I\left(p_j\,(v_i)\right)\right)} \tag{13}$$

$$Pint(v_i) = Eint_j\,(v_i) \to min; \tag{14}$$

$$\overline{vect_i} = p\,(middle_i) - p\,(v_i)\,; \tag{15}$$

$$p\,(middle_i) = \frac{(p\,(v_{i-1}) + p\,(v_{i-1}))}{2}; \tag{16}$$

$$\overline{vect_i} = \frac{\overline{vect_i}}{length(\overline{vect_i})} \tag{17}$$

where $I(p_j(v_i))$—image value (after threshold processing) at a point along the vector;
$p(v_i)$—point position;
$length(\overline{vect_i})$—long vector $\overline{vect_i}$;
$Pint(v_i)$—new position for point, relative to the energy of the action of internal strength.

The classic calculation of external energy was modified by a grouping of several parameters at the same time: expansion energy, gradient energy, image energy, and combination with the method of adding an additional contour. External energy consists of the impact of force expansion relative to the normal vector and relative to the vector to the center of the mass of the figure. The position relative to the normal vector is calculated by calculating the position with a minimum of energy during the normal vector (18)–(19). The normal vector is calculated relative to the neighboring, for the point of the contour of the points (20) and is directed inside the contour, or outward (depending on the value of the image), and normalized by (19). The minimum of energy is by finding the positions, the closest and the most gradient of the image. Where K—Image Processing Threshold.

$$Enorm_j(v_i) = \frac{1}{\left(\overline{vect_i} \cdot I\left(p_j(v_i)\right) \cdot \left\|G_\sigma \cdot I\left(p_j(v_i)\right)\right\|\right)} \tag{18}$$

$$Pnorm(v_i) = Enorm_j(v_i) \to min; \tag{19}$$

$$\overline{vect_i(x,y)} = \begin{cases} p(v_{i+1})(y,-x) - p(v_{i-1})(y,-x), \; if \; I\left(p_j(v_i)\right) > K; \\ p(v_{i+1})(-y,x) - p(v_{i-1})(-y,x), \; if \; I\left(p_j(v_i)\right) < K. \end{cases} \tag{20}$$

We calculate the relative position of the figure's center of mass using the minimum energy position vector (21)–(24). The center of the mass of the contour is calculated by searching the center of the masses for the supporting points of the contour (21). The vector is calculated against the point of the contour of the v and the center of the masses (24), and is directed inside the contour, or outward (depending on the value of the image), and then normalized by formula (24). The minimum of energy is by finding the positions, the closest and the most gradient of the image.

$$Ecenter_j(v_i) = \frac{1}{\left(\overline{vect_i} \cdot I\left(p_j(v_i)\right) \cdot \left\|G_\sigma \cdot I\left(p_j(v_i)\right)\right\|\right)}; \tag{21}$$

$$Pcenter(v_i) = Ecenter_j(v_i) \to min; \tag{22}$$

$$p(center) = \frac{1}{N}\sum_{i=1}^{N} p(v_i); \tag{23}$$

$$\overline{vect_i} = \begin{cases} p(center) - p(v_i), & I(p_j(v_i)) > K; \\ p(v_i) - p(center), & I(p_j(v_i)) < K. \end{cases} \tag{24}$$

After calculating the positions of points with a minimum energy value of forces, relative to the point of the contour, a new position is sought for the point for which the amount of energies of forces directing this point of the outline of calculated positions will be minimal (25).

$$p\left(v_i^{'}\right) = \frac{(a \cdot Pint(v_i) + b \cdot Pnorm(v_i) + c \cdot Pcener(v_i))}{a + b + c}, \tag{25}$$

where a, b, c—action factors affecting the contour of forces.

The algorithm calculation consists of next stages:

(1) The stage of calculating the position of minimum energy for the force that moves the point of the circuit relative to the neighbor points.
(2) The stage of calculating the position of minimum energy for the force that moves the point of the contour relative to the normal vector to the contour at this point.
(3) The stage of calculating the position of minimum energy for the force that moves the point of the circuit relative to the center of the mass of the contour.
(4) The final stage at which a new position for the point, relative to the calculated positions for each of the active forces and the influence of these forces is calculated.

The sequence of the algorithm steps consists in calculating new positions for all points of the contour (see Fig. 3).

The peculiarity of this modification is that: calculations in two-dimensional arrays are replaced by calculations in one-dimensional, which adds speed to the calculation of new positions for the points of the contour; vectors are used to direct the change in point position. The basic principles of active contours are preserved: dynamic deformation of the contour and preserved, but modified, the basic principle of calculations and search of the energy of the contour.

4 The Practical Results and Mechanical Examples

In this section, we will consider the practical application of the active contour method for detecting chaotic regions in Poincare' sections. So, in the framework of constructing a general concept for studying dynamic systems, software and algorithmic support for an intelligent system has been developed that automatically detects regular and chaotic regions of a dynamic system by Poincare' sections [2].

Also in [2], the problem of detecting regular cases on Poincaré sections was considered using an example of a mechanical system.

The area detection program performs the following steps.

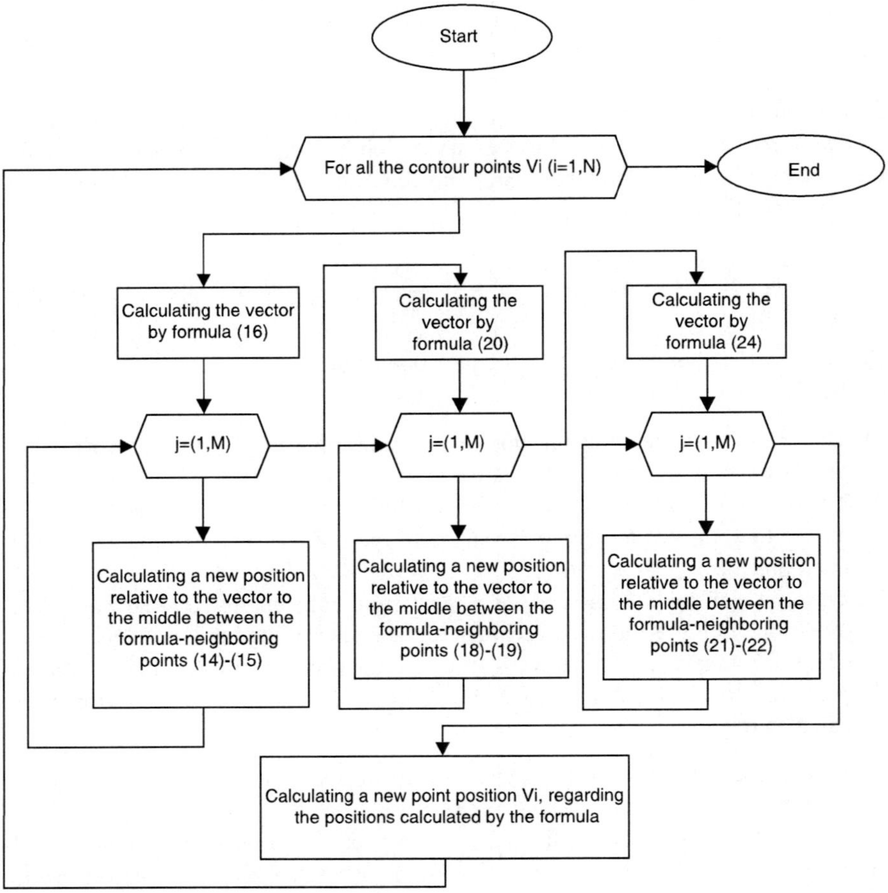

Fig. 3 Algorithm for a modified active contour method

(1) Firstly, we select the image and selects the σ values for the Gaussian smoothing.
(2) Then we select the initial position of the snake by clicking on the image and selecting control points which are later interpolated into a contour.
(3) As the initial approximation, a set of points close to the regular case is chosen. The more accurately you can select the starting points, the more accurately the boundary of the chaotic region will be detected.
(4) We specifies various control parameters for the snake. These include α—specifies the elasticity of the snake; this controls the tension in the contour by combining with the first derivative term; β—specifies the rigidity in the contour by combining with the second derivative term; γ—specifies the step size; κ - acts as the scaling factor for the energy term; $Eline$—weighing factor for intensity based potential term. $Eedge$—weighing factor for edge based potential term. $Eterm$—Weighing factor for termination potential term.

(5) Then we specifies the number of iterations for which contour's position is to be computed.
(6) For the standard test cases provided with the assignment, predefined values for all the variables are hard-coded for convenience, though they can be changed at any point.

The algorithm proposed in the previous section requires a fairly complex setup, the main points of which are as follows.

If the snake is initialized "too far" from the object boundary, it is possible that the contour may not be able to converge onto object boundary. An increase in the amount of smoothing (σ value) increases the range from which a snake can converge onto an object smoothing. This can be seen as that blurring of edges increases there "enactment" area. If the energy scaling factor is "too big" for a given image than though active contour can converge onto image boundary but keeps on wiggling along the object boundary. In most cases it ends up losing track on object boundary. The $Eedge$ value must be carefully chosen in case of binary images which has very high gradient values. The $Eterm$ defines the contribution of curvatures to the overall energy term. When weight associated with this component in dominant in the overall energy, the snakes seems to be attracted to corners first before converging onto the object boundary. In cases where its weight is not dominant, edges are traced before the corners by the snake.

From the KAM theory, we know that regular cases border on chaotic cases, therefore, highlighting the boundary of a chaotic region will allow us to distinguish the boundary of the region of regular cases.

A computer study was carried out for the problem considered in [1, 2]. For a given Hamiltonian system, chaotic regions can have a complex shape, be single or multiply connected. The studies performed allowed us to identify typical cases of chaotic regions.

(1) The chaotic region is simply connected and inside located (almost simply connected). Outside the area is a regular occurrence and inside is a chaos area (Fig. 4).
(2) The chaotic region is simply connected and outside located. Outside is chaos area and inside is a regular region (Fig. 5).
(3) The chaotic region is multiply connected and outside. Outside, it borders on the regular case, and on the inside is both chaos and the regular case (Fig. 6).

Figures 4a), 4b), 4c) show the process of constructing chaotic and regular regions using a computer program with different time intervals for case 1). The longer the time interval, the more accurate the image areas will be. Figures 4d), 4e), 4f) show the process of detecting chaotic regions. So the original image (Figure 4d) is converted to shades of gray and filtered by a Gaussian filter. On the Figure 4e)—the initial position of the contour is determined; on the Figure 4f)—shows the final contour, which is boundary the chaotic region.

Figures 5a), 5b), 5c) show the process of constructing chaotic and regular regions using a computer program with different time intervals for case 2). The chaotic

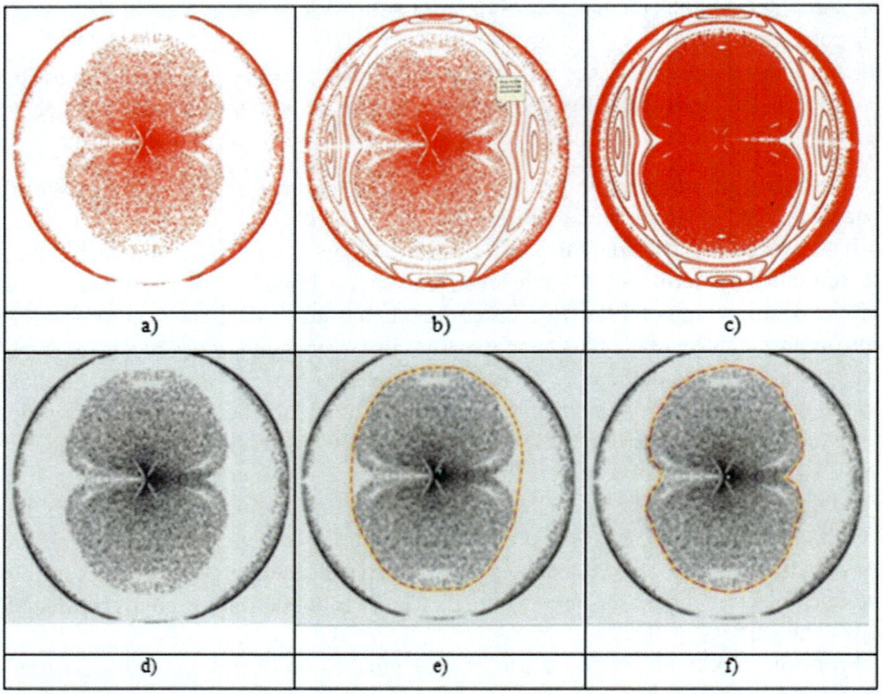

Fig. 4 The chaotic region is simply connected and inside located

region is simply connected and outside located. Outside is chaos area and inside is a regular region (Figure 5). The longer the time interval—the more accurate the image areas. For detection of chaotic regions from these images, Figure 5a) was selected. Figures 5d), 5e), 5f) show the process of detecting chaotic regions. So the original image (Figure 5e) is shown converted to shades of gray and processed by a Gaussian filter. On the Figure 5e) the initial position of the contour is determined. On the Figure 5f) the final contour is shown. It is bounder the chaotic region. Figures 5g), 5h), 5i) demonstrate the process of detecting next chaotic regions. This detection is more accurate.

On the Figures 6a), 6b), 6c) are shown the process of constructing chaotic and regular regions using a computer program with different time intervals for case 3). The chaotic region is multiply connected and outside. Outside, it borders on the regular case, and on the inside is both chaos and the regular case. Detection of chaotic regions for these case image was selected Figure 6b). Figures 6d), 6e), 6f) show the process of detecting chaotic regions. So the original image (Figure 6d) is shown converted to shades of gray and processed by a Gaussian filter. The image (Figure 6e) shows the initial position of the contour. On the image (Figure 6f) shows the final contour, which is bounder the chaotic region. Figure 6b) was

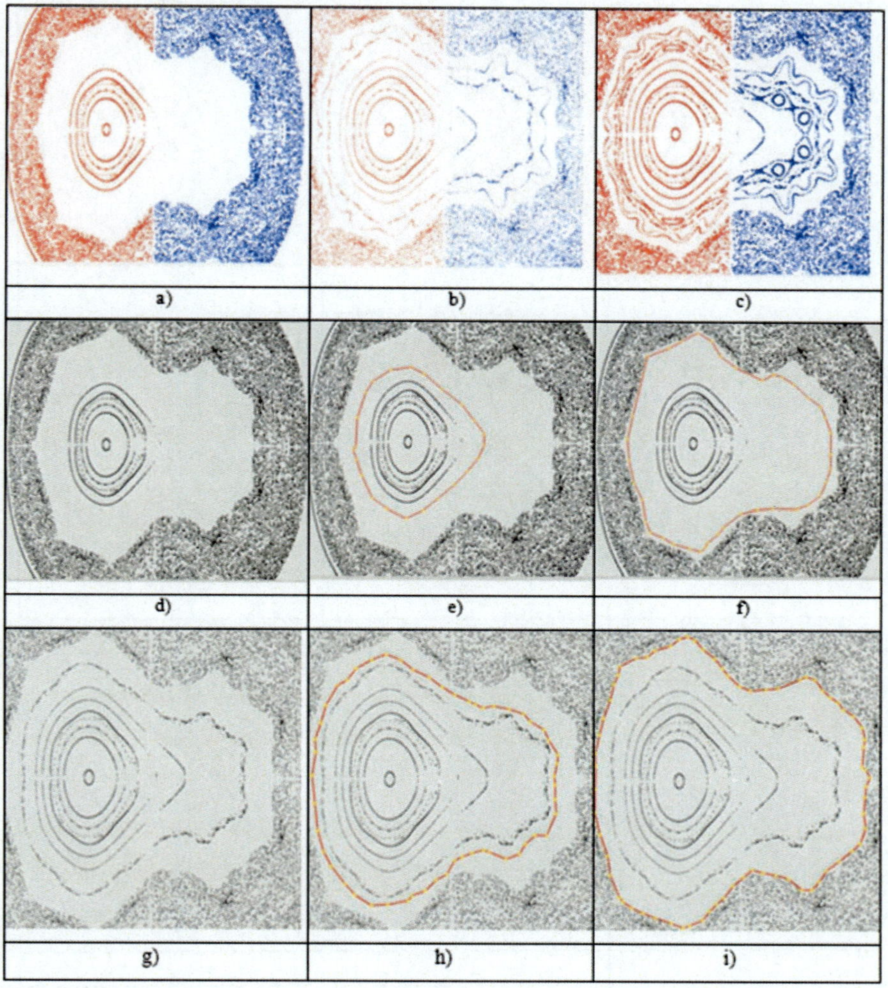

Fig. 5 The chaotic region is simply connected and outside located

selected Figures 6g), 6h), 6i) show the process of detecting two chaotic regions more accurate.

5 Conclusions

In this article we had investigated the problem of automatic detecting the chaotic behavior of the dynamical system. We had carried out a numerical study of the phase space of a dynamical system with a special Hamiltonian structure, represented on

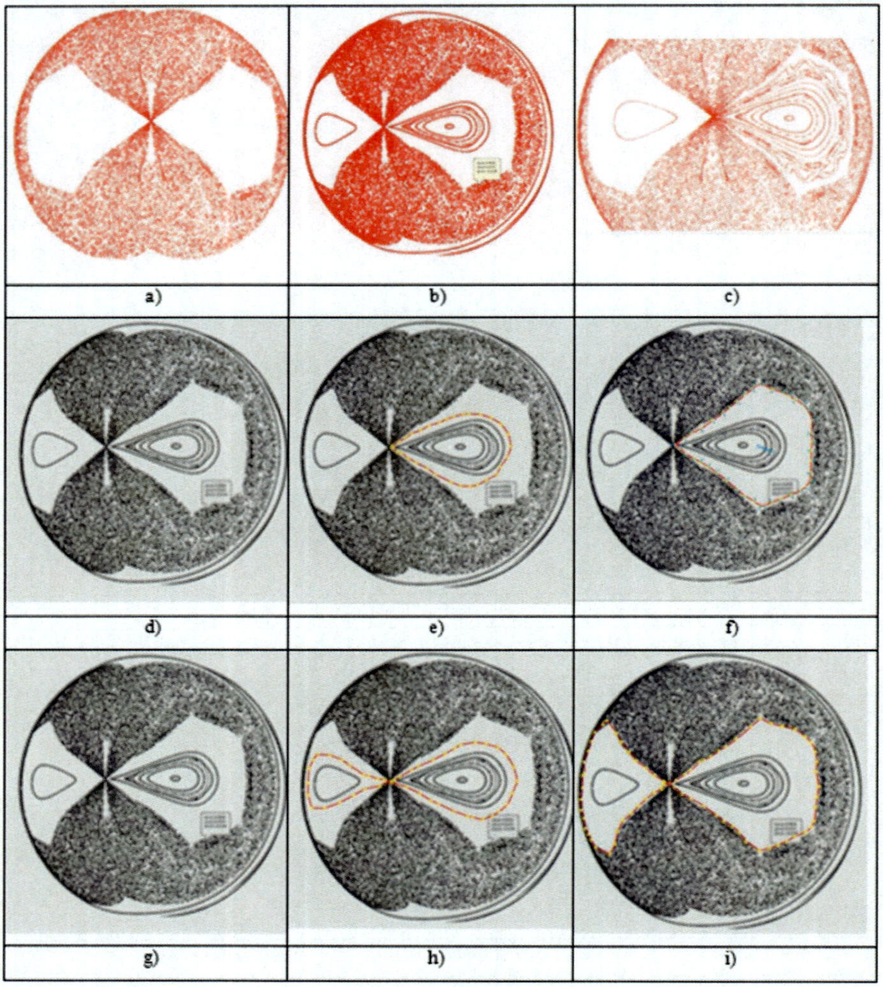

Fig. 6 The chaotic region is multiply connected and outside located

Poincare' sections of a point cloud in chaotic cases with help method active contour. We investigated various forms of chaotic regions and developed an algorithm for recognizing their in some special cases. We considered the practical application of the active contour method for detecting chaotic regions in Poincare' sections of the mechanical system.

References

1. Ruchkin, C.: The general conception of the intellectual investigation of the regular and chaotic behavior of the dynamical system hamiltonian structure. Applied Non-Linear Dynamical Systems. Springer Proceedings in Mathematics and Statistics, vol. 93. Springer, Cham (2014). https://doi.org/10.1007/978-3-319-08266-0_17
2. Ruchkin, K.A.: Development of computer system for analysis of Poincare' sections / K.A. Ruchkin. "Artificial intelligence". - Donetsk: NAS Ukraine. - 2009. - No 1., pp. 83–87 (2009)
3. Neumark, Yu.I.: The multidimensional geometry and image recognition. Soros Educational Magazine Number 7, pp. 119–123 (1996)
4. Nejmark Yu.I.: New the approach to numerical research of specific dynamic systems by pattern recognition and statistical modeling methods / Yu.I. Nejmark, I.V. Kotelnikov, L.G. Teklina. News High Schools "Application-oriented nonlinear dynamics". - 2010. - 18. - No2, pp. 3–15 (2010)
5. Neimark, Yu.I., Teklina, L.G.: On possibilities of using pattern recognition methods to study mathematical models. Pattern Recognit. Image Anal. **22**(1), 144–149 (2012). https://doi.org/10.1134/S1054661812010282
6. Neimark, Yu.I., Kotelnikov, I.V., Teklina, L.G.: Investigation of the structure of the phase space of a dynamical system as a pattern recognition problem. In: Conference on MMRO-12, pp. 177–180. M. MaksPress (2005)
7. Gottwald, G.A., Melbourne, I.: A new test for chaos in deterministic systems. Proc. Roy. Soc. A **460**, 603–611 (2004)
8. Gottwald, G.A., Melbourne, I.: Testing for chaos in deterministic systems with noise. Physica D **212**(1–2), 100–110 (2005)
9. Gottwald, G.A., Melbourne, I.: Comment on "Reliability of the 0–1 test for chaos". Phys. Rev. E **77**, 028201 (2008)
10. Gottwald, G.A., Melbourne, I.: On the implementation of the 0–1test for chaos. SIAM J. Appl. Dyn. **8**, 129–145 (2009)
11. Gottwald, G.A., Melbourne, I.: On the validity of the 0–1 test for chaos. Nonlinearity **22**, 1367–1382 (2009)
12. Gottwald, G.A., Melbourne, I.: A Huygens principle for diffusion and anomalous diffusion in spatially extended systems. Proc. Natl. Acad. Sci. USA **110**, 8411–8416 (2013)
13. Gottwald, G.A., Melbourne, I.: Central limit theorems and suppression of anomalous diffusion for systems with symmetry. submitted. (2013). https://arxiv.org/abs/1404.0770
14. Gottwald, G.A., Melbourne, I.: A test for a conjecture on the nature of attractors for smooth dynamical systems. Chaos **24**, 024403 (2014). https://doi.org/10.1063/1.4868984
15. Zachilas, L., Psarianos I.: Examining the chaotic behavior in dynamical systems by means of the 0-1 Test. J. Appl. Math. **2012**, 681296 (2012). https://doi.org/10.1155/2012/681296
16. Kyriakopoulos, N., Koukouloyannis, V., Skokos, Ch., Kevrekidis, P.: Chaotic behavior of three interacting vortices in a confined bose-einstein condensate. Chaos **24**, 024410 (2013). https://doi.org/10.1063/1.4882169
17. Manos, Th., Skokos, Ch., Athanassoula, E., Bountis T.: Studying the global dynamics of conservative dynamical systems using the SALI chaos detection method. Nonlinear Phenomena Complex Syst. **11**, 171–176 (2008)
18. Skokos, C., Parsopoulos, K.E., Patsis, P.A., Vrahatis, M.N.: Particle swarm optimization: an efficient method for tracing periodic orbits in three-dimensional galactic potentials. Monthly Notices Roy. Astron. Soc. **359**(1), 251–260 (2005). https://doi.org/10.1111/j.1365-2966.2005.08892
19. Petalas, Y., Parsopoulos, K., Vrahatis, M.: Stochastic optimization for detecting periodic orbits of nonlinear mappings. Nonlinear Phenom. Complex Syst. **11**, 285–291 (2008)

20. Petalas, Y., Antonopoulos, Ch., Bountis, T., Vrahatis, M.: Detecting Resonances using Evolutionary Algorithms. In: roceedings of the International Conference of "Computational Methods in Sciences and Engineering"(ICCMSE 2006). CRC Press, Boca Raton (2019). https://doi.org/10.1201/9780429070655-106
21. Strijov, V., Shakin, V.: An algorithm for clustering of the phase trajectory of a dynamic system. Math. Commun. Suppl. **1**, 159–165 (2001)
22. Kass, M., Witkin, A., Terzopoulos, D.: snakes: active contour models. Int. J. Comput. Vis. **1**(4), 321–331 (1988)
23. Cootes, T.F., Taylor, C.J., Cooper, D.H., Graham, J.: Active shape models - their training and application. Comput. Vis. Image Understand. **61**(1), 38–59 (1995)

Dynamics of Sensing Element of Micro- and Nano-Electromechanical Sensors as Anisotropic Size-Dependent Plate

Marina Barulina, Alexey Golikov, and Sofia Galkina

Abstract Micromechanical electronics sensors are widely used in various fields of science and technology. The automotive industry, marine and space techniques, navigation equipment, military vehicle, consumer electronics, robotics, smart systems—this is not a complete list of areas, in which these sensors are used. Nanoelectromechanical sensors (NEMS) are the next step of evolution of micromechanical electronics sensors. Due to the nanosized of sensing elements and other components of NEMS, they need non-classical approaches for the study of their dynamics. One more problem of NEMSs is the dependence of their characteristics on material anisotropy or orthotropy. In the article, equations of motion of a sensing element of micro- and nanoelectromechanical sensors as an anisotropic size-dependent plate were obtained based on the modified couple-stress theory. The sensing element was considered as rectangular console plate under the distributed force at the bottom of the sensing element. The dynamic version of the principle of virtual displacements and the third-order theory of laminated composite plates and shells were used for obtaining the differential equations of motion and natural boundary conditions.

Keywords MEMS · NEMS · Size-dependent plate · Mathematical model · Modified couple-stress theory · Nanoelectromechanical sensors · Sensing element

1 Introduction

The development of microelectronics technology, methods of selective and ultra-deep etching, three-dimensional shaping led to the creation of micromechanical electronics systems (MEMS)—small-sized and cheap micromechanical devices, which made by using microelectronics industry technologies [1–8].

M. Barulina (✉) · A. Golikov · S. Galkina
Institute of Precision Mechanics and Control Problems, Russian Academy of Sciences, Saratov, Russia
e-mail: marina@barulina.ru

J. Awrejcewicz (ed.), *Perspectives in Dynamical Systems II: Mathematical and Numerical Approaches*, Springer Proceedings in Mathematics & Statistics 363,
https://doi.org/10.1007/978-3-030-77310-6_14

The modern world is already difficult to imagine without a wide variety of inertial information sensors and devices based on them. The automotive industry, marine and space techniques, navigation equipment, military vehicle, consumer electronics, robotics, smart systems—this is not a complete list of areas, in which inertial information sensors and devices are used.

MEMS sensors and devices are widely spread in consumer electronics—cell phones, cars, quadcopters, radio-controlled toys—flipping cars, helicopters, and even flying fairies.

Nowadays, the processes and performance characteristics of MEMS sensors, and micro dimensional devices are relatively well understood in the frameworks of classical mechanics of deformable bodies, and also in the frameworks of solid-state physics, and using such methods like oscillation theory method, heat factors methods, theory of elasticity and thermoelectricity methods, theory of deterministic chaos and others [9–13].

However, the development of MEMS sensors still going. As follows from literature analysis, the development of MEMS sensors goes in several directions:

1. Increasing the efficiency of the inertial information sensors, systems, and devices based on them, improving the design of the sensor and the device [14, 15].
2. Increasing the efficiency by means of compensation of influence and mutual influence various disturbing factors on device characteristics in the supporting software [16–18].
3. In reducing of MEMS sizes and occurrence of a new devices type—nanoelectromechanical sensors (NEMS) [19].

The first and second points can be achieved by using non-classical approaches to the study of the dynamics of MEMS elements [20–22], or by a more in-depth study of the influence of material options on sensor characteristics [23, 24].

At the same time, for NEMS, besides specific issues due to their extremely small sizes, the same issues remain relevant, as for the MEMS-sensors, for example, temperature, electric and mechanical noises influence, etc. But, in the case of small sizes of NEMS, classical approaches to resolving these issues may not be suitable.

Thus, it was shown in [25–29] that the size effects play an important role in microstructures (microbeams, crystals, plates). For example, in [30] it is experimentally proved that the dimensionless natural frequencies increase to about 2.1 times with the beam thickness decreasing from 15 to 2.1 μm. There are many works devoted to the study of the influence of size effects on the dynamics of sensitive elements of nanosensors [30–34]. However, the results obtained in them relate either to beams or to isotropic plates. A large contribution to the study of anisotropic plates with size effects was made by Chen [35–40]. In [40] he proposed the new modified couple stress theory for anisotropic elasticity which contains three material length scale parameters. Based on this model, he developed composite laminated Kirchhoff plate model. But, as the authors know, there are no publications, in which a detailed variational formulation was provided for anisotropic nano-plates based on the new modified couple stress theory and the third-order plate theory.

2 Theoretical Formulations

Let's consider a rectangular anisotropic nanoplate of uniform thickness h (see Fig. 1) under the distributed force at the top ($x_3 = -h/2$) of the plate. The origin of the coordinate system is located at the midpoint of the left side of the nanoplate's midplane. Axis x_1, x_2 are taken along the length and width directions, respectively. The positive direction of the axis x_3 is downward from the midplane along the thickness direction. So, the coordinate of a midplane's point is $(x_1, x_2, 0)$. The density ρ_0 of the nanoplate is uniform.

3 Displacement Field

According to the third-order plate theory [41], the displacement field (u_1, u_2, u_3) can be expressed as:

$$\begin{aligned}u_1(x_1, x_2, x_3, t) &= u_0(x_1, x_2, t) + x_3\phi_1(x_1, x_2, t)\\&\quad - \frac{4}{3h^2}x_3^3\left(\phi_1(x_1, x_2, t) + \frac{\partial w_0(x_1, x_2, t)}{\partial x_1}\right)\\u_2(x_1, x_2, x_3, t) &= v_0(x_1, x_2, t) + x_3\phi_2(x_1, x_2, t)\\&\quad - \frac{4}{3h^2}x_3^3\left(\phi_2(x_1, x_2, t) + \frac{\partial w_0(x_1, x_2, t)}{\partial x_2}\right)\\u_3(x_1, x_2, x_3, t) &= w_0(x_1, x_2, t)\end{aligned} \tag{1}$$

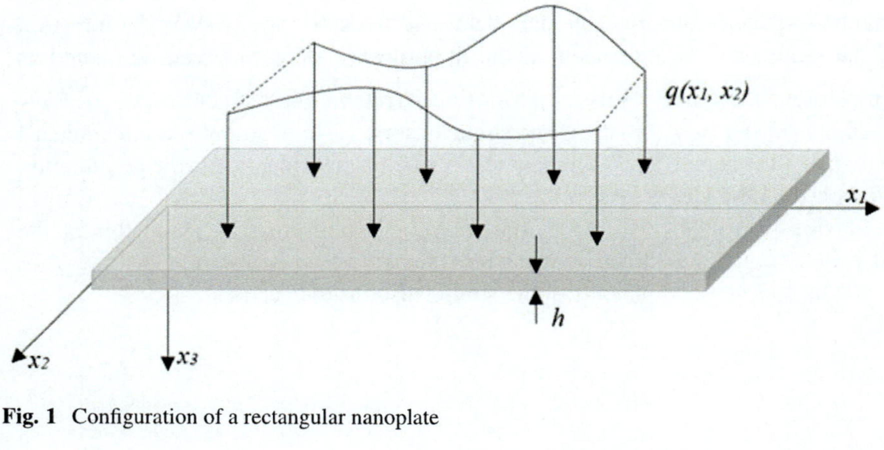

Fig. 1 Configuration of a rectangular nanoplate

where (u_0, v_0, w_0) are the displacement components of a midplane's point along (x_1, x_2, x_3) coordinate axis, ϕ_1 and ϕ_2 are the angles of rotation about the x_2- and x_1-axes, respectively.

4 The constitutive Relations

The new modified couple stress theory for anisotropic elasticity was proposed by Chen and Li in [40]. In this theory, three material length scale parameters are involved.

The constitutive relations have the following form according to the new modified couple stress theory [40]

$$\sigma_{ij} = \widetilde{C}_{ijkl}\varepsilon_{kl} \tag{2a}$$

$$m_{ij} = l_i^2 G_i \chi_{ij} + l_j^2 G_j \chi_{ji} \tag{2b}$$

$$\varepsilon_{ij} = \frac{1}{2}\left(u_{i,j} + u_{j,}\right) \tag{2c}$$

$$\chi_{ij} = \omega_{i,j} \tag{2d}$$

$$\omega_i = \frac{1}{2} e_{ijk} u_{k,j} \tag{2e}$$

where l_i—material length scale parameter and the subscript i means the direction of the shapes and arrangements of the impurities or defects, l_i can be treated as a measurement of sizes of the impurities or defects in microstructures; $\widetilde{C}_{ijkl}$, G_i—elasticity constants; σ, ε—stress and strain tensors; χ—curvature (rotation gradient) tensor; m—the couple stress moment tensor; u—displacements; e—the permutation symbol (the Levi-Civita symbol).

Obviously, σ_{ij}, ε_{ij}, m_{ij} are symmetric. χ_{ij} is nonsymmetric, and this is the difference from the modified couple stress theory.

Let us consider components of the strain tensor as the vector:

$$\begin{pmatrix} \varepsilon_1 \\ \varepsilon_2 \\ \gamma_{12} \\ \gamma_{23} \\ \gamma_{13} \end{pmatrix} = \begin{pmatrix} \varepsilon_{11} \\ \varepsilon_{22} \\ 2\varepsilon_{12} \\ 2\varepsilon_{23} \\ 2\varepsilon_{13} \end{pmatrix} \tag{3}$$

where ε_{ij} are defined by (2c).

The following relations are obtained by substituting (1) to (2c)–(3):

$$\begin{pmatrix} \varepsilon_1 \\ \varepsilon_2 \\ \gamma_{12} \end{pmatrix} = \begin{pmatrix} \varepsilon_1^0 \\ \varepsilon_2^0 \\ \gamma_{12}^0 \end{pmatrix} + x_3 \begin{pmatrix} \varepsilon_1^1 \\ \varepsilon_2^1 \\ \gamma_{12}^1 \end{pmatrix} + x_3^3 \begin{pmatrix} \varepsilon_1^3 \\ \varepsilon_2^3 \\ \gamma_{12}^3 \end{pmatrix}, \begin{pmatrix} \gamma_{23} \\ \gamma_{13} \end{pmatrix} = \begin{pmatrix} \gamma_{23}^{(0)} \\ \gamma_{13}^{(0)} \end{pmatrix} + x_3^2 \begin{pmatrix} \gamma_{23}^{(2)} \\ \gamma_{13}^{(2)} \end{pmatrix} \tag{4a}$$

where

$$\begin{pmatrix} \varepsilon_1^{(0)} \\ \varepsilon_2^{(0)} \\ \gamma_{12}^{(0)} \end{pmatrix} = \begin{pmatrix} u_{0,1} \\ v_{0,2} \\ u_{0,2} + v_{0,1} \end{pmatrix}, \begin{pmatrix} \varepsilon_1^{(1)} \\ \varepsilon_2^{(1)} \\ \gamma_{12}^{(1)} \end{pmatrix} = \begin{pmatrix} \phi_{1,1} \\ \phi_{2,2} \\ \phi_{1,2} + \phi_{2,1} \end{pmatrix} \tag{4b}$$

$$\begin{pmatrix} \varepsilon_1^{(3)} \\ \varepsilon_2^{(3)} \\ \gamma_{12}^{(3)} \end{pmatrix} = -c_1 \begin{pmatrix} \phi_{1,1} + w_{0,11} \\ \phi_{2,2} + w_{0,22} \\ \phi_{1,2} + \phi_{2,1} + 2w_{0,12} \end{pmatrix}$$

$$\begin{pmatrix} \gamma_{23}^{(0)} \\ \gamma_{13}^{(0)} \end{pmatrix} = \begin{pmatrix} w_{0,2} + \phi_2 \\ w_{0,1} + \phi_1 \end{pmatrix}, \begin{pmatrix} \gamma_{23}^{(2)} \\ \gamma_{13}^{(2)} \end{pmatrix} = -c_2 \begin{pmatrix} \gamma_{23}^{(0)} \\ \gamma_{13}^{(0)} \end{pmatrix} \tag{4c}$$

$$c_1 = \frac{4}{3h^2}, c_2 = \frac{4}{h^2} = 3c_1.$$

Acting similarly with the χ_{ij} components and according to (1), (2d), (2e), we obtain:

$$\begin{pmatrix} \chi_{11} \\ \chi_{22} \\ \chi_{33} \\ \chi_{12} \\ \chi_{21} \end{pmatrix} = \frac{1}{2} \begin{pmatrix} \chi_{11}^{(0)} \\ \chi_{22}^{(0)} \\ \chi_{33}^{(0)} \\ \chi_{12}^{(0)} \\ \chi_{21}^{(0)} \end{pmatrix} + \frac{1}{2} x_3^2 c_2 \begin{pmatrix} \chi_{11}^{(2)} \\ \chi_{22}^{(2)} \\ \chi_{33}^{(2)} \\ \chi_{12}^{(2)} \\ \chi_{21}^{(2)} \end{pmatrix} \tag{5a}$$

$$\begin{pmatrix} \chi_{13} \\ \chi_{23} \end{pmatrix} = c_2 x_3 \begin{pmatrix} \chi_{13}^{(1)} \\ -\chi_{23}^{(1)} \end{pmatrix} \tag{5b}$$

$$\begin{pmatrix} \chi_{31} \\ \chi_{32} \end{pmatrix} = \frac{1}{2} \begin{pmatrix} \chi_{31}^{(0)} \\ \chi_{32}^{(0)} \end{pmatrix} + \frac{1}{2} x_3 \begin{pmatrix} \chi_{31}^{(1)} \\ \chi_{32}^{(1)} \end{pmatrix} + \frac{1}{2} c_1 x_3^3 \begin{pmatrix} \chi_{31}^{(3)} \\ \chi_{32}^{(3)} \end{pmatrix} \tag{5c}$$

where

$$\begin{pmatrix} \chi_{11}^{(0)} \\ \chi_{22}^{(0)} \\ \chi_{33}^{(0)} \\ \chi_{12}^{(0)} \\ \chi_{21}^{(0)} \end{pmatrix} = \begin{pmatrix} w_{0,12} - \phi_{2,1} \\ -w_{0,12} + \phi_{1,2} \\ \phi_{2,1} - \phi_{1,2} \\ w_{0,22} - \phi_{2,2} \\ -w_{0,11} + \phi_{1,1} \end{pmatrix}$$

$$\begin{pmatrix} \chi_{13}^{(1)} \\ \chi_{23}^{(1)} \end{pmatrix} = \begin{pmatrix} w_{0,2} + \phi_2 \\ w_{0,1} + \phi_1 \end{pmatrix} \tag{5d}$$

$$\begin{pmatrix} \chi_{11}^{(2)} \\ \chi_{22}^{(2)} \\ \chi_{33}^{(2)} \\ \chi_{12}^{(2)} \\ \chi_{21}^{(2)} \end{pmatrix} = \begin{pmatrix} w_{0,12} + \phi_{2,1} \\ -w_{0,12} - \phi_{1,2} \\ -\phi_{2,1} + \phi_{1,2} \\ w_{0,22} + \phi_{2,2} \\ -w_{0,11} - \phi_{1,1} \end{pmatrix}$$

$$\begin{pmatrix} \chi_{31}^{(0)} \\ \chi_{32}^{(0)} \end{pmatrix} = \begin{pmatrix} v_{0,11} - u_{0,12} \\ -u_{0,22} + v_{0,12} \end{pmatrix}, \begin{pmatrix} \chi_{31}^{(1)} \\ \chi_{32}^{(1)} \end{pmatrix} = \begin{pmatrix} \phi_{2,11} - \phi_{1,12} \\ \phi_{2,12} - \phi_{1,22} \end{pmatrix}, \begin{pmatrix} \chi_{31}^{(3)} \\ \chi_{32}^{(3)} \end{pmatrix} = \begin{pmatrix} -\chi_{31}^{(1)} \\ -\chi_{32}^{(1)} \end{pmatrix} \tag{5e}$$

5 Principle of Virtual Displacements

The strain energy U in the region V occupied by the elastically deformed material is written as

$$U = U_\sigma + U \tag{6}$$

where U_σ—"classical" part of the strain energy, U_χ—size-dependent part of the strain energy:

$$U_\sigma = \int_V \sigma_{ij}\varepsilon_{ij} dV, U_\chi = \int_V m_{ij}\chi_{ij} dV$$

The work done by the external forces is

$$W = \int_\Omega q dx_1 dx_2 \tag{7}$$

The kinematic energy K can be written as

$$K = \frac{1}{2}\int_V \rho_0 \left(\dot{u}_i^2\right) d \tag{8}$$

6 Governing Equations

The expression of the dynamic version of the principle of virtual displacements:

$$\int_{t_1}^{t_2} \delta U - \delta K - \delta W = \int_{t_1}^{t_2} \delta U_\sigma + \delta U_\chi - \delta K - \delta W = 0 \tag{9a}$$

where δU, δK, δW are the variation of strain energy, kinetic energy, and work done by external applied forces:

$$\delta U_\sigma = \int_\Omega \int_{-\frac{h}{2}}^{\frac{h}{2}} (\sigma_{11}\delta\varepsilon_{11} + \sigma_{22}\delta\varepsilon_{22} + \sigma_{12}\delta\gamma_{12} + \sigma_{13}\delta\gamma_{13} + \sigma_{23}\delta\gamma_{23}) dx_3 dx_1 dx_2, \tag{9b}$$

$$\begin{aligned}\delta U_\chi = \int_\Omega \int_{-\frac{h}{2}}^{\frac{h}{2}} &(m_{11}\delta\chi_{11} + m_{22}\delta\chi_{22} + m_{33}\delta\chi_{33} + m_{12}(\delta\chi_{12} + \delta\chi_{21}) \\ &+ m_{13}(\delta\chi_{13} + \delta\chi_{31}) + m_{23}(\delta\chi_{23} + \delta\chi_{32})) dx_3 dx_1 dx_2,\end{aligned} \tag{9c}$$

$$\delta K = \int_\Omega \int_{-\frac{h}{2}}^{\frac{h}{2}} \rho_0 (\dot{u}_i \delta\dot{u}_i) dx_3 dx_1 dx_2, \tag{9d}$$

$$\delta W = \int_{\Omega} q\delta w dx_1 dx_2, \tag{9e}$$

After substituting (4a) in (9b), (5a)–(5c) in (9c), (1) in (9d) we obtain:

$$\begin{aligned}\delta U_\sigma = \int_{\Omega} \Big[& N_{11}\delta\varepsilon_1^{(0)} + M_{11}\delta\varepsilon_1^{(1)} + P_{11}\delta\varepsilon_1^{(3)} + N_{22}\delta\varepsilon_2^{(0)} + M_{22}\delta\varepsilon_2^{(1)} + P_{22}\delta\varepsilon_2^{(3)} \\ & + N_{12}\delta\gamma_{12}^{(0)} + M_{12}\delta\gamma_{12}^{(1)} + P_{12}\delta\gamma_{12}^{(3)} + N_{13}\delta\gamma_{13}^{(0)} - R_{13}c_2\delta\gamma_{13}^{(0)} + \\ & + N_{23}\delta\gamma_{23}^{(0)} - R_{23}c_2\delta\gamma_{23}^{(0)} \Big] dx_1 dx_2\end{aligned} \tag{10}$$

$$\begin{aligned}\delta U_\chi = \int_{\Omega} \Big[& A_{11}\delta\chi_{11}^{(0)} + c_2C_{11}\delta\chi_{11}^{(2)} + A_{22}\delta\chi_{22}^{(0)} + c_2C_{22}\delta\chi_{22}^{(2)} + A_{33}\delta\chi_{33}^{(0)} + c_2C_{33}\delta\chi_{33}^{(2)} \\ & + A_{12}\left(\delta\chi_{12}^{(0)} + \delta\chi_{21}^{(0)}\right) + c_2C_{12} + A_{13}\delta\chi_{31}^{(0)} + B_{13}\delta\chi_{31}^{(1)} + 2c_2B_{13}\delta\chi_{13}^{(1)} \\ & - c_1D_{31}\delta\chi_{31}^{(1)} + A_{23}\delta\chi_{32}^{(0)} + B_{23}\delta\chi_{32}^{(1)} - 2c_2B_{23}\delta\chi_{23}^{(1)} - c_1D_{23}\delta\chi_{32}^{(1)} \Big] dx_1 dx_2\end{aligned} \tag{11}$$

$$\begin{aligned}\delta K = \int_{\Omega} & \left[\left(I_0\dot{u}_0 + I_1\dot{\phi}_1 - c_1I_3\dot{\varphi}_1\right)\delta\dot{u}_0 + \left(I_1\dot{u}_0 + I_2\dot{\phi}_1 - c_1I_4\dot{\varphi}_1\right)\delta\dot{\phi}_1 \right. \\ & + c_1\left(-I_3\dot{u}_0 - I_4\dot{\phi}_1 + c_1I_6\dot{\varphi}_1\right)\delta\dot{\varphi}_1 + \left(I_0\dot{v}_0 + I_1\dot{\phi}_2 - c_1I_3\dot{\varphi}_2\right)\delta\dot{v}_0 \\ & + \left(I_1\dot{v}_0 + I_2\dot{\phi}_2 - c_1I_4\dot{\varphi}_2\right)\delta\dot{\phi}_2 + c_1\left(-I_3\dot{v}_0 - I_4\dot{\phi}_2 + c_1I_6\dot{\varphi}_2\right)\delta\dot{\varphi}_2 \\ & \left. + I_0\dot{w}_0\delta\dot{w}_0 \right] dx_1 dx_2\end{aligned} \tag{12}$$

where

$$N_{ij} = \int_{-\frac{h}{2}}^{\frac{h}{2}} \sigma_{ij} dx_3,\ M_{ij} = \int_{-\frac{h}{2}}^{\frac{h}{2}} x_3\sigma_{ij} dx_3,\ R_{ij} = \int_{-\frac{h}{2}}^{\frac{h}{2}} x_3^2\sigma_{ij} dx_3,\ P_{ij} = \int_{-\frac{h}{2}}^{\frac{h}{2}} x_3^3\sigma_{ij} dx_3,$$

$$\mathrm{A_{ij}} = \int_{-\frac{h}{2}}^{\frac{h}{2}} \mathrm{m_{ij}dx_3},\ \mathrm{B_{ij}} = \int_{-\frac{h}{2}}^{\frac{h}{2}} \mathrm{x_3m_{ij}dx_3},\ C_{ij} = \int_{-\frac{h}{2}}^{\frac{h}{2}} x_3^2 m_{ij} dx_3,\ D_{ij} = \int_{-\frac{h}{2}}^{\frac{h}{2}} x_3^3 m_{ij} dx_3,$$

$$\varphi_1 = \phi_1 + w_{0,1},\ \varphi_2 = \phi_2 + w_{0,2},\ I_i = \int_{-\frac{h}{2}}^{\frac{h}{2}} \rho_0 x_3^i dx_3.$$

After substituting (4b), (4c) in (10), (5d), (5e) in (11), and integration (9a) by parts in respect (10)–(12), and collecting the coefficients for δu_0, δv_0, δw_0, $\delta\phi_1$, $\delta\phi_2$, the following system of equations of motion will be obtained:

$$\delta u_0: \quad N_{11,1} + N_{12,2} + \frac{1}{2}A_{23,22} + \frac{1}{2}A_{13,12} = I_0\ddot{u}_0 + J_1\ddot{\phi}_1 - c_1 I_3 \ddot{w}_{0,1}$$

$$\delta v_0: \quad N_{22,2} + N_{12,1} - \frac{1}{2}A_{13,11} - \frac{1}{2}A_{23,12} = I_0\ddot{v}_0 + J_1\ddot{\phi}_2 - c_1 I_3 \ddot{w}_{0,2} \qquad (13)$$

$$\begin{aligned}\delta w_0: \quad & (N_{13} - c_2 R_{13})_{,1} + (N_{23} - c_2 R_{23})_{,2} + c_1 \left(P_{11,11} + 2P_{12,12} + P_{22,22}\right) \\ & -\frac{1}{2}K_{1,12} + \frac{1}{2}K_{2,11} - \frac{1}{2}K_{2,22} + c_2 B_{13,2} - c_2 B_{23,11} + q = I_0 \ddot{w}_0 \\ & + c_1 I_3 \left(\ddot{u}_{0,1} + \ddot{v}_{0,2}\right) + c_1 J_4 \left(\ddot{\phi}_{1,1} + \ddot{\phi}_{2,2}\right) - c_1^2 I_6 \left(\ddot{w}_{0,12} + \ddot{w}_{0,22}\right)\end{aligned}$$

$$\begin{aligned}\delta \phi_1: \quad & (M_{11} - c_1 P_{11})_{,1} + (M_{12} - c_1 P_{12})_{,2} - (N_{13} - c_2 R_{13}) \\ & -\frac{1}{2}K_{3,12} - \frac{1}{2}K_{4,22} + \frac{1}{2}K_{5,1} + \frac{1}{2}K_{6,2} + c_2 B_{23} = J_1 \ddot{u}_0 + \kappa \ddot{\phi}_1 - c_1 J_4 \ddot{w}_{0,1}\end{aligned}$$

$$\begin{aligned}\delta \phi_2: \quad & (M_{22} - c_1 P_{22})_{,2} + (M_{12} - c_1 P_{12})_{,1} - (N_{23} - c_2 R_{23}) \\ & +\frac{1}{2}K_{3,11} + \frac{1}{2}K_{4,12} + \frac{1}{2}K_{7,1} - \frac{1}{2}K_{5,2} - c_2 B_{13} = J_1 \ddot{v}_0 + \kappa \ddot{\phi}_2 - c_1 J_4 \ddot{w}_{0,2}\end{aligned}$$

where

$K_1 = A_{11} - A_{22} + c_2(C_{11} - C_{22})$, $K_2 = A_{12} + c_2 C_{12}$, $K_3 = -B_{13} + c_1 D_{13}$, $K_4 = -B_{23} + c_1 D_{23}$, $K_5 = A_{12} - c_2 C_{12}$, $K_6 = A_{22} - A_{33} + c_2(C_{33} - C_{22})$, $K_7 = A_{33} - A_{11} + c_2(C_{11} - C_{33})$, $J_i = I_i - c_1 I_{i+2}$, $\kappa = I_2 - 2c_1 I_4 + c_1^2 I_6$.

Natural boundary conditions can be obtained from the following relation:

$$\begin{aligned}\int\limits_{\partial\Omega} \Bigg[& \mathcal{H}_1 \delta u_0 + \mathcal{H}_2 \delta v_0 + \mathcal{H}_3 \frac{\partial \delta u_0}{\partial x_2} - \mathcal{H}_3 \frac{\partial \delta v_0}{\partial x_1} + \mathcal{H}_4 \delta w_0 + \mathcal{H}_5 \frac{\partial \delta w_0}{\partial x_1} + \mathcal{H}_6 \frac{\partial \delta w_0}{\partial x_2} \\ & + \mathcal{H}_7 \delta \phi_1 + \mathcal{H}_8 \delta \phi_2 + \mathcal{H}_9 \frac{\partial \delta \phi_1}{\partial x_2} - \mathcal{H}_9 \frac{\partial \delta \phi_2}{\partial x_1} \Bigg] d\sigma = 0\end{aligned}$$

where $\partial\Omega$—the piecewise smooth boundary curve of Ω, (n_1, n_2)—coordinates of the normal n to $\partial\Omega$,

$$\mathcal{H}_1 = N_{11} n_1 + N_{12} n_2 + \frac{1}{2}\left(A_{13,1} n_2 + A_{23,2} n_2\right)$$

$$\mathcal{H}_2 = N_{12} n_1 + N_{22} n_2 + \frac{1}{2}\left(-A_{13,1} n_1 - A_{23,2} n_1\right)$$

$$\mathcal{H}_3 = \frac{1}{2}\left(-A_{13} n_1 - A_{23} n_2\right)$$

$$\mathcal{H}_4 = \left(c_1\left(P_{11,1} + \mathrm{P}_{12,2}\right) + N_{13} - c_2\left(R_{13} + B_{23}\right) - \frac{1}{4}\left(K_{1,2} - 2K_{2,1}\right)\right)n_1 + \left(c_1\left(P_{22,2} + \mathrm{P}_{12,1}\right) + N_{23} - c_2\left(R_{23} - B_{13}\right) - \frac{1}{4}\left(K_{1,1} + 2K_{2,2}\right)\right)n_2 - \left(c_1 I_3 \ddot{u}_0 + c_1 J_4 \ddot{\phi}_1 - c_1^2 I_6 \ddot{w}_{0,1}\right)n_1 - \left(c_1 I_3 \ddot{v}_0 + c_1 J_4 \ddot{\phi}_2 - c_1^2 I_6 \ddot{w}_{0,2}\right)n_2$$

$$\mathcal{H}_5 = \left(-\left(\mathrm{c}_1 P_{11} + \frac{1}{2}K_2\right)n_1 - \left(\mathrm{c}_1 P_{12} - \frac{1}{4}K_1\right)n_2\right)$$

$$\mathcal{H}_6 = \left(-\left(\mathrm{c}_1 P_{12} + \frac{1}{4}K_1\right)n_1 - \left(\mathrm{c}_1 P_{22} - \frac{1}{2}K_2\right)n_2\right)$$

$$\mathcal{H}_7 = \left(M_{11} - c_1 P_{11}\right)n_1 + \left(M_{12} - c_1 P_{12}\right)n_2 + \frac{1}{2}\left(K_5 n_1 + \left(-K_{3,1} - K_{4,2} + K_6\right)n_2\right)$$

$$\mathcal{H}_8 = \left(M_{22} - c_1 P_{22}\right)n_2 + \left(M_{12} - c_1 P_{12}\right)n_1 + \frac{1}{2}\left(\left(K_{3,1} + K_{4,2} + K_7\right)n_1 - K_5 n_2\right)$$

$$\mathcal{H}_9 = \frac{1}{2}\left(K_3 n_1 + K_4 n_2\right).$$

7 Discussion

Let's compare Eqs. (13) with the equations of motion of plates obtained by Reddy according to the third-order theory [41]:

$$N_{11,1} + N_{12,2} + \mathcal{X}_u = I_0 \ddot{u}_0 + J_1 \ddot{\phi}_1 - c_1 I_3 \ddot{w}_{0,1}$$

$$N_{22,2} + N_{12,1} + \mathcal{X}_v = I_0 \ddot{v}_0 + J_1 \ddot{\phi}_2 - c_1 I_3 \ddot{w}_{0,2} \tag{14}$$

$$\left(N_{13} - c_2 R_{13}\right)_{,1} + \left(N_{23} - c_2 R_{23}\right)_{,2} + c_1\left(P_{11,11} + 2P_{12,12} + P_{22,22}\right) + q + \mathcal{X}_w = I_0 \ddot{w}_0 + c_1 I_3\left(\ddot{u}_{0,1} + \ddot{v}_{0,2}\right) + c_1 J_4\left(\ddot{\phi}_{1,1} + \ddot{\phi}_{2,2}\right) - c_1^2 I_6\left(\ddot{w}_{0,12} + \ddot{w}_{0,22}\right)$$

$$(M_{11}-c_1P_{11})_{,1}+(M_{12}-c_1P_{12})_{,2}-(N_{13}-c_2R_{13})+\mathcal{X}_{\phi_1}=J_1\ddot{u}_0+\kappa\ddot{\phi}_1-+c_1J_4\ddot{w}_{0,1}$$

$$(M_{22}-c_1P_{22})_{,2}+(M_{12}-c_1P_{12})_{,1}-(N_{23}-c_2R_{23})+\mathcal{X}_{\phi_2}=J_1\ddot{v}_0+\kappa\ddot{\phi}_2-+c_1J_4\ddot{w}_{0,2}$$

where $\mathcal{X}_u$, $\mathcal{X}_v$, $\mathcal{X}_w$, $\mathcal{X}_{\phi_1}$, $\mathcal{X}_{\phi_2}$—addendums that distinguish Reddy's equations from the equations obtained in this article. They are defined by the following expressions:

$$\mathcal{X}_u=\frac{1}{2}A_{23,22}+\frac{1}{2}A_{13,12},\ \mathcal{X}_v=-\left(\frac{1}{2}A_{13,11}+\frac{1}{2}A_{23,12}\right)$$

$$\mathcal{X}_w=\frac{1}{2}\left(-K_{1,12}+K_{2,11}-K_{2,22}\right)+c_2\left(B_{13,2}-B_{23,1}\right)$$

$$\mathcal{X}_{\phi_1}=-\frac{1}{2}K_{3,12}-\frac{1}{2}K_{4,22}+\frac{1}{2}K_{5,1}+\frac{1}{2}K_{6,2}+c_2B_{23}$$

$$\mathcal{X}_{\phi_2}=\frac{1}{2}K_{3,11}+\frac{1}{2}K_{4,12}+\frac{1}{2}K_{7,1}-\frac{1}{2}K_{5,2}-c_2B_{13}$$

Equations (14) show that the right part of the equations did not change after taking into account size-dependent effects. Also, only the left part of equations has some addendums due to using the new modified couple stress theory. Moreover, the natural boundary conditions consist of nine equations whereas the "classical" third-order theory leads to only six natural boundary conditions.

Thus, we can use for numerical simulation of dynamics of plates the same functions, which are used for in the framework of the third-order theory. However, if we have a deal, for example, with a laminated composite plate, the problem of study of its dynamics looks more complicated.

8 Conclusions

The equations of motion of a NEMS sensing element as anisotropic rectangular size-dependent nanoplate were obtained by using the dynamic version of the principle of virtual displacements, the modified couple-stress theory and the third-order theory of laminated composite plates and shells.

The obtained equations were compared with the equations of the plate's motion based on the third-order theory obtained by Reddy. This comparison showed that taking into account the size-dependent effects according to the modified couple-stress theory leads to additional terms on the left side of the equations and does not lead to a change in the right "dynamic" part.

Acknowledgments The work was supported by RFBR grant 19-08-00807.

References

1. Kraft, M., White, N.: Mems for Automotive and Aerospace Applications. Woodhead Publishing, Oxford (2013)
2. Nihtianov, S., Luque, A.: Smart Sensors and MEMS. Intelligent Sensing Devices and Microsystems for Industrial Applications. Woodhead Publishing, Duxford (2018)
3. Dussy, S., Durrant, D., Moy, T., Perriault, N., Célerier B.: MEMS gyro for space applications. Overview of European activities. In: AIAA Guidance, Navigation, and Control Conference and Exhibit, San Francisco, CA (2005)
4. Lei, X., Liqing, F., Ruikun, H., Xu, L.: Research on MEMS technology application in fuse. IOP Conf. Series Earth Environmental Sci. **252**, 022011 (2019)
5. Lin, Y.-C., Wang, W.-S., Chen, L.Y., Chen, M.W., Gessner, T., Esashi, M.: Nanoporous gold for MEMS packing applications. IEEJ Trans. Sensors Micromach. **133**(2), 31–36 (2013)
6. Wei, X.L., Liu, J.Q., Liu, H.F., Wu, W.J., Fan, J., Tu, L.C.: Electroplating of three-dimensional Sn-rich solder for MEMS packaging applications. J. Micromech. Microeng. **29**, 4 (2019)
7. Hiroyuki, F.: MEMS/MOEMS application to optical communication. Proc. SPIE. **4560**, 11–17 (2001)
8. Barbour, N.M.: Inertial navigation sensors. In: NATO RTO Lecture Series 232. Advances in Navigation Sensors and Integration Technology (2004)
9. Baranova, E.A., Evstifeev, M.I., Eliseev, D.P.: Simulation of translational vibrations effect on torque-to-balance RR-type MEMS gyroscope. Gyroscopy Navig. **9**(1), 50–56 (2018)
10. Barulina, M.A.: Mathematical proof why crystallographic plane (111) is better for micro-electromechanical sensors sensing elements. In: 19th SGEM International Multidisciplinary Scientific GeoConference EXPO Proceedings (2019)
11. Zhang, Y., Zhou, B., Song, M., Hou, B., Xing, H., Zhang, R.A.: Novel MEMS gyro north finder design based on the rotation modulation technique. Sensors (Basel). **5**, 17 (2017)
12. Cropp, A., Collingwood, Ch., Dussy, S.: The characterization and testing of MEMS gyros for GIOVE-A. In: AIAA Guidance, Navigation, and Control Conference and Exhibit, 21–24 Aug, Keystone, Colorado (2006)
13. Kausinis, S., Barauskas, R.: Parametric sensitivity of MEMS-gyro. Solid State Phenom. **113**, 495–499 (2006)
14. Evstifeev, M.I., Untilov, A.A.: The requirements to manufacturing accuracy for elastic suspension of a micromechanical gyro. Gyroscopy Navig. **41**(2), 24–31 (2003) (in Russ.)
15. Shkel, A., Howe, R., Horowrtz, R.: Micromachined gyroscopes; challenges, design solutions and opportunities. In: LARP International Workshop on Micro Robots, Micro Machines and Systems, pp. 27–34. Russian Academy of Sciences, Moscow (1999)
16. Barulina, M.A., Dzhashitov, V.E., Pankratov, V.M.: Mathematical models of thermal control systems of micromechanical gyroscopes. Gyroscopy Navig. **38**(3), 48–60 (2002) (in Russ.)
17. Tsukamoto, T., Tanaka, S.: MEMS rate integrating gyroscope with temperature corrected virtual rotation. In: IEEE International Symposium on Inertial Sensors and Systems (INERTIAL) (2019)
18. Sun, B.: The similarity laws for MEMS gyro with temperature changes. In: The 21st International Congress on Sound and Vibration, Beijing, China (2014)
19. Wilson, J.: Sensor Technology Handbook. Newnes, Amsterdam (2004)

20. Barulina, M.A.: Frequency equations and self-induced vibrations of the elements of the vibratory micromechanical gyroscopes based on Timoshenko shift theory. J. Nano Microsyst. Techn. "Nano i mikrosistemnaya tehnika". **177**(4), 27–38 (2015)
21. Awrejcewicz, J., Krysko, A.V., Erofeev, N.P., Dobriyan, V., Barulina, M.A., Krysko, V.A.: Quantifying chaos by various computational methods. Part 1: simple systems. Entropy. **3**(20), 175 (2018)
22. Awrejcewicz, J., Krysko, A.V., Erofeev, N.P., Dobriyan, V., Barulina, M.A., Krysko, V.A.: Quantifying chaos by various computational methods. Part 2: vibrations of the Bernoulli–Euler beam subjected to periodic and colored noise. Entropy. **3**(20), 170 (2018)
23. Christensen, R.M.: Mechanics of cellular and other low-density materials. Int. J. Solids Struct. **37**(1–2), 93–104 (2000)
24. Hopcroft, M., Nix, W., Kenny, T.: What is the Young's modulus of silicon? J. Microelectromech. Syst. **19**(2), 229–238 (2010)
25. Fleck, N.A., Muller, G.M., Ashby, M.F., Hutchinson, J.W.: Strain gradient plasticity: theory and experiment. Acta Metall. Mater. **42**, 475–487 (1994)
26. Ma, Q., Clarke, D.R.: Size dependent hardness of silver single crystals. J. Mater. Res. **10**, 853–863 (1995)
27. McFarland, A.W., Colton, J.S.: Role of material microstructure in plate stiffness with relevance to microcantilever sensors. J. Micromech. Microeng. **15**, 1060–1067 (2005)
28. Lei, J., He, Y., Guo, S., Li, Z., Liu, D.: Size-dependent vibration of nickel cantilever microbeams: experiment and gradient elasticity. AIP Adv. **6**, 105202 (2016)
29. Li, Z., He, Y., Lei, J., Guo, S., Liu, D., Wang, L.: A standard experimental method for determining the material length scale based on modified couple stress theory. Int. J. Mech. Sci. **141**, 198–205 (2018)
30. Lu, L., Guo, X., Zhao, J.: A unified size-dependent plate model based on nonlocal strain gradient theory including surface effects. Appl. Math. Model. **68**, 583–602 (2019)
31. Li, L., Hu, Y.: Nonlinear bending and free vibration analyses of nonlocal strain gradient beams made of functionally graded material. Int. J. Eng. Sci. **107**, 77–97 (2016)
32. Ebrahimi, F., Barati, M.R.: Hygrothermal effects on vibration characteristics of viscoelastic FG nanobeams based on nonlocal strain gradient theory. Compos. Struct. **159**, 433–444 (2017)
33. Tsiatas, G.C., Yiotis, A.J.: Size effect on the static, dynamic and buckling analysis of orthotropic Kirchhoff-type skew micro-plates based on a modified couple stress theory: comparison with the nonlocal elasticity theory. Acta Mech. **4**(226), 1267–1281 (2014)
34. Yang, Z., He, D.: Vibration and buckling of functionally graded sandwich micro-plates based on a new size-dependent model. Int. J. Appl. Mech. **11**, 1950004 (2019)
35. Chen, W.J., Li, L., Ma, X.: A modified couple stress model for bending analysis of composite laminated beams with first order shear deformation. Compos. Struct. **93**, 2723–2732 (2011)
36. Chen, W.J., Chen, W., Sze, K.Y.: A model of composite laminated Reddy beam based on a modified couple stress theory. Compos. Struct. **94**, 2599–2609 (2012)
37. Chen, W., Ma, X., Li, L.: A model of composite laminated Reddy plate based on new modified couple stress theory. Compos. Struct. **94**, 2143–2156 (2012)
38. Chen, W.J., Li, X.: Size-dependent free vibration analysis of composite laminated Timoshenko beam based on new modified couple stress theory. Arch. Appl. Mech. **83**, 431–444 (2013)
39. Chen, W.J., Si, J.: A model of composite laminated beam based on the global–local theory and new modified couple stress theory. Compos. Struct. **9**, 99–107 (2013)
40. Chen, W., Li, X.: A new modified couple stress theory for anisotropic elasticity and microscale laminated Kirchhoff plate model. Arch. Appl. Mech. **3**(84), 323–341 (2013)
41. Reddy, J.N.: Mechanics of Laminated Composite Plates and Shells, Theory and Analysis, 2nd edn. CRC Press, Boca Raton (2006)

Dynamic Analysis and Damage of Composite Layered Plates Reinforced by Unidirectional Fibers Subjected Low Velocity Impact

Josef Soukup, Milan Zmindak, Pavol Novak, Frantisek Klimenda, Michal Kaco, and Lenka Rychlikova

Abstract Currently, for dynamic modeling in composite structures at low and high speeds are used mainly Finite Element Method (FEM). For these analyzes commercial FEM software ABAQUS/explicit, LS-DYNA, AUTODYN and PAM CRASH, etc., are used in practice. In the present study, low-velocity impact response of composite laminates was studied using ABAQUS/Explicit code (FEM) to investigate damage by employing various damage criteria. The basic material properties in and transverse to the fiber directions, such as the elastic moduli, strains at failure, and plastic moduli among others are determined by simple tests in tension, compression, and shear. The material properties AS4/PEEK was used in numerical simulations and have been taken from the literature. Layer is considered as homogeneous transversely isotropic and layer stacking sequence is symmetrical or unsymmetrical. The solution in the form of time integration can be, depending on the problem, accomplished via implicit or explicit methods. For many of dynamic problems explicit methods have shown more suitable, cause they do not require stiffness, mass and damping matrix decomposition. In the plates examined, von Mises's stress and damage caused shear stress in the matrix and fiber were evaluated. From the results obtained, it was found that the von Mises stress was approximately the same for all types layer stacking sequence.

Keywords FEM · ABAQUS · Composite material

J. Soukup (✉) · F. Klimenda · L. Rychlikova
Faculty of Mechanical Engineering, University of J. E. Purkyne in Usti nad Labem, Usti nad Labem, Czech Republic
e-mail: josef.soukup@ujep.cz

M. Zmindak · P. Novak · M. Kaco
Faculty of Mechanical Engineering, University of Zilina, Zilina, Slovak Republic

J. Awrejcewicz (ed.), *Perspectives in Dynamical Systems II: Mathematical and Numerical Approaches*, Springer Proceedings in Mathematics & Statistics 363,
https://doi.org/10.1007/978-3-030-77310-6_15

1 Introduction

Fiber reinforced polymers (FRP) are most commonly used materials in various fields of industry. In the recent years quickly developing industries such as aerospace, ship and car industry almost completely rely on composite materials, especially on layered polymers reinforced with glass, aramid or carbon fibers and sandwich constructions consisting of FRP coatings with a foam core. Such constructions offer high strength at low weight, which considerably improve their performance (higher loading capacity, lower fuel consumption, etc.) especially in ship and aerospace industry [1, 2]. These materials also have good antiballistic properties, for example modern bulletproof vests are made from aramid fibers.

The most important characteristic of the composite materials is that they can be layered, with the fibers in each layer running in a different direction. This allows an engineer to design structures with unique properties, furthermore a structure can be designed so that it will bend in one direction, but not another. Impact damage is one of the main problems that composite structures face, there needs to be a way of reducing that damage when it occurs, reducing it enough so that the integrity of the structure is not comprised.

Today, typically, numerical models based on lamina-level failure criteria are used to simulate the damage of the fiber-reinforced composite material, although with well-accepted limitations. In this constitutive models, composite, are modelled as orthotropic linear elastic materials within the failure surface. The failure surface is defined by the failure criterion as maximum stress/strain criterion, Hashin's criterion, Christensen's criterion, Chang-Chang's criterion, Puck's criterion, LARC, etc. [3, 4].

They are many definition of low velocity or low energy impact due of the great number of parameters that should be study such as the velocity, the shape and the mass of the impactor [5]. The dynamical response of the structure depends therefore on the duration of the contact between the structure and the impactor. Cantwell and Morton [6] have proposed that every dynamic solicitation corresponding to an impact speed below 10–20 m/s can be considered as a low velocity impact. On other hand, Abrate [7] considers that the impact speed limit defining a low velocity impact is five to ten time greater than the one proposed by Cantwell and Morton (100 m/s). Liu et al. [8] uses a different approach based on the internal damage of the impacted structure. They postulate that a high velocity impact leads to fibres rupture where as low velocity impact leads to internal delamination and matrix cracking.

2 Theory Background and Solution Method

The contact between two components or bodies is a static phenomenon if the two bodies are static equilibrium. Otherwise the contact is a dynamic phenomenon. A dynamic contact is often much complicated than static one. The term 'contact-impact' is often used to stress the dynamic effects in contact phenomena [9, 10]. By nature, contact phenomena always involve friction phenomena. However, friction effects may be neglected in situations where frictional forces are sufficiently small. Therefore, we may have a frictionless contact, which is a special case of general contact. Mechanical problems involving contact are inherently non-linear and contact problems involve unknown boundary conditions.

2.1 Transient Stress Analysis

Figure 1 shows the transient analysis model of laminate and punch. At the t moment, the equilibrium equation can be deduced as:

$$\sigma_{ij,j}^{t} = \rho^{t}\ddot{u}_{i} + \mu^{t}\dot{u}_{i} \tag{1}$$

where ρ^t and μ^t are the density and dynamic friction coefficient of laminate at t moment, respectively.

$$\sigma_{ij}^{t} n_{j} - \overline{T}_{i}^{t} = 0 \ (\text{on } S_{\sigma}) \tag{2}$$

where, S_σ represents the stress boundary.

The equivalent integration of the equilibrium equation and the load boundary condition can be expressed, as follows:

$$\int_{V} \delta u_{i}\left(\sigma_{ij,j}^{t} - \rho^{t}\ddot{u}_{i} - \mu^{t}\dot{u}_{i}\right) dV - \int_{S_{\sigma}} \delta u_{i}\left(\sigma_{ij}^{t}\, n_{j} - \overline{T}_{i}^{t}\right) dS = 0 \tag{3}$$

As there will be geometric nonlinearity during the deformation of composite under the low-velocity impact load, strain tensor at the t moment can be expressed, as follows [11, 12]:

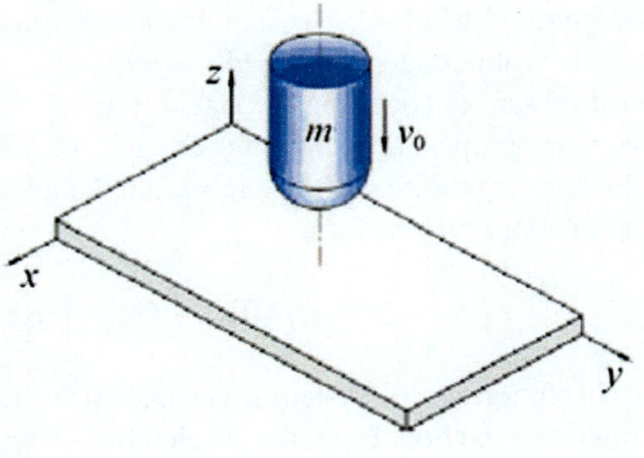

Fig. 1 Transient analysis model of laminate and punch

$$\varepsilon_{ij}^{t} = \frac{1}{2}\left(u_{i,j}^{t} + u_{j,i}^{t} + u_{k,i}^{t}\, u_{k,j}^{t}\right), (i, j, k = x, y, z) \tag{4}$$

Decompose the above equation in to linear and nonlinear terms

$$\varepsilon = \varepsilon_{L} + \varepsilon_{NL} \tag{5}$$

There is the following relationship of σ_{ij}^{t} and ε_{kl}^{t} at t moment

$$\sigma_{ij}^{t} = \overline{Q}_{ijkl}^{t-\Delta t} + \varepsilon_{kl}^{t} \tag{6}$$

where $\overline{Q}_{ijkl}^{t-\Delta t}$ represents material elasticity matrix at $t - \Delta t$ moment, and it can be obtained by coordinate transformation [13]

$$\overline{Q}_{ijkl}^{t-\Delta t} = [T]\ Q^{t-\Delta t}[T]^{T} \tag{7}$$

Using Eqs. (3) and (6), the stress equilibrium equation at each moment can finally be deduced as

$$\int \left(\delta\varepsilon_{ij}\, \overline{Q}_{ijkl}^{t-\Delta t}\, \varepsilon_{kl} + \delta u_{i}\, \rho^{t} \ddot{u}_{i} + \delta u_{i}\, \mu^{t} \dot{u}_{i}\right) dV + \int_{V_{n-1}} \sigma_{ij}^{n-1} \delta\left(\Delta\eta_{ij}\right) dV = \int_{S_{\sigma}} \overline{T}_{i}^{n}\, \delta u_{i}\, dS \tag{8}$$

where $\overline{T}_{i}^{t}$ and $\overline{T}_{i}^{n}$ are the surface force at t moment and nth step in numerical analysis respectively; $\delta\varepsilon_{ij}$ represents the strain at t moment; and, $\Delta\eta_{ij}$ is the nonlinear term of strain increment.

2.2 Solution Methods

Recently the most successful method for modeling the dynamic response of a structure is FEM [14]. The solution in the form of time integration can be, depending on the problem, accomplished via implicit or explicit methods. Although implicit methods are unconditionally stable (they are not dependent on the time step size), for wave propagation problems explicit methods have shown more suitable, cause they do not require stiffness, mass and damping matrix decomposition. The system of equations has the form

$$\mathbf{M}\ddot{\mathbf{u}}_{(\mathrm{t})} + \mathbf{C}\dot{\mathbf{u}}_{(\mathrm{t})} + \mathbf{K}\mathbf{u}_{(\mathrm{t})} = \mathbf{F}_{(\mathrm{t})}^{\mathrm{ext}} \tag{9}$$

The solution of this system is carried out for each time step via the explicit central difference method. Here, the acceleration in time t has the form

$$\ddot{\mathbf{u}}_{(t)} = \mathbf{M}^{-1}\left[\mathbf{F}_{(t)}^{ext} - (\mathbf{C}\dot{\mathbf{u}}_{(t)} + \mathbf{K}\mathbf{u}_{(t)})\right] = \mathbf{M}^{-1}\left[\mathbf{F}_{(t)}^{ext} - \mathbf{F}_{(t)}^{\text{int}}\right] \tag{10}$$

where $\boldsymbol{F}_t^{ext}$ is the vector of external forces and $\boldsymbol{F}_t^{int}$ is the vector of internal forces gives as

$$\mathbf{F}_t^{\text{int}} = \sum\left(\int_{\Omega}\left(\mathbf{B}^T\sigma_n d\Omega + F^{hg}\right)\right) + F^{cont} \tag{11}$$

Velocities and accelerations have the form

$$\Delta t^2\ddot{\mathbf{u}}_{(\mathrm{t})} = \mathbf{u}_{(\mathrm{t}-\Delta\mathrm{t})} - 2\mathbf{u}_{(\mathrm{t})} + \mathbf{u}_{(\mathrm{t}+\Delta\mathrm{t})} \tag{12}$$

$$2\Delta t\ \dot{\mathbf{u}}_{(\mathrm{t})} = \mathbf{u}_{(\mathrm{t}+\Delta\mathrm{t})} - \mathbf{u}_{(\mathrm{t}-\Delta\mathrm{t})} \tag{13}$$

The starting procedure has the form

$$\mathbf{u}_{(\mathrm{t}-\Delta\mathrm{t})} = \mathbf{u}_{(0)} - \Delta t\dot{\mathbf{u}}_{(0)} + \frac{\Delta t^2}{2}\ddot{\mathbf{u}}_{(0)} \tag{14}$$

By applying zero initial conditions to the displacements and velocities, the starting procedure has the form

$$\ddot{\mathbf{u}}_{(t-\Delta t)} = \mathbf{M}^{-1}\mathbf{F}_{(0)}^{\mathbf{ext}} \tag{15}$$

The stability of the central difference method depends on the length of the time step, which has to be divided into the shortest natural domains in the finite element mesh. The critical time step is computed by following relation

$$\Delta t^{crit} = \frac{2}{\omega_{\max}} \tag{16}$$

where ω_{max} is the maximum natural circular frequency. The calculation is based on Courant-Friedrichs-Lewy condition (CFL condition) for solving partial differential equations numerically by the method of finite differences

$$\omega_{\max} = 2\frac{\mathrm{c}}{1} \tag{17}$$

where c is the wave speed in the material and l is the characteristic length. By substitution (Eq. 9) into (Eq. 7) we obtain relation for critical time step

$$\Delta t = \frac{l}{c} \tag{18}$$

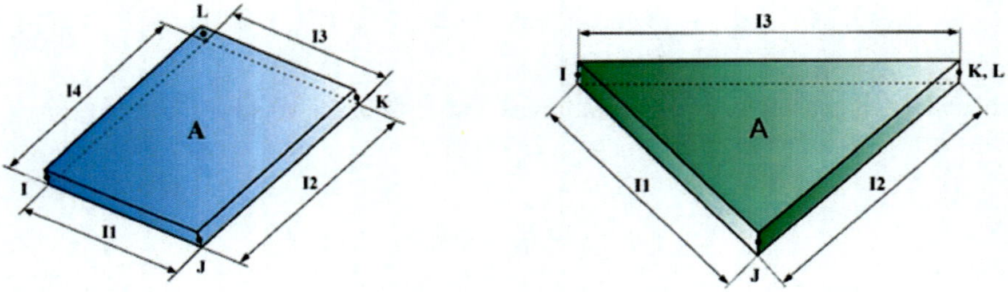

Fig. 2 Quadrilateral and triangular shell element

where Δt is time required for wave propagation in rod with length l. During time step calculation, ABAQUS/explicit program check size of all finite elements. For the numerical stability of calculation was used coefficient 0.9 for time step reduction

$$\Delta t = 0.9\frac{l}{c}x \tag{19}$$

Characteristic length of a shell element is given as

$$l = \frac{A}{\max\left(l_1, l_2, l_3, l_4\right)} \tag{20}$$

where A is the element area, l_i are lengths sides of Fig. 2. For triangular shell element the relation has the form

$$l = \frac{2A}{\max\left(l_1, l_2, l_3\right)} \tag{21}$$

Wave propagation velocity in a shell element is given by relation

$$c = \sqrt{\frac{E}{\rho\left(1 - \mu^2\right)}}x \tag{22}$$

where E is the Young modulus, ρ is mass density and μ is the Poisson number.

3 Description of Problem and Modelling Approach

Today, typically, finite element numerical models based on lamina-level failure criteria are used to simulate the damage of the fiber-reinforced composite material. Damage modeling usually encompasses two phases: damage initiation and damage evolution.

3.1 Materials

In the present paper, Hashin's criterion is implemented to identify fiber and matrix failure initiation. This criterion involves four damage modes, namely, fiber tension, fiber compression, matrix tension and matrix compression modes according to the following equations:

1. Fiber tensile failure: $(\hat{\sigma}_{11} \geq 0)$:

$$\left(\frac{\hat{\sigma}_{11}}{X_T}\right)^2 + \frac{\hat{\sigma}_{12}^2 + \hat{\sigma}_{13}^2}{S_{12}^2} = \begin{cases} \geq 1 \ \text{failure} \\ < 1 \ \text{no failure} \end{cases} \tag{23}$$

2. Fiber compressive failure $(\hat{\sigma}_{11} < 0)$:

$$\left(\frac{\sigma_{11}}{X_C}\right)^2 = \begin{cases} \geq 1 \ \text{failure} \\ < 1 \ \text{no failure} \end{cases} \tag{24}$$

3. Matrix tensile failure $(\hat{\sigma}_{22} \geq 0)$:

$$F_{mt} = \left(\frac{\hat{\sigma}_{22}}{Y^T}\right)^2 + \left(\frac{\hat{\sigma}_{12}}{S_{12}}\right)^2 = 1 \tag{25}$$

4. Matrix compressive failure $(\hat{\sigma}_{22} < 0)$:

$$F_{mc} = \left(\frac{\hat{\sigma}_{22}}{2S_{23}}\right)^2 + \left[\left(\frac{Y^C}{2S_{23}}\right)^2 - 1\right]\frac{\hat{\sigma}_{22}}{Y^C} + \left(\frac{\hat{\sigma}_{12}}{S_{12}}\right)^2 = 1 \tag{26}$$

where, σ_{ij} $\widehat{\sigma}_{ij}$ are effective stress, X^T and X^C are tensile and compressive strength of composite laminate in fiber direction, Y^T and X^C are tensile and compressive strength in transverse direction, S_{12} and S_{23} are longitudinal and transverse shear strength of the composite, respectively. The coefficient α is for shear stress contribution on the fiber tensile failure.

The material of present composite is an AS4/PEEK quasi-isotropic laminate. For the simulation of impact damage has been used four types of orientation layers (layup):

$$\begin{bmatrix} 0 \\ \hline 0 \\ \hline 0 \\ \hline 0 \end{bmatrix}_S, \begin{bmatrix} 0 \\ \hline 0 \\ \hline 90 \\ \hline 90 \end{bmatrix}_S, \begin{bmatrix} 45 \\ \hline 45 \\ \hline 45 \\ \hline 45 \end{bmatrix}_S \tag{27}$$

Material parameters of the laminate plate are listed in Tables 1 and 2.

Table 1 Material properties of AS4/PEEK

E_{11} (GPa)	E_{22} (GPa)	ν_{12} (–)	G_{12} (GPa)	G_{13} (GPa)	G_{23} (GPa)	ρ (kg/m^3)	X_T (MPa)	X_C (MPa)
138	10.2	0.3	5.7	5.7	3.7	1570	2070	1360
Y_T (MPa)	Y_C (MPa)	S_L (MPa)	S_T (MPa)					
86	230	186	86					

Table 2 Fracture energy of laminate (course of damage) [15]

	F_f^t	F_f^c	F_m^t	F_m^{tc}
Fracture energy (N/mm)	12.5	12.5	1.0	1.0

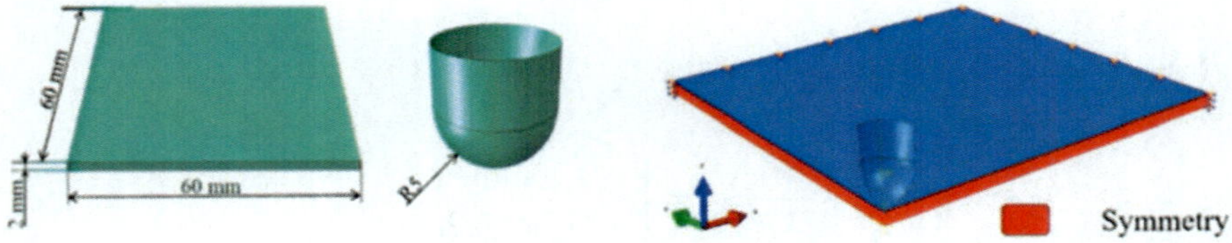

Fig. 3 Geometry of composite plate and impactor (to the left), boundary conditions (to the right)

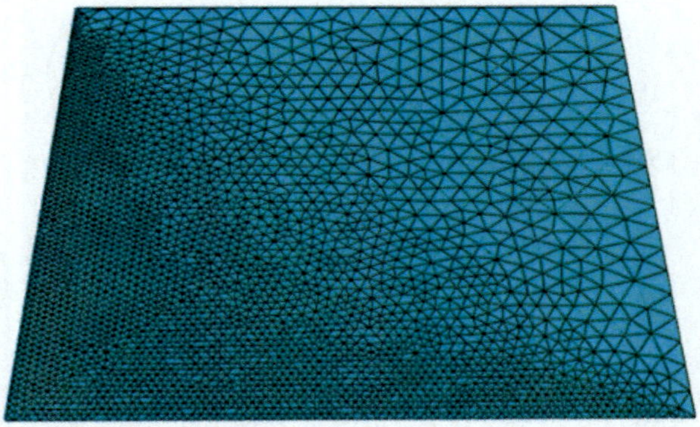

Fig. 4 Finite element mesh of conventional shells

3.2 Finite Element Modeling

In the next simulations were considered composite plate with dimensions 120 × 120 × 2 mm composed from eight layers. Due to the symmetry, only quarter of the geometry was modelled to save the computational cost (Fig. 3). The composite plate structure was created in ABAQUS/explicit using the composite module [16]. This module involves the formation of conventional and volumetric shells (Figs. 4 and 5). This module defines the individual layers of the composite structure, the type of integration rule, symmetry, material properties, thickness, orientation and the number of integration points of the layer.

The composite plate is composed from the eight-node brick hexahedral elements with one integration point (C3D8R) and 50,000 elements were used in the simula-

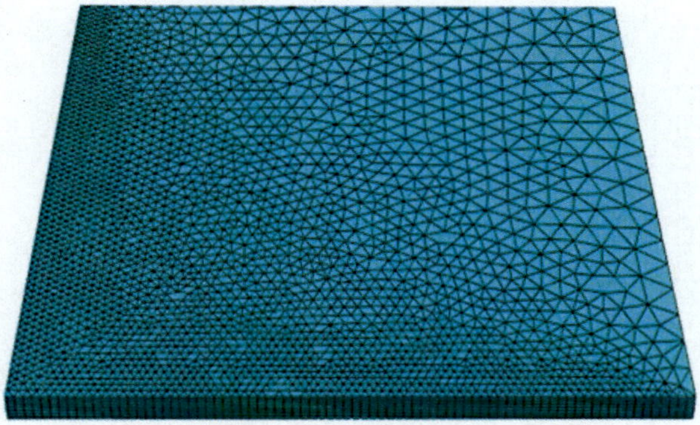

Fig. 5 Finite element mesh of volumetric shells

tion. A refined, uniform mesh was used in the impact region. The ABAQUS/Explicit simulations presented here examined the penetration of plate specimens.

The projectile has cylindrical shape with semi-spherical fillet with a radius $R = 5$ mm. Since results from ballistic experiments showed negligible deformation, plastic deformation of the projectile is not considered. The projectile impacts the plate perpendicularly, right is center of the plate with a defined initial speed $v_i = 100$ m/s and the plate was supported on all edges. FE mesh for the shell geometry was created using 5450 linear triangular elements S3R and for solid geometry has been used 39.520 linear SC6R brick elements (Fig. 5).

The composite structure consists of 8 layers, one layer having a thickness of 0.25 mm. The number of integration points has been set by default. The following four types of layer orientation were used to simulate the impact damage (27).

4 Result

The ABAQUS/Explicit simulations presented here examined the penetration of composite plate samples impacted with steel rod with a hemispherical. Figures 6 and 7 show the dependence of acceleration and velocity on time at the node where is maximum displacement. The maximum value of acceleration is $5.2e^{+08}$ m/s^2 and maximum value of velocity is 139 m/s.

In the volumetric shell geometry, the maximum stress was von Mises 2090 MPa and the shear damage occurred only in the area close to the impactor impact (Fig. 8). In conventional shell geometry, damage propagate from the impact point to the edge of tested plate, and shear damage also occurred at the edges of the plate. The maximum von Mises stress reached 2100 (Fig. 9).

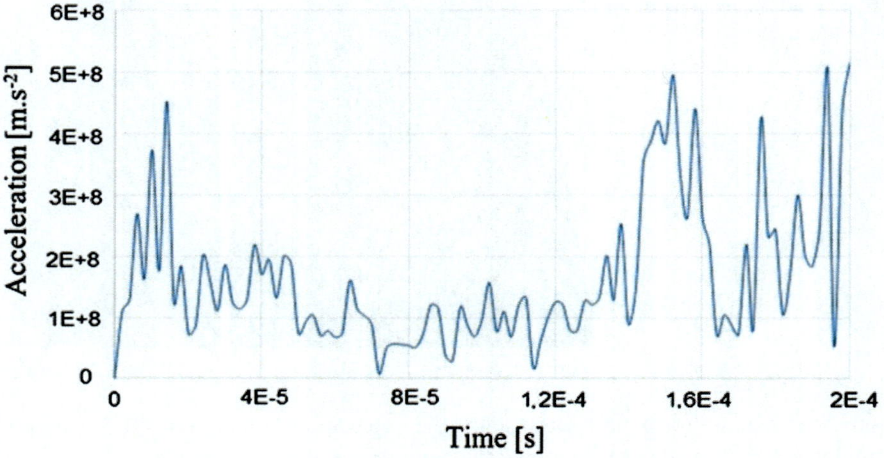

Fig. 6 Time history of acceleration at node with maximum displacement

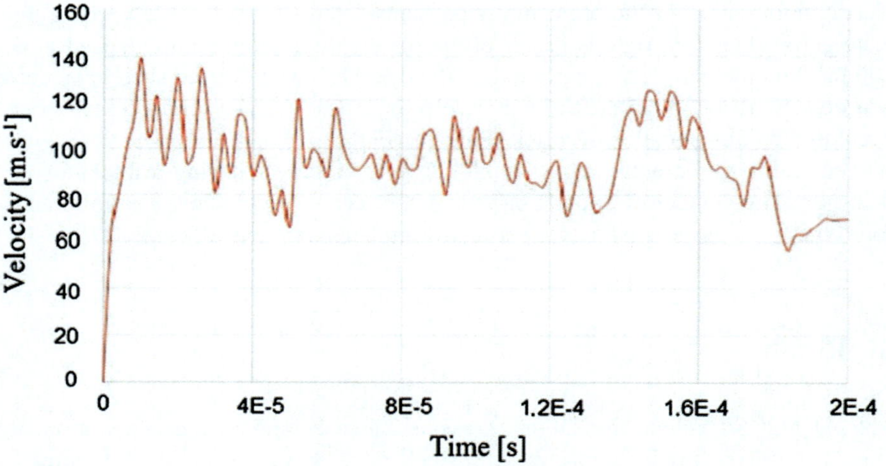

Fig. 7 Time history of velocity at maximum displacement

5 Conclusion

In this paper for the analysis of laminate composite plates two models are used. The first is the solid based model and other is shell based model. There were also compared four different arrangement of the layers of the composite. As a criterion damage the composite plate was used Hashin damage model. The results obtained show that the von Mises stress have approximately the same value for all types of arrangements of the layers. For solid model, and also for the shell model was the largest von Mises stress in the arrangement of layers $[0/0/90/90]_s$

Fig. 8 Course of von Mises stress (to the left) and shear stress (to the right) for volumetric shell geometry—$[0/0/90/90]_s$

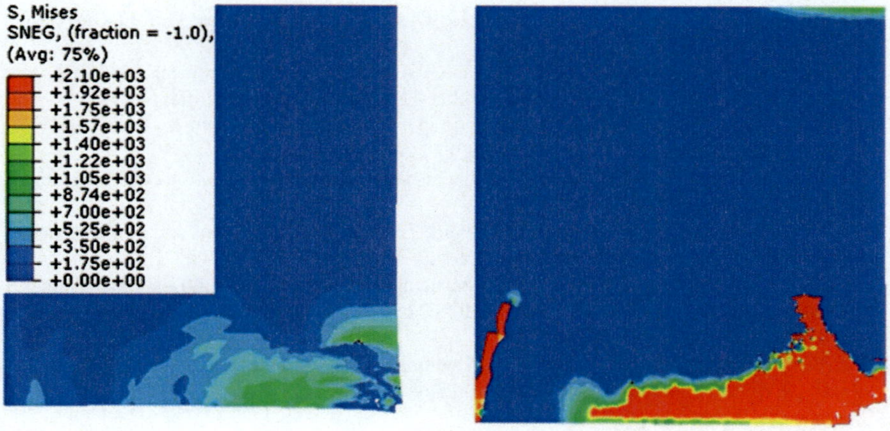

Fig. 9 Course of von Mises stress (to the left) and shear stress (to the right) for conventional shell geometry—$[0/0/90/90]_s$

the lowest von Mises stress was in the arrangement of the layers $[90/0/0/0]_S$. For the arrangement of layers $[90/0/0/0]_S$, $[45/45/45/45]_s$ is the lowest von Mises stress for shell based model. The largest shear stresses were in the arrangement of the layers $[45/45/45/45]_s$ for solid as well as shell based model. The largest deformation was at the area of impact, which gradually propagate to the depth of the material. From the results we can see that the orientation of the layers in the composite structures can have a significant effect on the behavior of the structure.

Acknowledgments This work was supported by internal grant of Jan Evangelista Purkyně, Faculty of Mechanical Engineering (UJEP-IGS-2018-48-002-1 and UJEP-SGS-2018-48-002-2).

References

1. Barbero, E.J.: Introduction to Composite Materials Design. CRC Press, Boca Raton (2011)
2. Marulo, F., Guida, M., Maio, L., Ricci, F.: Numerical simulation and experimental experiences of impact on composite structures. In: Dynamic Response and Failure of Composite Materials and Structures. Woodhead Publishing, Duxford (2017)
3. Barbero, E.J.: Finite Element Analysis of Composite Materials Using ABAQUS. CRC Press, Boca Raton (2013)
4. Zhang, H., Wang, M., Wen, W., Xu, Y., Cuiu, H., Chen, J.: A full-process numerical analyzing method of low velocity impact damage and residual strength for stitched composites. Appl. Sci. **8**, 2698 (2018). https://doi.org/10.3390/app8122698
5. Dong, S., Sheldon, A., Carney, K.: Modeling of carbon-fiber-reinforced polymer (CFRP) composites in LS-dyna with optimization of material and failure parameters in LS-OPT. In: 15th International LS-DYNA Users Conference, 10–12 June 2018, Detroit, MI, USA
6. Cantwell, W.J., Morton, J.: The impact resistance of composite materials—a review. Composites. **22**, 347–362 (1991)
7. Abrate, S.: Impact Engineering of Composite Structures. Springer, Udine (2011)
8. Liu, D., Malvern, L.E.: Matrix cracking in impacted glass/epoxy plates. J. Compos. Mater. **21**, 594–609 (1987)
9. Zhong, Z.-H.: Finite Element Procedures for Contact-Impact Problems. Oxford University Press, Oxford (1993)
10. Wriggers, P.: Computational Contact Mechanics. Antony Rowe Ltd., Chippenham (2002)
11. Bathe, K.J.: Finite Element Procedures. Prentice-Hall Inc., Englewood Cliffs, NJ (1995)
12. Crisfield, M.A.: Non-linear Finite Element Analysis of Solids and Structures. Essentials, vol. 1. John Wiley & Sons Ltd., Chichester (2000)
13. Belytschko, T., Liu, W.K., Moran, B.: Nonlinear Finite Elements for Continua and Structures. John Wiley & Sons Ltd., Chichester (2000)
14. Ibrahimbegovic, A.: Nonlinear Solid Mechanics, Theoretical Formulations and Finite Element Solution Methods. Springer, New York (2009)
15. Nanderi, M., Khonsari, M.M.: Ch. 4: Stochastic analysis of inter- and intra-laminar damage in notched PEEK laminates. In: EBSCO HOST Connection. Online version, vol. 7, p. 383. Louisiana State University (2013)
16. ABAQUS. Theory manual version 6.10. Documentation [online]. 2015. [cit. 10.03.2015]. Available from: <http://abaqusdoc.ucalgary.ca/books/stm/default.htm> (2010)

Identification of Nonlinear Joint Interface Parameters Using Instantaneous Power Flow Balance Approach

R. Anish and K. Shankar

Abstract Joints in assembled structures can affect the dynamic behaviour of mechanical structures under dynamic loading conditions. Mathematical modelling of such structures, one need to consider the joint interface effects accurately. In this paper a bolted lap joint is modelled with a nonlinear spring and a damper to simulate the nonlinear effects like softening phenomena due to slip, associated with the joint structures. The known parametric model of the assembled beam structure with joint interface non-linearity was simulated 'experimentally' under a harmonic external excitation to find the responses. The parameter identification was formulated as an inverse problem using Particle Swarm Optimization algorithm. The error between experimentally measured and numerically predicted response matching and a novel Instantaneous Power Flow Balance criterion based objective functions are used for the identification of nonlinear parameters. The identified nonlinear parameters show the accuracy of the current method over other time domain methods.

Keywords Structural identification · Power flow · Substructure · Particle swarm optimization

1 Introduction

Structural identification problems are inverse analysis problems, which concerned with system modeling from input output information. Most commonly used parameter identification problems are vibration-based identification techniques. In recent years researchers are biased towards the area of nonlinear structural parameter identification with the objective to develop more accurate mathematical model. Nonlinear parameter identification problems are more generic in nature and there

---

R. Anish (✉) · K. Shankar
Machine Design Section, Department of Mechanical Engineering,
Indian Institute of Technology Madras, Chennai, India
e-mail: skris@iitm.ac.in

J. Awrejcewicz (ed.), *Perspectives in Dynamical Systems II: Mathematical and Numerical Approaches*, Springer Proceedings in Mathematics & Statistics 363,
https://doi.org/10.1007/978-3-030-77310-6_16

exist no common analysis procedure that can be applied to all nonlinear problems at all instances. Kerschen et al. [1], Noel and Kerschen [2] reviewed the state of art nonlinear structural parameter identification techniques. Koh et al. [3, 4] used the substructure technique to decomposes a large system to small manageable subsystems to improve the convergence of structural parameter identification. Varghese and Shankar [5] introduced an inverse identification technique, which investigates the application of substructural power flow to linear structural parameter estimation. Kapania and Park [6] studied a two-step identification process using Time Finite element Method (TFM) for the structural parameter identification. Kumar and Shankar [7] worked on nonlinear parametric identification by formulating an inverse problem using substructure acceleration matching objective function with Genetic Algorithms.

Bowden and Dugundji [8] explain the global dynamics of jointed structures by considering both the linear and nonlinear analysis separately. Li et al. [9] proposed a general analytical method for predicting the vibrations and power flow between two coupled beams. Ahmadian and Jalali [10] presented an accurate lumped parameter model for bolted lap joints and the responses were found using the method of multiple scales. Ahmadian and Jalali [11] developed a nonlinear generic joint element for Finite Element modelling of bolted lap joints. Ma et al. [12] presented a technique for constructing a non-parametric model for identifying the dynamic effects of bolted joints using laser vibrometry analysis.

Several researchers revealed the superiority of the Particle Swarm Optimization (PSO) algorithm (proposed by Eberhart and Kennedy [13]) over other non-classical algorithms. In this work a multi objective optimization problem was formulated using a novel concept of instantaneous power flow balance criteria based objective function along with substructure acceleration matching criteria as a weighted aggregation approach for the nonlinear structural parameter identification.

2 Joint Interface Model

Generally the nonlinearities associated with joints are functions of relative displacements and the direct measurement of relative displacements between two contacting surfaces is not possible to get in practice. In this paper the parameters of a nonlinear joint are identified through an inverse problem approach using acceleration response matching and instantaneous power flow balance objective functions. A Finite Element based lumped joint interface model is used here to simulate the experiment on assembled structure of steel beams. The lumped model consists of a linear translational stiffness (K_L) and torsional Stiffness (K_θ), nonlinear translational spring (K_{NL}) and a linear damper (C) and is shown in Fig. 1. A parametric model with joint interface coefficients, $K_L = 8.089 \times 10^8$ N/m; $C = 0.281$ Ns/m; $K_\theta = 3264$ N/rad; $K_{NL} = 3.722 \times 10^6$ N/m^3, was selected for the analysis. A finite element modal analysis of the nonlinear joint system was carried

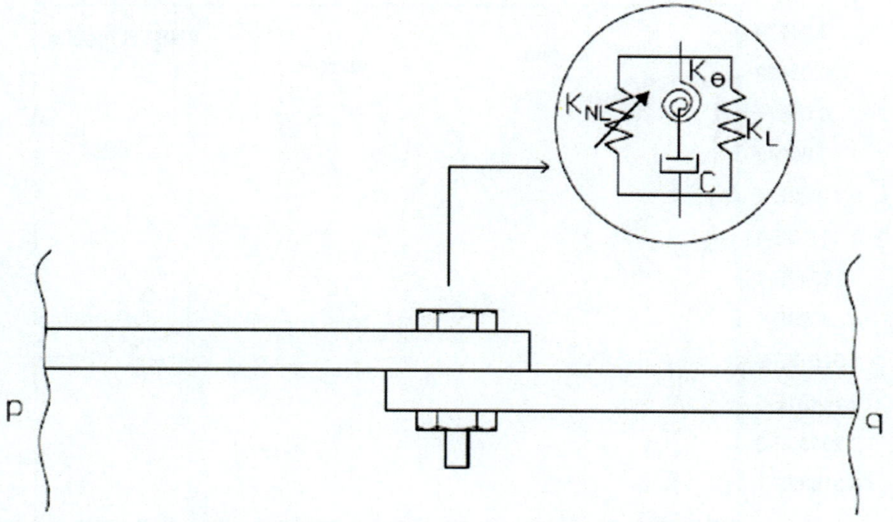

Fig. 1 Lumped joint interface model

out in ANSYS® using BEAM188, MASS21, COMBIN14 and nonlinear spring COMBIN39 elements and found the first three frequencies to be 9.8306, 69.07 and 226.92 Hz. Figure 2 shows the frequency response plots of the model using ANSYS software. The dynamic responses of the model in terms of acceleration, velocity and displacement are generated using a single frequency harmonic excitation applied very close to fixed end of the assembled beam.

3 Substructure Formulation

Substructure without overlapping is studied here. The equations of motion for the substructure considered may be extracted from the system of partitioned equations following the method described in Koh et al. [3–5].

$$[M_{rf} M_{rr} M_{rg}] \begin{Bmatrix} \ddot{u}_f \\ \ddot{u}_r \\ \ddot{u}_g \end{Bmatrix} + [C_{rf} C_{rr} C_{rg}] \begin{Bmatrix} \dot{u}_f \\ \dot{u}_r \\ \dot{u}_g \end{Bmatrix} + [K_{rf} K_{rr} K_{rg}] \begin{Bmatrix} u_f \\ u_r \\ u_g \end{Bmatrix} = \{\mathrm{P_r}(t)\} \tag{1}$$

where, the subscript '*r*' denotes internal DOFs of the concerned substructure, subscripts '*f*' and '*g*' represents the interface DOFs. Let subscript '*j*' denote all interface DOFs (i.e. '*f*' and '*g*' included) for concise presentation the above equation can be written as

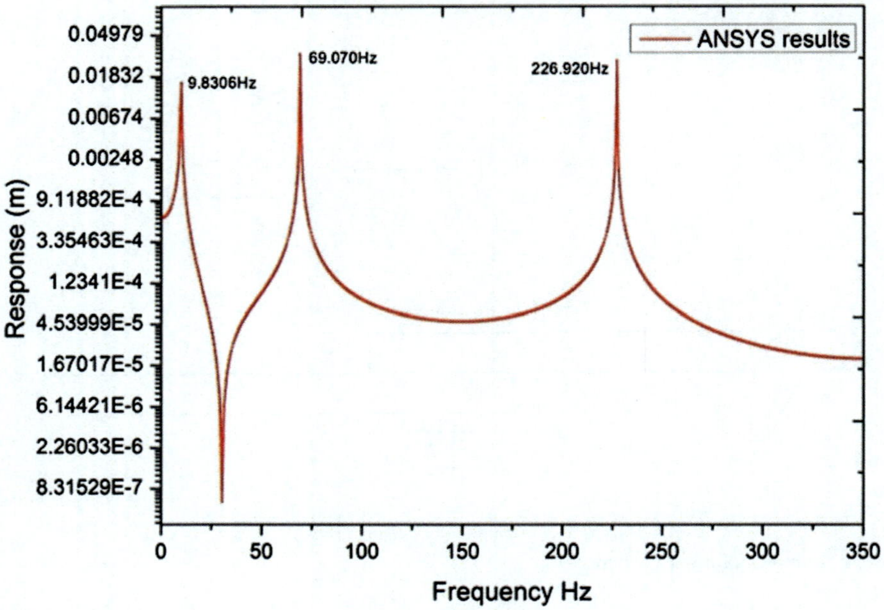

Fig. 2 Frequency response plots of the model using ANSYS

$$\left[M_{rj}M_{rr}\right]\left\{\begin{matrix}\ddot{u}_j\\ \ddot{u}_r\end{matrix}\right\}+\left[C_{rj}C_{rr}\right]\left\{\begin{matrix}\dot{u}_j\\ \dot{u}_r\end{matrix}\right\}+\left[K_{rj}K_{rr}\right]\left\{\begin{matrix}u_j\\ u_r\end{matrix}\right\}=\{\mathrm{P_r}(t)\} \tag{2}$$

This can be rearranged to bring the interior partitions to the left and interface effects in the form of a force on to the right as,

$$M_{rr}\ddot{u}_r(t)+C_{rr}\dot{u}_r(t)+K_{rr}u_r(t)=\mathrm{P_r}(t)-M_{rj}\ddot{u}_j(t)-C_{rj}\dot{u}_j(t)-K_{rj}u_j(t) \tag{3}$$

$P_r(t)$ is the excitation force applied on the interior node(s). In the absence of force excitation within the substructure, then $P_r(t)$ is taken zero that means the force application is outside the substructure. The left side of Eq. (3) represents the output from the substructure while right side is treated as input to the substructure.

4 Objective Functions

Two identification methods have been reported in this paper: one based on acceleration (response/modal data) matching objective function and the other based on Instantaneous Power Flow Balance Objective function, both uses the inverse problem strategy for identification. A multi objective problem was formulated by combining both the conventional acceleration matching and instantaneous power flow balance approach. The substructure identification method used in this work, allows the researchers to concentrate, the sensor measurements to a smaller zone of interest and thereby reducing the computational time.

4.1 Substructure Acceleration Matching

The concept of sub structuring is the 'divide and conquer' rule, in which the global structure is sub divided into substructures. The first fitness function is formulated by comparing the error between measured and estimated acceleration as

$$f_1 = \sum_{i=1}^{M} \sum_{j=1}^{T} \frac{(\ddot{x}_m - \ddot{x}_e)^2}{T * M} \tag{4}$$

where, the subscripts 'm' and 'e' represents the measured and estimated acceleration response for fitness evaluation. 'M' is the number of measurement points or sensor locations and 'L' is the number of time steps. Here the measured acceleration response was simulated numerically and Fig. 3 represents the acceleration matching between global and substructure accelerations of the joint assembly.

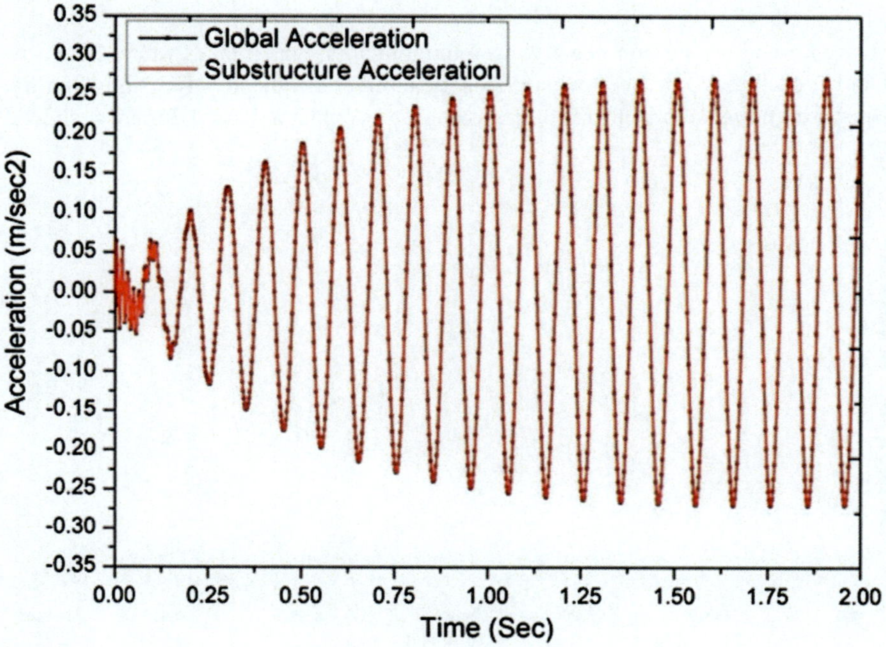

Fig. 3 Acceleration matching global structure and substructure accelerations

4.2 Instantaneous Power Flow Balance

Power flow balance is another form of law of conservation of energy. The power balance criteria states that within a substructure the net sum of power that is net sum of input, damping and transferred power is zero. The main purpose of instantaneous power study is to formulate an objective function in terms of instantaneous power flow balance as

$$f_2 = \frac{1}{T}\sum_{i=1}^{T}\left(IP_e^b\right)^2 \tag{5}$$

where, 'T' is the number of time steps and superscript 'e' denotes the estimated instantaneous power balance for objective function evaluation.

4.3 Weighted Aggregation Approach

The two objective functions, Eqs. (4) and (5) are combined together to form a multi objective optimization problem using weighted aggregation approach. The combined objective function can be represented as

$$f = w_1 f_1 + w_2 f_2 \tag{6}$$

where, w_1 and w_2 are non-negative weighting factors, which takes values between 0 and 1, such that $\sum w_i = 1$, where 'i' is the number of objective functions. In this paper, weighting factors of {0.5, 0.5} was chosen as explained in [5]. Figure 4, shows

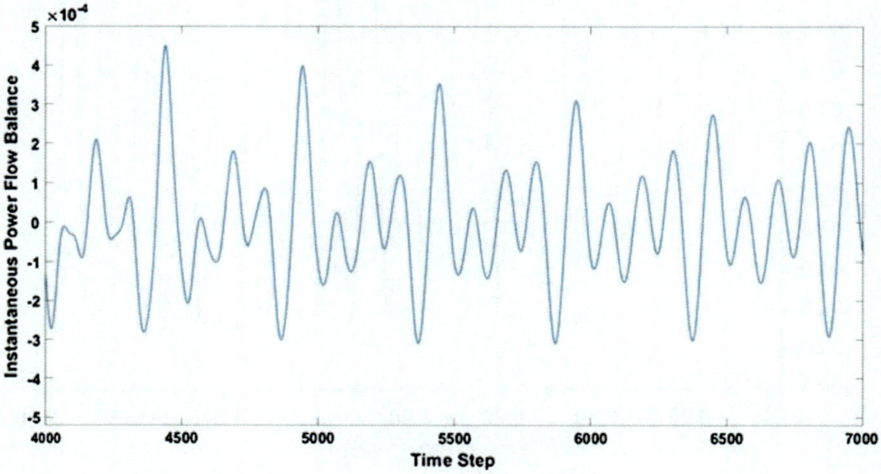

Fig. 4 Instantaneous power flow balance plot

the Instantaneous Power Flow Balance (IPFB) plot corresponding to each time step in the time span 0.8–1.4 s. From Fig. 4 it is found that, IPFB is of the order of $1e^{-4}$ which is close to zero. Now the combined objective function is passed to Particle Swarm Optimization algorithm for minimization and thereby structural parameter identification.

5 Identification of Joint Interface Parameters

The nonlinear joint interface parameter identification is explained through a numerical model of two beam connected by a single bolted lap joint as in Ahmadian and Jalali [10] was selected. The assembly consists of two identical steel beams connected by a bolted lap joint of same material. Length of each beam is taken as 280 mm with and width 25 and 5 mm thickness. One end of the beam assembly is fixed and the other end carries a steel block of 0.1739 kg as shown in Fig. 5. Experiments show the nonlinearities associated with bolted assembly varies with bolt tightness due to softening phenomena, which are modelled using a nonlinear flexibility at the joint interface. In this particular simulation study, it is assumed that the bolt is tightened with a constant torque (2 Nm), which remains constant during the analysis. For the simulation study, a finite element model of the coupled beam structure is developed by discretizing the assembly into 28-Euler Bernoulli beam elements. The lap joint interface is modelled using a lumped joint model of linear and nonlinear springs and a linear damper. The structure is excited with a harmonic force of magnitude 0.5 N and a frequency 20 Hz, at a distance 40 mm, close to fixed end of the beam assembly.

The substructure method without overlap is chosen for parameter identification study. This will allow the researcher to concentrate on a small region of interest, here the joint interface portion. In this case study, 60 mm to both sides of the bolt axis was selected for the generation of substructure. The response for internal

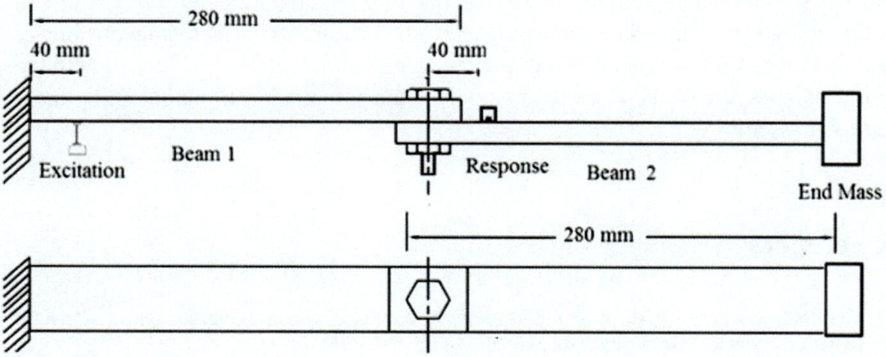

Fig. 5 Geometrical diagram of simulated model

Table 1 Actual values and identified results of lap joint interface parameters

	Joint parameters				Damping constants	
Substructure parameter	K_L (N/m)	K_θ (N/rad)	K_{NL} (N/m^3)	C (Ns/m)	α	β
Actual value	8.089×10^8	3264	3.722×10^6	0.281	3.244	1.209×10^{-4}
Identified value	7.948×10^8	3263	2.986×10^6	0.299	3.189	1.210×10^{-4}
Mean absolute % error	1.740	0.002	19.778	6.646	1.690	0.004

DOF of the substructures is simulated using the time span of 0–2 s and sampling time step of 0.0002 s using the Newmark beta numerical integration method. In the substructure parameter identification procedure, the coefficients of linear (both translational and rotary) springs, nonlinear spring and linear damper terms of the lumped joint model are assumed as unknown parameters. The PSO with population of 50, generation of 300, acceleration coefficients two and an inertia weight 0.9 has been used for identification. The identification results are shown in Table 1. The range of mean absolute error was found to be 0.002–19.778% the higher value in the nonlinear parameter is due to the magnifying effect of the relative displacement in the mathematical formulations.

6 Conclusions

A novel concept of nonlinear joint parameter identification using combined acceleration matching and instantaneous power flow balance as objective functions in time domain has been introduced. The proposed method was successfully implemented on parameter estimation of a lap joints assembly as a nonlinear system. The error in identification using the proposed method was found to be less than 6% for the linear parameters and 19.778% for the non-linear parameters which are in acceptable limits. The identified results show the feasibility of instantaneous power flow balance objective function along with acceleration matching approach with the aid of substructure concept, a valuable tool for the parameter estimation of nonlinear lap joint assembly.

References

1. Kerschen, G., et al.: Past, present and future of nonlinear system identification in structural dynamics. Mech. Syst. Signal Process. **20**(3), 505–592 (2006)
2. Noel, J.-P., Kerschen, G.: Nonlinear system identification in structural dynamics: 10 more years of progress. Mech. Syst. Signal Process. **83**, 2–35 (2017)

3. Koh, C.G., See, L.M., Balendra, T.: Estimation of structural parameters in time domain: a substructure approach. Earthquake Eng. Struct. Dyn. **20**(8), 787–801 (1991)
4. Koh, C.G., Hong, B., Liaw, C.Y.: Sub structural and progressive structural identification methods. Eng. Struct. **25**(12), 1551–1563 (2003)
5. Varghese, C.K., Shankar, K.: Identification of structural parameters using combined power flow and acceleration approach in a substructure. Int. J. Eng. Technol. Innov. **1**(1), 65–79 (2016)
6. Kapania, P., S.: parametric identification of nonlinear structural dynamic systems using time finite element method. AIAA J. **35**(4), 719–726 (1997)
7. Kumar, R.K., Shankar, K.: Parametric identification of structures with nonlinearities using global and substructure approaches in the time domain. Adv. Struct. Eng. **12**(2), 195–210 (2009)
8. Bowden, M., Dugundji, J.: Joint damping and nonlinearity in dynamics of space structures. AIAA J. **28**(4), 740–749 (1990)
9. Li, W.L., Bonilha, M.W., Xiao, J.: Vibrations and power flows in a coupled beam system. J. Vib. Acoust. **129**(5), 616–622 (2007)
10. Ahmadian, H., Jalali, H.: Identification of bolted lap joints parameters in assembled structures. Mech. Syst. Signal Process. **21**(2), 1041–1050 (2007)
11. Ahmadian, H., Jalali, H.: Generic element formulation for modelling bolted lap joints. Mech. Syst. Signal Process. **21**(5), 2318–2334 (2007)
12. Ma, X., Bergman, L., Vakakis, A.: Identification of bolted joints through laser vibrometry. J. Sound Vib. **246**(3), 441–460 (2001)
13. Eberhart, R., Kennedy, J.: A new optimizer using particle swarm theory. In: Proceedings of the Sixth International Symposium on Micro Machine and Human Science, 1995 (MHS'95). IEEE (1995)

Numerical Procedure for the Sensitivity Analysis of Hybrid Systems

Radosław Pytlak, Damian Suski, and Tomasz Tarnawski

Abstract The paper presents the numerical procedure for the evaluation of adjoint equations of hybrid systems for the purpose of gradients calculations of functions dependent on states of hybrid systems. It is assumed that hybrid systems can have a finite number of discrete states, in each discrete state the hybrid system is described by a set of ordinary differential equations. In particular, the proposed procedure can be used in optimization procedures for solving optimal control problems with hybrid systems. For this reasons the presented procedure is based on the implementation of a Runge–Kutta method which is advocated as the most suitable numerical procedures for integration of differential equations with controls represented by piecewise constant functions (Hager, Numer Math 87:247–282, 2000; Pytlak, Numerical Methods for Optimal Control Problems with State Constraints. Lecture Notes in Mathematics, vol. 1707. Springer, Berlin, 1999; Schwartz and Polak, SIAM J Control Optim 34:1235–1269, 1996). Since we are dealing with hybrid systems our numerical procedure is equipped with the procedure for locating switching points which determine the change of a discrete state. The evaluation of adjoint equations is consistent with the system equation discretization. We show the effectiveness of our procedure on the example of planning a haemodialysis process.

Keywords Hybrid systems · Optimal control · Adjoint equations

R. Pytlak (✉)
Warsaw University of Technology, Faculty of Mathematics and Information Science, Warsaw, Poland
e-mail: r.pytlak@mini.pw.edu.pl

D. Suski
Warsaw University of Technology, Institute of Automatic Control and Robotics, Warsaw, Poland
e-mail: d.suski@mchtr.pw.edu.pl

T. Tarnawski
Military Institute of Armament Technology, Zielonka, Poland
e-mail: tarnawskit@witu.mil.pl

J. Awrejcewicz (ed.), *Perspectives in Dynamical Systems II: Mathematical and Numerical Approaches*, Springer Proceedings in Mathematics & Statistics 363,
https://doi.org/10.1007/978-3-030-77310-6_17

1 Introduction

Hybrid systems are systems with mixed discrete–continuous dynamics [11]. In this work we use a definition of a hybrid which we used in our previous paper [7]. We restrict our analysis to systems with autonomous transitions and without state jumps during transitions

Definition 1 A hybrid system $\mathcal{H}$ is a tuple

$$\mathcal{H} = (\mathcal{Q}, \mathcal{U}, \mathcal{I}, \mathcal{F}, \mathcal{T}, \mathcal{G}) \tag{1}$$

where

- $\mathcal{Q}$ is a finite set of discrete states. Its elements are denoted by q.
- $\mathcal{U}$ is a set of admissible controls. The elements of $\mathcal{U}$ are measurable functions $u : I \to U$, where I can be any closed interval of $\mathbb{R}$ and U is a fixed subset of $\mathbb{R}^m$.
- $\mathcal{I}$ is a function which assigns to every discrete state q a set

$$\mathcal{I}(q) = \left\{x \in \mathbb{R}^n : g_q(x) \leq 0\right\}, \ g_q : \mathbb{R}^n \to \mathbb{R}^{n_{\mathcal{I}_q}} \tag{2}$$

  such that as long as a hybrid systems is in a discrete state q the continuous state trajectory x stays in $\mathcal{I}(q)$. We therefore say that $\mathcal{I}(q)$ is an *invariant set* for a discrete state q.
- $\mathcal{F}$ is a function which assigns to every discrete state q a function $f^q : \mathcal{I}(q) \times U \to \mathbb{R}^n$ such that in a discrete state q the continuous state evolves according to a differential equation

$$x' = f^q(x, u) \tag{3}$$

- $\mathcal{T}$ is a subset of $\mathcal{Q} \times \mathcal{Q}$, which collects all pairs of discrete states (q, q') such that the transition from a state q to a state q' is possible.
- $\mathcal{G}$ assigns to each pair $(q, q') \in \mathcal{T}$ a subset of $\mathcal{I}(q)$ boundary such that when a continuous state trajectory is about to leave $\mathcal{I}(q)$ through its boundary at a point $x_t \in \mathcal{G}(q, q') \subset \partial\mathcal{I}(q)$ a discrete state changes from q to q'. We call such an event a *transition* and $\mathcal{G}$ plays a role of a *transition guard*.

Taking into account the considerations and the presented definition we now put down the optimal control problem of interest, as follows:

$$\min_{u \in U} \phi(x(t_1)) \tag{4}$$

$$x(t_0) = x_0 \tag{5}$$

$$x' = \begin{cases} f^1(x, u) & \text{if} g(x) < 0 \\ f^2(x, u) & \text{if} g(x) > 0 \end{cases}$$

$$\psi_i^1(x(t_1)) = 0, \text{ for } i \in E \tag{6}$$

$$\psi_j^2(x(t_1)) \le 0, \text{ for } j \in I. \tag{7}$$

The admissible control is a function $u : [t_0, t_1] \to U$. In the paper we do not consider the system evolution in a sliding mode as we did in our paper [7] not because the analysis prsented here is not applicable to that case but for the simplicity of presentation. For the same reason we assume that the system evolve according to ordinary differential equations, the presented results could be extended to a quite broad class of differential–algebraic equations with index 3.

To calculate the optimal solution of the considered optimal control problem we need gradients of all functionals defining the problem. They can be determined in the similar way thus we only consider the gradient of the functional ϕ. Since x depends uniquely on u the value $\phi(x_{t_f})$ depends on u.

We assume that at time t_t state trajectory crosses the hyperplane $g(x) = 0$, and up to time t_t the system evolves according to the equation $x' = f^1(x, u)$ and then according to the equation $x' = f^2(x, u)$. Under that assumption the adjoint equations for the functional ϕ are (here t_t^-, t_t^+ are latest and earliest times of system being in states 1 and 2 respectively):

$$\lambda' = \begin{cases} -\left(f^1\right)_x^T \lambda \; t \in (t^+, t_f) \\ -\left(f^2\right)_x^T \lambda \; t \in [0, t^-) \end{cases}$$

with the endpoint condition

$$\lambda(t_f) = -\phi_x(x(t_f))^T$$

and the jump condition

$$\lambda(t_t^+) - \lambda(t_t^-) = \pi_t g_x^T(x(t_t))$$
$$f^1(x(t_t^-), u(t_t^-))^T \lambda(t_t^-) = f^2(x(t_t^+), u(t_t^+))^T \lambda(t_t^+)$$

where ν_t is some number.

Having a solution to the adjoint equations we can provide the estimate of the functional ϕ change due to the control variable variations d:

$$\delta\phi(d) = \int_{t_0}^{t_t} -\lambda^T \left(f^1\right)_u (x, u) d dt - \int_{t_t}^{t_f} \lambda^T \left(f^2\right)_u (x, u) d dt. \tag{8}$$

One issue which must be analysed is the extent of the jump in adjoint variables. To this end we solve the jump conditions equations getting

$$\pi_t = -\frac{\lambda(t_t^+)^T \left(f^2(x(t_t^+), u(t_t^+)) - f^1(x(t_t^-), u(t_t^-))\right)}{g_x(x(t_t)) f^1(x(t_t^-), u(t_t^-))}$$

$$\lambda(t_t^-) = \lambda(t_t^+) - \frac{\lambda(t_t^+)^T \left(f^2(x(t_t^+), u(t_t^+)) - f^1(x(t_t^-), u(t_t^-))\right)}{g_x(x(t_t)) f^1(x(t_t^-), u(t_t^-))} g_x^T(x(t_t)).$$

However numerical integration of system and adjoint equations introduce errors. One of the goals of the paper is to analyse how these errors influence accuracy of the evaluated functionals gradients.

Between transitions, a system of differential equations is integrated with the help of an appropriate numerical integration scheme. The numerical integration scheme can be presented in a from of an equation

$$X(k+1) = \varphi(X(k), u(k), h(k))) \tag{9}$$

which describes the relation between the augmented state at the next step $X(k+1)$, the augmented state at the present step $X(k)$, the actual control $u(k)$ and the actual stepsize $h(k)$ [5]. The discrete step k corresponds to a time instant $t(k)$.

During the numerical integration a possible violation of invariant set conditions has to be monitored. This task is realized by checking the sign changes of $g(x(k))$. in subsequent steps, where $x(k) = x(t(k))$ at time $t(k)$ corresponding to discrete time k. When a sign change of $g(x(k))$ between discrete steps k and $k+1$ is detected, the following problem is solved

$$\text{find } t_t \in [t(k), t(k+1)], \text{ s.t. } \tilde{g}(t_t) = 0 \tag{10}$$

where $\tilde{g}(\cdot)$ is a function, which approximates $g(x(\cdot))$ on a time interval $[t(k), t(k+1)]$. When a transition time t_t is found, the actual iteration of numerical integration is repeated but with a stepsize $h(k) = t_t - t(k)$ instead of $h(k) = t(k+1) - t(k)$. We denote by k_t the discrete time at which the transition takes place.

If we define adjoint variables as a solution of *adjoint equations* (for the simplicity of presentation we denote, e.g., $\varphi_X^2(X(k), u(k), h(k))$ by $\varphi_X^2(k)$, here $\hat{\phi}$ and $\hat{g}$ are functions which are based on ϕ and g but are built to take into account the extended state X)

$$\Lambda(N) = -\hat{\phi}_X(N)^T \tag{11a}$$

for $k = N-1, \ldots, k_t+1$

$$\Lambda(k) = \left(\varphi_X^2(k)\right)^T \Lambda(k+1) \tag{11b}$$

$$\Lambda^+(k_t) = \left(\varphi_X^2(k_t)\right)^T \Lambda(k_t+1) \tag{11c}$$

$$\Lambda^-(k_t) + \pi(k_t)\left(\hat{g}_X(k_t)\right)^T = \Lambda^+(k_i) \tag{11d}$$

$$\left(\varphi_h^1(k_t-1)\right)^T \Lambda^-(k_t) = \left(\varphi_h^2(k_t)\right)^T \Lambda(k_t+1) \tag{11e}$$

$$\Lambda(k_t - 1) = \left(\varphi_X^1(k_t - 1)\right)^T \Lambda^-(k_t) \tag{11f}$$

for $k = k_t - 2, \ldots, 1$

$$\Lambda(k) = \left(\varphi_X^1(k)\right)^T \Lambda(k+1) \tag{11g}$$

then the derivative of $\hat{\phi}(u) = \tilde{\phi}(X^u(N))$ with respect to a control $u(k)$ is equal to

$$\phi_{u(k)}(u) = -\Lambda(k+1)^T \varphi_u(k). \tag{12}$$

2 Numerical Procedure

The numerical procedure described in this section was introduced in [5] and later extended in [6] by adding to it the procedure for consistent initialization of higher index DAEs.

We start our considerations from a control system described by ODEs

$$x'(t) = f(x(t), u(t)),\ x(t_0) = x_0. \tag{13}$$

The system equations (13) are numerically integrated over the time interval $[t_0, t_f]$ using the Runge–Kutta scheme [2]

$$x_i(k+1) = x(k) + h(k) \sum_{j=1}^{s} a_{ij} f\left(x_j(k+1),\ u(t(k) + c_i h(k))\right) \tag{14}$$

for $i = 1, \ldots, s$ and

$$x(k+1) = x(k) + h(k) \sum_{i=1}^{s} b_i f\left(x_i(k+1),\ u(t(k) + c_i h(k))\right). \tag{15}$$

$x(k) \simeq x(t(k))$ is the numerical approximation of the state at $t(k)$. The constant coefficients a_{ij}, b_i, c_i for $i, j = 1, \ldots, s$ define the Runge–Kutta scheme. At each step of the Runge–Kutta scheme the nonlinear system (14) is first solved for variables $x_i(k+1)$, $i = 1, \ldots, s$ and then (15) is used to calculate $x(k+1)$. The scheme is repeated for steps $k = 0, \ldots, K-1$, where $x(0) = x_0$, $t(0) = t_0$ and $t(K) = t_f$. In our code we assume the control variable is constant along the single integration step, so we have

$$u(t(k) + c_i h(k)) = u(k). \tag{16}$$

Let us rewrite the numerical integration scheme (14)–(15) in a vector form

$$\begin{pmatrix} x_1 - x - h\sum_{j=1}^{s} a_{1j} f(x_j, u) \\ \vdots \\ x_s - x - h\sum_{j=1}^{s} a_{sj} f(x_j, u) \\ x^+ - x - h\sum_{i=1}^{s} b_i f(x_i, u) \end{pmatrix} = \begin{pmatrix} 0 \\ \vdots \\ 0 \\ 0 \end{pmatrix}. \tag{17}$$

For the sake of a more compact notation we omitted the discrete step argument and introduced the symbols $x_i = x_i(k+1)$, $x = x(k)$, $x^+ = x(k+1)$, $h = h(k)$, $u = u(k)$. If we now define the augmented state vector $X(k)$ as

$$X(k) = \left(x_1(k)^T, \ldots, x_s(k)^T, x(k)^T\right)^T, \tag{18}$$

then (17) can be presented in a form of the implicit discrete time state equation

$$F\left(X(k+1), X(k), u(k)\right) = 0. \tag{19}$$

System (14)–(15) is fully implicit discrete time and, under some nonsingularity assumption, can be expressed as explicit. If the Jacobian of $\hat{F}$ with respect to $X(k+1)$, denoted by F_{X^+}, exists and is nonsingular for all $k = 0, \ldots, N-1$, then from the Implicit Function Theorem there exists an unique function φ such that

$$\begin{aligned} X(k+1) &= \varphi(X(k), u(k), h(k)),\ k = 0, \ldots, N-1, \\ F(\varphi(X(k), u(k), k), X(k), u(k), h(k)) &= 0,\ k = 0, \ldots, N-1. \end{aligned} \tag{20}$$

The adjoint equations (11b) for the discrete state equation (19) become

$$\Lambda(k) = -F_X^T(k)\left(F_{X^+}^T(k)\right)^{-1} \Lambda(k+1), \tag{21}$$

where $\Lambda(k)$ is the discrete adjoint variable at a discrete step k,

$$F_{X^+}(k) = \frac{\partial F(X(k+1), X(k), u(k))}{\partial X(k+1)},\quad F_X(k) = \frac{\partial F(X(k+1), X(k), u(k))}{\partial X(k)}. \tag{22}$$

$$\Lambda(k) = \left(l_1(k)^T, \ldots, l_s(k)^T, \lambda(k)^T\right)^T,$$

It can be shown that $l_1(k) = 0, \ldots, l_s(k) = 0$ for steps $k = 0, \ldots, K-1$.

The Eqs. (11d)–(11e), can be transformed into equations

$$\Lambda^-(k_t) + \pi(k_t)\hat{g}_X(k_t)^T = -F_X^2(k_t)^T\left[F_{X^+}^2(k_t)\right]^{-T} \Lambda(k_t+1)$$

$$\Lambda^-(k_t)^T\left[F^1_{X^+}(k_t-1)\right]^{-1}F^1_h(k_t-1)=\Lambda(k_t+1)^T\left[F^2_{X^+}(k_t)\right]^{-1}F^2_h(k_t).$$

and, by taking into account the partial derivatives matrices (we apply the notation $f_{xi}=f_x(x_i,u)$)

$$F_{X^+}(k)=\begin{pmatrix} I-ha_{11}f_{x1} & \dots & -ha_{1s}f_{xs} & 0\\ \vdots & & \vdots & \vdots\\ -ha_{s1}f_{x1} & \dots & I-ha_{ss}f_{xs} & 0\\ -hb_1f_{x1} & \dots & -hb_sf_{xs} & I\end{pmatrix},\quad F_X(k)=\begin{pmatrix}0 & \dots & 0 & -I\\ \vdots & & \vdots & \vdots\\ 0 & \dots & 0 & -I\\ 0 & \dots & 0 & -I\end{pmatrix},$$

one can solve these equations to obtain

$$\pi(k_t)=\left[\lambda(k_t+1)\left(-h(k_t)\sum_{i=1}^s b_if^2_{xi}(k_t)z_i(k_t)+\sum_{i=1}^s b_if^2(x_i(k_t+1),u(k_t))\right)-\right.$$

$$\left.\lambda^+(k_t)^T\left(-h(k_t-1)\sum_{i=1}^s b_if^1_{xi}(k_t-1)z_i(k_t-1)+\sum_{i=1}^s b_if^1(x_i(k_t),u(k_t-1))\right)\right]$$

$$\Big/\, g_x(k_t)\left(h(k_t-1)\sum_{i=1}^s b_if^1_{xi}(k_t-1)z_i(k_t-1)-\sum_{i=1}^s b_if^1(x_i(k_t),u(k_t-1))\right),$$

where z is the solution to the equations

$$\begin{pmatrix} I-ha_{11}f^2_{x1} & \dots & -ha_{1s}f^2_{xs} & 0\\ \vdots & & \vdots & \vdots\\ -ha_{s1}f^2_{x1} & \dots & I-ha^2_{ss}f_{xs} & 0\\ -hb_1f^2_{x1} & \dots & -hb_sf^2_{xs} & I\end{pmatrix}\begin{pmatrix}z_1\\ \vdots\\ z_s\\ z^+\end{pmatrix}=\begin{pmatrix}-\sum_{j=1}^s a_{1j}f^2(x_j,u)\\ \vdots\\ -\sum_{j=1}^s a_{sj}f^2(x_s,u)\\ -\sum_{i=1}^s b_if^2(x_i,u)\end{pmatrix}$$

We can show that z is bounded thus we will also have

$$\pi(t_k)\to\pi_t \tag{23}$$

provided that $h(k)\to 0$ and the applied Runge–Kutta method has the property that $\sum_{i=1}^s b_i=1$.

The relation (23) is not the only one we have to show in order to establish the convergence of the numerical integration of adjoint equations of hybrid systems to their continuous time counterparts. The second issue which has to be to investigated is the influence of event time perturbation, due to the perturbation of the solution x, on the order of convergence of the numerical procedure (for more details see [3, 9]).

Suppose that on an open interval (a,b), such that $t_t\in(a,b)$, the state trajectory $x(t)$ is pertubed by $\delta x(t)$. To that state perturbation corresponds the perturbation of the switching time, $t_t + \delta t_t$. We will have

$$g(x(t_t + \delta t_t) + \delta x(t_t + \delta t_t)) = 0. \tag{24}$$

Evaluating g around $x(t_t + \delta t_t) + \delta x(t_t + \delta t_t)$ will result in

$$\begin{aligned} 0 = g(x(t_t + \delta t_t) + \delta x(t_t + \delta t_t)) = g_x(x(t_t))\delta x(t_t) + \\ g_x(x(t_t)) f^1(x(t_t), u(t_t))\delta t_t + o(\delta t_t, \delta x). \end{aligned} \tag{25}$$

Eventually, we have

$$\delta t_t = -\frac{1}{g_x(x(t_t)) f^1(x(t_t), u(t_t))} g_x(x(t_t))\delta x(t_t) + o(\delta t_t, \delta x). \tag{26}$$

This implies that as $\delta x(t_t) \to 0$ (what is ensured by the numerical integration scheme) also $\delta t_t \to 0$ and the numerically determined transition time converges to the actual transition time. The convergence of the discrete adjoint equations can be justified by showing the resemblance of continuous and discrete adjoint schemes, as it was done in [1, 8].

3 Implementation Details

Building the numerical procedure for sensitivity analysis of hybrid systems requires solving several nontrivial issues:

(1) the choice of numerical procedure for integrating system equations
(2) creating efficient numerical procedure for the switching points location
(3) the choice of a numerical scheme for evaluating sensitivity information.

As far as the first issue is concerned we have to take into account kind of parameters with respect to which sensitivity analysis is going to be performed. If parameters in question are coefficients of system equations one could consider multistep methods such as Adams method or BDF—in that case functions f^1 and f^2 will be smooth enough over the whole horizon so building polynomials of appropriate orders for approximating solutions would be effective. On the other hand, if parameters in question are coefficients of piecewise constant (linear) approximations of control functions influencing solutions of system equations, then one–step methods such as Runge–Kutta methods would be more appropriate. In that case at each point of time where controls exhibit discontinuity an integration procedure must be restarted. Since one–step methods are restarted at every integration step they are more suitable for systems with discontinuous (or not differentiable) control functions. Furthermore, if an integration method is going to be applied to systems described by stiff differential equations then we are limited to implicit methods.

Our aim is to develop numerical software for sensitivity analysis of hybrid systems which we would like to control. Therefore, for the reasons outlined above, the sensible choice for the numerical integration method is that based on an implicit Runge–Kutta method. We decided to use the RADAU5 procedure in its C++ implementation. At the current state of the development of our software we rely on linear algebra numerical procedures for factorizing matrices and for solving linear equations present in the original RADAU5 procedure (Fortran, or C++ implementations), however we plan to replace these procedures by the BLAS linear algebra subroutines and Harwell software for solving sparse sets of linear equations.

The evaluation of switching points could be time consuming if it followed the path of calculating a step in our integration procedure. Suppose that we have

$$g(x(k)) < 0, \;\; g(x(k+1)) > 0, \tag{27}$$

where $x(k+1)$ is determined according to (15). This implies that the switching point t_t lies in the subinterval $(t(k), t(k+1))$, and a kind of the secant procedure could be used to approximate t_t. The procedure will need several evaluations of x at intermediate points before reaching the approximation of $x(t_t)$ at which $g(x(t_t)) = 0$. But that would require performing the same linear algebra calculations as we have in the step size setting, i.e., solving linear equations (and possibly costly matrices factorizations). Since we need to determine t_t with high accuracy the outlined secant procedure would significantly deteriorate the performance of the integration procedure for hybrid systems.

Therefore, in order to determine t_t, we use a polynomial approximation to x. According to the applied Runge–Kutta method we have state values at the intermediate points $t(k)$, $t(k) + c_i h(k)$, $i = 1, \ldots, s$, which according to our nomenclature, are $x_i(k)$, $i = 1, \ldots, s$. Having c_i and $x_i(k)$ we build the polynomial approximation $\tilde{x}$ as follows

$$\tilde{x}(t) = x_1(k) + \tau \left(x_2(k) + (\tau - c_1)\left(x_2(k) + \ldots + (\tau - c_{s-1})x_s(k)\right)\cdots\right), \tag{28}$$

where $\tau = (t - t(k))/h(k)$. It can be shown that this approximation gives at points $t \in [t(k), t(k) + h(k)]$, $k = 1, \ldots, K$ the same accuracy of x approximation as in times $t(k)$, $k = 1, \ldots, K$ the Runge–Kutta scheme.

As far as gathering the sensitivity information is concerned we have two approaches. We can either use sensitivity equations, or we can solve adjoint equations and then calculate gradients on the basis of adjoint variables. The first approach is effective when we have few parameters with respect to which we perform sensitivity analysis. Since our software is aimed at optimal control problems in which control functions are approximated by piecewise constant functions (with many parameters) that approach is ruled out.

When we have to solve adjoint equations we can either apply continuous model of these equations, or we can refer to discrete time adjoint equations derived from a discrete time system equations being the result of the numerical integration of system equations (how it can be achieved is sketched in the previous section).

Suppose that the sensitivity analysis of hybrid equations is based on continuous time adjoint equations. We need to integrate them as accurately as system equations and, if system equations are integrated by an implicit method, an implicit method will have to be applied to adjoint equations. This means that step sizes will vary and jacobians will have to be evaluated at points of time at which forward integration procedure has not called for jacobians. Furthermore, these jacobians will have to be factorized in order to solve corresponding linear equations. Eventually, the integration of adjoint equations can be time consuming, and we have to bear in mind that for each functional we have to solve adjoint equations with different boundary conditions. We avoid these efficiency problems when we base our sensitivity analysis procedure on discrete time adjoint equations.

4 Numerical Results

The primary goal of our work on sensitivity analysis of hybrid system is the potential application of the developed numerical tools to solving optimal control problems with hybrid systems.

In [7] we presented an example of optimal control problems solved with the help of our software. Here, we discuss the application of our approach to the problem of planning a haemodialysis process. The problem is fully described in [10], in this paper we present some results obtained for a variant of the problem. We do that in order to show that our application for sensitivit analysis can contribute to high accuracy of obtained solutions of optimal control problems with hybrid systems.

The system equations that determine the concentrations of urea and phosphorus concentrations in intracellural fluid— C_{IC}^{urea}, $C_{IC}^{PO_4}$ C_{EC}^{urea}, urea and phosphorus concentations in extracellural fluid—C_{EC}^{urea}, $C_{EC}^{PO_4}$ and ultrafiltration volume—UFR are as follows

$$\frac{dC_{EC}^{urea}}{dt} = \frac{K_{IE}^{urea} \cdot (C_{IC}^{urea} - C_{EC}^{urea}) - C_{EC}^{urea} \cdot (K_D^{urea} + K_r^{urea} + K^{ufr})}{0.34 \cdot V(0) - UFR} \tag{29}$$

$$\frac{dC_{IC}^{urea}}{dt} = \frac{K_{IE}^{urea} \cdot (C_{EC}^{urea} - C_{IC}^{urea}) + G^{urea}}{0.66 \cdot V(0)} \tag{30}$$

$$\frac{dC_{EC}^{PO_4}}{dt} = \frac{K_{IE}^{PO_4} \cdot (C_{IC}^{PO_4} - C_{EC}^{PO_4}) - K_D^{PO_4} \cdot C_{EC}^{PO_4}}{0.34 \cdot V(0) - UFR} + K_3^{PO_4} + K_4^{PO_4} \tag{31}$$

$$\frac{dC_{IC}^{PO_4}}{dt} = \frac{K_{IE}^{PO_4} \cdot (C_{EC}^{PO_4} - C_{IC}^{PO_4})}{0.66 \cdot V(0)} \tag{32}$$

$$\frac{dUFR}{dt} = U_{ufr} \tag{33}$$

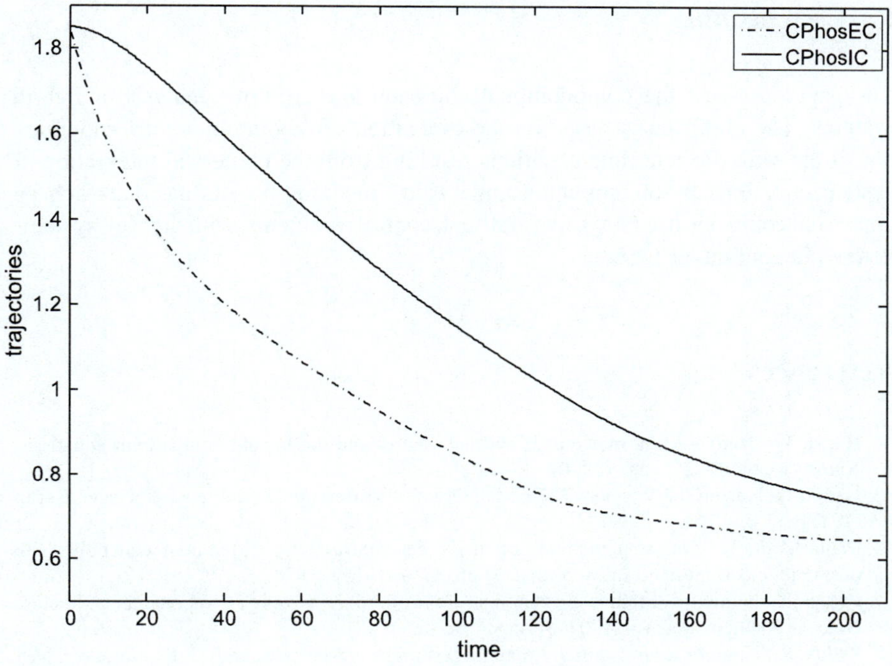

Fig. 1 State trajectories for the phosphorus $C_{IC}^{PO_4}$ and $C_{EC}^{PO_4}$ (rebound of $C_{EC}^{PO_4}$ can be seen in 190 min of haemodialysis)

$$K_4^{PO_4} = \alpha \cdot \max\left(C_{min}^{PO_4} - C_{IC}^{PO_4}, 0\right) \tag{34}$$

$$K_3^{PO_4} = \beta \cdot \max\left(C_{max}^{PO_4} - C_{IC}^{PO_4}, 0\right) \tag{35}$$

The model coeficients are explained in [10]. The algebraic equations are responsible for hybrid behaviour of the system equations.

Having combined kinetic models of urea and phosphorus we look for proper concentrations of urea and phosphorus at the end of the haemodialysis process by controlling the parameters Q_B, Q_D and U_{ufr} (see [10] for details). In other words, by solving the optimal control problem we want to choose a proper dialysis membrane in order to achieve final parameters of haemodialysis.

Some of the optimal trajectories are shown in Fig. 1. One can observe that from the time $t \approx 190$ the system is in the sliding mode. The optimization procedure (essentially the procedure presented in [4]) equipped with our sensiivity evaluation module needed 31 iteriations to find the solution with the accuracy 10^{-8}. The system equations have been integrated with absolute and relative accuracies equal to 10^{-9}.

5 Conclusions

The paper presents the computational approach to sensitivity analysis of hybrid systems. The approach is based on the evaluation of adjoint equations which are consistent with discrete time equations resulting from the numerical integration of system equations by an implicit Runge–Kutta method. We outline that such an approach could be used to solve optimal control problems with hybrid systems defined in continuous time.

References

1. Hager, W.: Runge–Kutta methods in optimal control and the transfmored adjoint equations. Numer. Math. **87**, 247–282 (2000)
2. Hairer, E., Norsett, S., Wanner, G.: Solving Ordinary Differential Equations I. Springer, Berlin (1993)
3. Mannshardt, R.: One-step methods of any order for ordinary differential equations with discontinuous right-hand sides. Numer. Math. **31**, 131–152 (1978)
4. Pytlak, R.: Numerical Methods for Optimal Control Problems with State Constraints. Lecture Notes in Mathematics, vol. 1707. Springer, Berlin (1999)
5. Pytlak, R.: Numerical procedure for optimal control of higher index daes. J. Discret. Dyn. Nat. Soc. **29**, 1–24 (2011)
6. Pytlak, R., Zawadzki, T.: On solving optimal control problems with higher index daes. Optim. Methods Softw. **32**, 1139–1162 (2014)
7. Pytlak, R., Suski, D., Tarnawski, T.: Optimal Control of Hybrid Systems with Sliding Modes. In: Springer Proceed in Mathematics & Statistics, vol. 248, pp. 283–293. Springer, Berlin (2018)
8. Schwartz, A., Polak, E.: Consistent approximations for optimal control problems based on Runge–Kutta integration. SIAM J. Control Optim. **34**, 1235–1269 (1996)
9. Shampine, L.F., Thompson, S.: Event location for ordinary differential equations. Comput. Math. Appl. **39**, 43–54 (2000)
10. Stecz, W., Pytlak, R., Rymarz, A., Niemczyk, S.: Application of dynamic optimisation for planning a haemodialysis process. BMC Nephrol. **20**, 236 (2019)
11. van der Schaft, A., Schumacher, H.: An Introduction to Hybrid Dynamical Systems. Springer, London (2000)

Asymptotic Stability of Fractional Variable-Order Discrete-Time Equations with Terms of Convolution Operators

Dorota Mozyrska, Małgorzata Wyrwas, and Piotr Oziablo

Abstract The stability of linear systems with the Caputo fractional-, variable-order difference operators of convolution type is investigated. We present the recurrence formula for the solution to linear initial value problems with the operator that is defined as the convex combination of two fractional-, variable-order difference operators. The conditions for asymptotic stability of considered equations are formulated and proven. Finally, some examples that illustrate our results are presented.

Keywords Fractional order difference system · Fractional variable-order operator · Stability

1 Introduction

Fractional calculus is a mathematical tool which has been recently employed to model some real life processes and to achieve (in some of the cases) more precise control procedures, see for instance [4, 10, 13]. It provides new possibilities in many fields of science such as thermodynamics, chaotic systems or biophysics, see for instance [8, 9]. Finding answers to some simple questions related to the fractional calculus as for instance a half order derivative concept, a new window of opportunity to mathematical and real world was opened. Additionally, many new questions and (sometimes intriguing) results arisen. It is known that for instance regarding the problems from biology and ecology that arise in population dynamical studies some species should be modeled by use of difference equations, see for instance [5, 6]. Many interesting results on discrete fractional calculus have been already published, see for instance [1–3, 12, 14].

D. Mozyrska (✉) · M. Wyrwas · P. Oziablo
Bialystok University of Technology, Faculty of Computer Science, Białystok, Poland
e-mail: d.mozyrska@pb.edu.pl; m.wyrwas@pb.edu.pl; p.oziablo@student.pb.edu.pl

J. Awrejcewicz (ed.), *Perspectives in Dynamical Systems II: Mathematical and Numerical Approaches*, Springer Proceedings in Mathematics & Statistics 363,
https://doi.org/10.1007/978-3-030-77310-6_18

Since the behaviour of some models provides to nonzero equilibrium points and systems with the Grünwald–Letnikov operator have only zero as stationary point, in the paper the linear discrete-time systems with the Caputo operators are considered in order to permit the existence of nonzero stationary points. The convex combination of the Caputo-type fractional difference operators with two variable-order functions is taken into account. If the variable-order functions coincide, then one gets the Caputo-type difference system with one variable-order operator of convolution type that was defined originally in [11]. It is worth to stress that in fact instead of convex combination one can take the arbitrary linear combination of fractional variable-order operators and the results will be similar, but we decided to take into account the convex one in order to have the system with one fractional variable-order operator in some cases.

The aim of this paper is to perform the stability analysis of equations and systems with the described combination of the Caputo-type fractional variable-order operators. In the sequel the descriptions of the considered linear Caputo-type difference fractional variable-order equations and their stability analysis are presented. Since we consider the convolution type operators, there is possible to solve an initial value problems for linear equations using $\mathcal{Z}$-transform method. We analyse linear system with constant coefficients and fractional variable-order differences with two variable order functions ν and μ such that $\nu(k), \mu(k) \in (0, 1]$. The sufficient and necessary condition for asymptotic stability of the considered linear equation is presented and an example that illustrates the result is given. In the case of the linear system with the considered convex combination of the Caputo-type fractional variable-order operators we formulate and prove the condition that implies the instability of the system. Finally, the regions of locations of eigenvalues of matrices associated to the systems in order to guarantee the asymptotic stability are presented in examples.

2 Preliminaries

Definition 1 For $k, l \in \mathbb{Z}$ and a given order function $\nu(\cdot)$ we define the oblivion function, as a discrete function of two variables, by its values $a^{[\nu(l)]}(k)$ given as

$$a^{[\nu(l)]}(k) = \begin{cases} 0 & \text{for } k < 0 \\ 1 & \text{for } k = 0 \\ (-1)^k \frac{\nu(l)[\nu(l)-1]\cdots[\nu(l)-k+1]}{k!} & \text{for } k > 0 \end{cases} . \tag{1}$$

Formula (1) in Definition 1 is equivalent to the following recurrence with respect to $k \in \mathbb{N}$:

$$\begin{aligned} a^{[\nu(l)]}(0) &= 1\,, \\ a^{[\nu(l)]}(k) &= a^{[\nu(l)]}(k-1)\left[1 - \frac{\nu(l)+1}{k}\right] \text{ for } k \geq 1\,. \end{aligned} \tag{2}$$

Therein the paper we use in systems those order functions $\nu(\cdot)$ with values only in $[0, 1]$. But definitions of fractional-, variable-order summations and differences we can state for any nonnegative functions $\nu(\cdot)$ well-defined on $\mathbb{Z}$.

For a function $y : \mathbb{Z} \to \mathbb{R}$ the forward difference operator is defined as $(\Delta y)(k) = y(k+1) - y(k)$, see [7]. Let $q \in \mathbb{N}_0 := \{0, 1, 2, \ldots\}$ and $\Delta^q := \Delta \circ \cdots \circ \Delta$ is q-fold application of operator Δ. Then, $(\Delta^q y)(k) = \sum_{i=0}^{q} (-1)^{q-i} \binom{q}{i} y(k+i)$.

Definition 2 Let $\nu : \mathbb{Z} \to \mathbb{R}_+ \cup \{0\}$. For a function $y : \mathbb{Z} \to \mathbb{R}$ the *fractional-, variable-order sum of convolution type* is given by

$$\left(\Delta^{-\nu(\cdot)} y\right)(k) := \left(a^{[-\nu(\cdot)]} * y\right)(k) = \sum_{i=0}^{k} a^{[-\nu(i)]}(i) y(k-i),$$

where $k \in \mathbb{N}_0$ and "$*$" denotes the convolution operator. Additionally, we define $\left(\Delta^0 y\right)(k) := y(k)$.

Observe that having convolution operator we can write that

$$\mathcal{Z}\left[\Delta^{-\nu(\cdot)} y\right](z) = Y(z) \mathcal{Z}\left[a^{[-\nu(\cdot)]}\right](z), \tag{3}$$

where $Y(z) := \mathcal{Z}[y](z)$ and $\mathcal{Z}\left[a^{[-\nu(\cdot)]}\right](z) = \sum_{i=0}^{\infty} (-1)^i \binom{-\nu(i)}{i} z^{-i}$.

(a) For $\nu(k) \equiv \alpha$, (3) can be shortly written as $\mathcal{Z}\left[\Delta^{-\nu(\cdot)} y\right](z) = \left(\frac{z}{z-1}\right)^{\alpha} Y(z)$.
(b) Let $\nu_1, \nu_2 : \mathbb{Z} \to \mathbb{R}_+ \cup \{0\}$. Then, the following equality holds
(c) $\left(\Delta^{-\nu_1(\cdot)} \Delta^{-\nu_2(\cdot)} y\right)(k) = \left(\Delta^{-\nu_2(\cdot)} \Delta^{-\nu_2(\cdot)} y\right)(k)$.

Here we define the Caputo fractional-, variable-order difference operator of convolution type.

Definition 3 Let $\nu : \mathbb{Z} \to (q-1, q]$, $q \in \mathbb{N}_1$. Then, *the Caputo fractional-, variable-order difference operator of convolution type* $\Delta^{\nu(\cdot)}$ with order function $\nu(\cdot)$ for a function $y : \mathbb{Z} \to \mathbb{R}$ is defined by

$$\left(\Delta^{\nu(\cdot)} y\right)(k) = \left(\Delta^{-(q-\nu(\cdot))} \Delta^q y\right)(k). \tag{4}$$

It is easy to see that

(a) For $q = 1$: $\left(\Delta^{\nu(\cdot)} y\right)(k) = \left(\Delta^{-(1-\nu(\cdot))} \Delta y\right)(k)$.
(b) For $\nu(k) \equiv q \in \mathbb{N}_1$, we have $\left(\Delta^{\nu(\cdot)} y\right)(k) = \left(\Delta^q y\right)(k)$.
(c) For $q = 1$: $\mathcal{Z}\left[\Delta^{\nu(\cdot)} y\right](z) = \left((z-1)Y(z) - zy(0)\right) \mathcal{A}_\nu(z)$, where $Y(z) = \mathcal{Z}[y](z)$ and a complex function $\mathcal{A}_\nu$ is defined by

$$\mathcal{A}_\nu(z) := \mathcal{Z}\left[a^{[\nu(\cdot)-1]}\right](z) = \sum_{i=0}^{\infty} (-1)^i \binom{\nu(i)-1}{i} z^{-i}. \tag{5}$$

Observe that the Caputo fractional-, variable-order difference operator of convolution type can be extended to vector functions in the componentwise manner, then one can study the difference systems with fractional variable order equations and Caputo operator.

3 Equations and Stability Issue

Let us consider two order functions $\nu : \mathbb{Z} \to (0, 1]$ and $\mu : \mathbb{Z} \to (0, 1]$. In the considered type of equations we use a convex combination of two operators to receive new one in te following form: $p\Delta^{\nu(\cdot)} + (1-p)\Delta^{\mu(\cdot)}$, where $p \in [0, 1]$. The main goal of this paper is to investigate behaviour of solutions and state conditions on their stability, of the following type of linear equations with constant coefficients

$$p\left(\Delta^{\nu(\cdot)}x\right)(k) + (1-p)\left(\Delta^{\mu(\cdot)}x\right)(k) + \omega x(k) = u(k)\,,\ k \geq 0, \tag{6}$$

with initial value $x(0) \in \mathbb{R}$, where $\nu, \mu : \mathbb{Z} \to (0, 1]$ are variable-order functions, $p \in [0, 1]$, $\omega \in \mathbb{R}$, function $u : \mathbb{N}_0 \to \mathbb{R}$ is an input function, $y : \mathbb{N}_0 \to \mathbb{R}$ is a response and $\omega > 0$ is given real constant.

For constant order function it could be also considered, namely for $\nu \equiv \alpha$ and $\mu \equiv \beta$ we have: $p\left(\Delta^{\alpha}x\right)(k) + (1-p)\left(\Delta^{\beta}x\right)(k) + \omega y(k) = u(k)$. For $\alpha = \beta$, we get the equation with only one Caputo operator, i.e. $\left(\Delta^{\alpha}x\right)(k) = -\omega x(k) + u(k)$.

Let $\mathfrak{a}(j, p, \nu, \mu) := p a^{[\nu(j)-1]}(j) + (1-p)a^{[\mu(j)-1]}(j)$, $j > 0$. Equation (6) can be rewritten in the following recurrence way:

$$\begin{aligned}
x(1) &= (1-\omega)\,x(0) + u(0)\,,\\
x(k) &= (1-\omega)\,x(k-1) - \sum_{i=1}^{k-1} \mathfrak{a}(k-i, p, \nu, \mu)\,(x(i) - x(i-1)) + u(k-1)\,,\ k \geq 2
\end{aligned} \tag{7}$$

and $x(0) = x_0 \in \mathbb{R}$ is given.

Then,

$$\mathcal{Z}\left[\left(p\Delta^{\nu(\cdot)} + (1-p)\Delta^{\mu(\cdot)}\right)x\right](z) = ((z-1)X(z) - zx(0))\left(p\mathcal{A}_{\nu}(z) + (1-p)\mathcal{A}_{\mu}(z)\right),$$

where $X(z) = \mathcal{Z}[x](z)$ and $\mathcal{A}_{\nu}$ is given by (5).

Taking $\mathcal{Z}$-transform of Eq. (6) we get

$$X(z) = ((z-1)\mathcal{A}_2(z) + \omega)^{-1}[z\mathcal{A}_2(z)x(0) + U(z)]\,, \tag{8}$$

where $U(z) = \mathcal{Z}[u](z)$, $\mathcal{A}_2(z) := p\mathcal{A}_{\nu}(z) + (1-p)\mathcal{A}_{\mu}(z)$. In order to get the solution of (6) one needs to take inverse $\mathcal{Z}$-transform of (8) Now, we are ready to state conditions for asymptotical stability and instability of system (6).

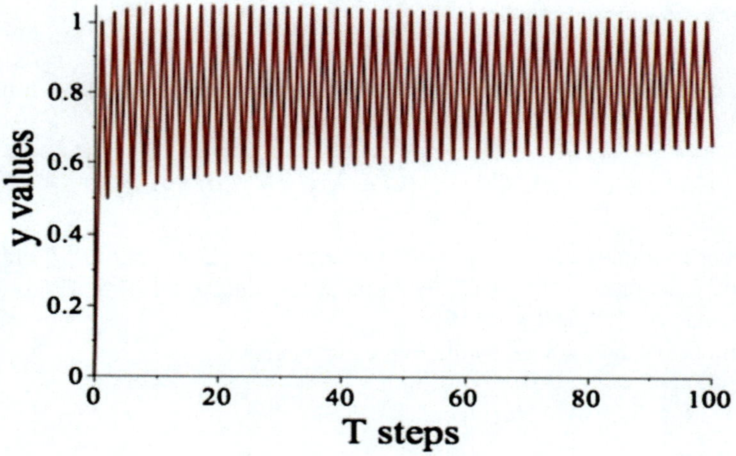

Fig. 1 The response (stable) to unit input with $\omega = 1.465$, $p = 0.5$, $\alpha = 0.1$, $\beta = 0.9$

We can formulate condition on the asymptotic stability of Eq. (6).

Proposition 1 *Equation (6) is asymptotically stable if and only if*

$$0 < \omega < 2\sum_{i=0}^{\infty}\left(p\binom{\nu(i)-1}{i} + (1-p)\binom{\mu(i)-1}{i}\right). \tag{9}$$

Proof Observe that the set of roots of the equation $(z-1)\mathcal{A}_2(z)+\omega = 0$ is related with the asymptotic stability of (6). Hence, we get $\omega = (1-z)\mathcal{A}_2(z)$. For the scalar equation (6) the points $z = -1$ and $z = 1$ correspond to the border of its asymptotic stability. Then for $z = 1$ one gets $\omega = 0$ and consequently, $\omega > 0$ if $z < 1$. Moreover, for $z > -1$ one gets $\omega < 2\sum_{i=0}^{\infty}\left(p\binom{\nu(i)-1}{i} + (1-p)\binom{\mu(i)-1}{i}\right)$. Therefore, the thesis holds.

For constant orders we have the following result.

Corollary 1 *Let $\alpha > \beta$ and $p \in [0, 1]$. Equation (6) for $\nu(k) \equiv \alpha$ and $\mu(k) \equiv \beta$ is asymptotically stable if and only if*

$$0 < \omega < p2^{\alpha} + (1-p)2^{\beta}. \tag{10}$$

Example 1 Let us take $\nu(k) = \alpha = 0.1$, and $\mu(k) = \beta = 0.9$ with different coefficients $p = 0.5$. Then the limit for $p2^{\alpha} + (1-p)2^{\beta} = 1.4691$. We take $\omega = 1.465$. In Fig. 1 we illustrate the response for unit input through equation $0.5\left(\Delta^{0.1}x\right)(k) + 0.5\left(\Delta^{0.9}x\right)(k) + \omega x(k) = u(k)$.

4 Systems

We now consider control systems with compound variable-orders of the following form:

$$\left(p\Delta^{\nu(\cdot)} + (1-p)\Delta^{\mu(\cdot)}\right)(x)(k) = Ax(k) + Bu(k)\,,\ k \geq 1, \tag{11}$$

with initial condition $x(0) = x_0 \in \mathbb{R}^n$, where $\nu, \mu : \mathbb{Z} \to \mathbb{R}_+ \cup \{0\}$ are order functions, function $u : \mathbb{N}_0 \to \mathbb{R}^m$ is an input function, $x : \mathbb{N}_0 \to \mathbb{R}^m$ is a state function and $A \in \mathbb{R}^{n\times n}$, $B \in \mathbb{R}^{n\times m}$.

System (11) can be solved by the following recursive

$$\begin{aligned} x(1) &= (I + A)\,x(0) + Bu(0)\,, \\ x(k) &= (I + A)\,x(k-1) \\ &\quad - \sum_{i=1}^{k-1} \mathfrak{a}(k-i, p, \nu, \mu)\,(x(i) - x(i-1)) + Bu(k-1)\,,\ k \geq 2 \end{aligned} \tag{12}$$

and $x(0) = x_0 \in \mathbb{R}^n$ is given and I denotes identity matrix.

Proposition 2 *Let* $\operatorname{spec}(A) = \{\lambda_i : i = 1, \dots, k\}$, $k \leq n$ *and*

$$\begin{aligned} w_i :=& 2\left|\sin\frac{\varphi_i}{2}\right| \left(\left(\sum_{k=0}^{\infty} \mathfrak{a}(k, p, \nu, \mu)\cos(k\varphi_i)\right)^2 \right. \\ &\left. + \left(\sum_{k=0}^{\infty} \mathfrak{a}(k, p, \nu, \mu)\sin(k\varphi_i)\right)^2\right)^{0.5}, \end{aligned} \tag{13}$$

where $\varphi_i = \arg(\lambda_i)$. *If there is* $\lambda_i \in \operatorname{spec}(A)$ *such that*

$$|\lambda_i| > w_i\,, \tag{14}$$

then system (11) *is unstable.*

Proof Observe that $w_i = |(e^{j\varphi} - 1)\mathcal{A}_2(e^{j\varphi})|$, where $\varphi \in \mathbb{R}$. Then by (14) we get $\lambda_i \notin \{(z-1)\mathcal{A}_2(z) : |z| \leq 1\}$. Hence the eigenvalue λ_i of the matrix A lies outside of the image of unit circle in the mapping $\mathcal{A}_2 : \mathbb{C} \to \mathbb{C}$ given by $\mathcal{A}_2 := p\mathcal{A}_\nu + (1-p)\mathcal{A}_\mu$.

From Proposition 2, we get that if system (11) is stable then $|\lambda_i| \leq w_i$ for all $\lambda_i \in \operatorname{spec}(A)$.

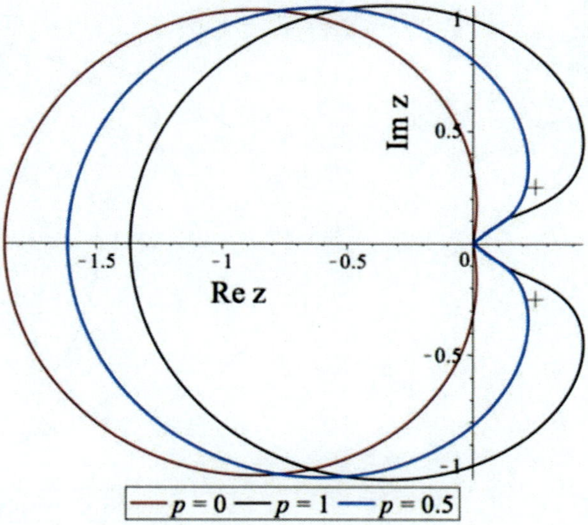

Fig. 2 Stability region for operators $p\Delta^{0.2} + (1-p)\Delta^{0.9}$, $p \in \{0, 0.5, 1\}$ and eigenvalues of A with $q = 0.25$-crosses

Example 2 Now let us study the system with a variable–order of the form

$$\left(p\Delta^{\nu(\cdot)} + (1-p)\Delta^{\mu(\cdot)}\right)(x)(k) = Ax(k-1)\,,\ k \geq 1,$$

with initial condition $x(0) = x_0 \in \mathbb{R}^2$ and matrix $A = q\begin{bmatrix} 1 & -1 \\ 1 & 1 \end{bmatrix}$. Because $\mathrm{spec}(A) = \{q - q \cdot j, q + q \cdot j\}$, so $|\lambda_i| = |q|\sqrt{2}$. Then

(a) for $q > 0$: $\varphi_i \in \{\frac{\pi}{4}, \frac{7\pi}{4}\}$;
(b) for $q < 0$: $\varphi_i \in \{\frac{3\pi}{4}, \frac{5\pi}{4}\}$.

– Let us take $\nu(k) = 0.2$, and $\mu(k) = 0.9$ with different coefficients $p \in \{0, 0.5, 1\}$. In Fig. 2 we illustrate by crosses eigenvalues of A for $q = 0.25$ (positive real parts of eigenvalues of A) and stability region

$$\mathrm{Int}\left\{(z-1)\mathcal{A}_2\left(e^{i\varphi}\right), \varphi \in [0, 2\pi]\right\}.$$

We see that the eigenvalues of matrix A are outside the region of stability, for $p = 0$ and $p = 0.5$, and inside region of the stability for $p = 1$ (for smaller value of order function).

– Let us take $\nu(k) = 1 - \frac{1}{k+1}$, and $\mu(k) = 1$ with different coefficients $p \in \{0, 0.5, 1\}$. In Fig. 3 we illustrate by crosses eigenvalues of A for $q = -0.25$ (negative real parts of eigenvalues of A) and stability region

$$\mathrm{Int}\left\{(z-1)\mathcal{A}_2\left(e^{i\varphi}\right), \varphi \in [0, 2\pi]\right\}.$$

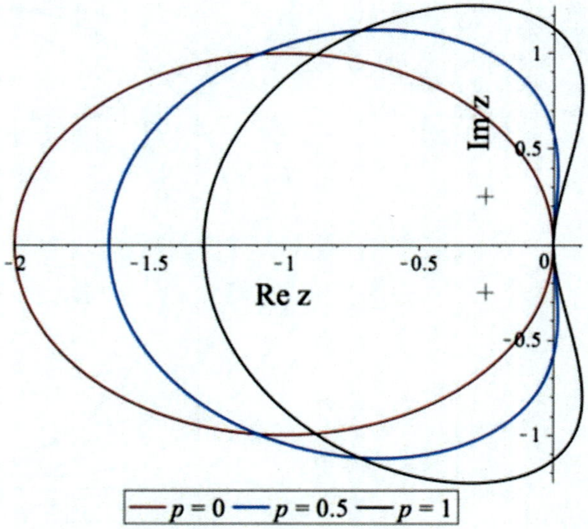

Fig. 3 Stability region for operators $p\Delta^{1-\frac{1}{k+1}} + (1-p)\Delta^1$, $p \in \{0, 0.5, 1\}$ and eigenvalues of A with $q = -0.25$-crosses

We see that the eigenvalues of matrix A are inside three regions of stability, for $p \in \{0, 0.5, 1\}$.

5 Conclusions

In the paper we studied the stability of the linear systems with the Caputo fractional-, variable-order difference operators of convolution type. Using the $\mathcal{Z}$-transform we formulated and proven the conditions for the asymptotic stability of the considered systems. In the example we determined the stability regions of location of eigenvalues of matrices that are associated to the systems as the sets of the images of some maps.

Acknowledgments The work was supported by Polish funds of National Science Center, granted on the basis of decision DEC-2016/23/B/ST7/03686.

References

1. Axtell, M., Bise, E.M.: Fractional calculus applications in control systems. In: Proceedings of the IEE 1990 International Aerospace and Electronics Conference, vol. 311, pp. 536–566. New York (1990). https://doi.org/10.1109/NAECON.1990.112826
2. Baranowski, J., Bauer, W., Zagórowska, M., Piątek, P.: On digital realizations of non-integer order filters. Circuits Syst. Signal Process. **35**(6), 2083–2107 (2016). https://doi.org/10.1007/s00034-016-0269-8

3. Bastos, N.R.O., Ferreira, R.A.C., Torres, D.F.M.: Discrete-time fractional variational problems. Signal Process. **91**(3), 513–524 (2011). https://doi.org/10.1016/j.sigpro.2010.05.001
4. Caponetto, R., Dongola, G., Fortuna, L., Petras, I.: Fractional Order Systems: Modeling and Control Applications. Series on Nonlinear Science, Series A, vol. 72. World Scientific, Singapure (2010)
5. Caswell, H.: Matrix Population Models: Construction, Analysis, and Interpretation. Sinauer Associates, Sunderland (2001)
6. Cushing, J.: An Introduction to Structured Population Dynamics. Society for Industrial and Applied Mathematics, Philadelphia (1998). https://doi.org/10.1137/1.9781611970005
7. Ferreira, R.A.C., Torres, D.F.M.: Fractional h-difference equations arising from the calculus of variations. Appl. Anal. Discrete Math. **5**(1), 110–121 (2011). https://doi.org/10.2298/AADM110131002F
8. Hilfer, R.: Applications of Fractional Calculus in Physics. World Scientific, Singapore (2000)
9. Machado, J.A.T., Silva, M.F., Barbosa, R.S., Jesus, I.S., Reis, C.M., Marcos, M.G., Galhano, A.F.: Some applications of fractional calculus in engineering. Math. Problems Eng. **2010**, 34pp. (2010). https://doi.org/10.1155/2010/639801. Article ID 639801
10. Matlob, M.A., Jamali, Y.: The concepts and applications of fractional order differential calculus in modeling of viscoelastic systems: a primer. Crit. Rev. Biomed. Eng. **47**(4), 249–276 (2019). https://doi.org/10.1615/CritRevBiomedEng.2018028368
11. Mozyrska, D., Wyrwas, M.: Stability of linear systems with Caputo fractional-, variable-order difference operator of convolution type. In: 2018 41st International Conference on Telecommunications and Signal Processing (TSP). IEEE, Athens (2018). https://doi.org/10.1109/TSP.2018.8441360
12. Sierociuk, D., Dzielinski, A.: Fractional Kalman filter algorithm for the states parameters and order of fractional system estimation. Int. J. Appl. Math. Comput. Sci. **16**(1), 129–140 (2006)
13. Sun, H., Zhang, Y., Baleanu, D., Chen, W., Chen, Y.: A new collection of real world applications of fractional calculus in science and engineering. Commun. Nonlinear Sci. Numer. Simul. **64**, 213–231 (2018). https://doi.org/10.1016/j.cnsns.2018.04.019
14. Vinagre, B.M., Monje, C.A., Caldero, A.J.: Fractional order systems and fractional order actions. In: 41st IEEE CDC (ed.) Tutorial Workshop#2: Fractional Calculus Applications in Automatic Control and Robotics. Las Vegas (2002)

Dynamics of Circular Plates Under Temperature and Mechanical Loadings

Simona Doneva, Jerzy Warminski, and Emil Manoach

Abstract The purpose of this paper is to study nonlinear oscillations of a heated plate subjected to dynamic loading. The response of moderately thick circular plate at elevated temperatures subjected to harmonic loading is analysed. A mathematical model of the plate is derived applying the geometrically nonlinear Reisner-Mindlin plate theory. A reduced model of the plate motion is created and the harmonic balance method is applied to the reduced model taking into account the first vibration mode. The numerical simulations are performed for the plate subjected to uniformly distributed harmonic loading and different temperatures. The model represented by a set of partial differential equations is reduced to one degree of freedom system by Galerkin orthogonalization method based on the first vibration mode. The obtained reduced nonlinear one degree of freedom model with cubic nonlinearity is studied by the harmonic balance method. The influence of the amplitude of the loading and the elevated temperature on the frequency response functions is studied.

Keywords Circular plate dynamics · Nonlinear oscillations · Heat loading

S. Doneva (✉)
Institute of Mechanics, Bulgarian Academy of Sciences, Sofia, Bulgaria

Department of Applied Mechanics, Lublin University of Technology, Lublin, Poland
e-mail: s.doneva@pollub.pl

J. Warminski
Department of Applied Mechanics, Lublin University of Technology, Lublin, Poland

E. Manoach
Institute of Mechanics, Bulgarian Academy of Sciences, Sofia, Bulgaria

J. Awrejcewicz (ed.), *Perspectives in Dynamical Systems II: Mathematical and Numerical Approaches*, Springer Proceedings in Mathematics & Statistics 363,
https://doi.org/10.1007/978-3-030-77310-6_19

1 Introduction

The circular plates are frequently used in mechanical and civil engineering structures. This specific shape of the plates and the corresponding type of support are imposed by the operation exploitation conditions. As structural components the plates are often subjected to mechanical and thermal loadings which lead to large amplitude vibrations. Mechanical and thermal loads usually occur simultaneously and as a result, the displacement and the temperature fields are created in close connection with each other. The two fields have to be defined simultaneously taking into account the relationship between them. The dynamic thermoelasticity allows the stresses generated by the temperature and mechanical fields, as well as the temperature distribution and propagation arising due to time-dependent mechanical and heat forces to be determined.

In 1956 Biot [1] developed the concept of coupled thermoelasticity in order to solve the paradox inherent in the classical uncoupled theory that elastic changes have no effect on the temperature. The equations of elasticity and heat conduction are coupled in this theory. Finally, when the deformation of the body is accounted for in the heat conduction equation, and the influence of changes in temperature appear in the equations of motion, the problem becomes coupled thermoelasticity one. In the pioneer books [2, 3], the basis of thermoelasticity and the influence of temperature coupling on the strain field are presented.

The temperature can change the reaction of mechanical structure and disguise a damage of the structure and thus provide incorrect data in a system dedicated for damage detection. The authors of paper [4] studied the geometrically nonlinear vibrations of Timoshenko beam at elevated temperature. The results show that short heat pulses with high magnitudes can lead to vibrations with quite large amplitudes.

The thermomechanical, geometrically nonlinear vibrations of rectangular plate model based on higher order shear deformation theory have been studied in [5]. The authors demonstrated that close to critical points, even small temperature variation can cause unexpected change in the reaction of the system. They noted highly nonlinear dynamic behaviour of the system, including periodic, quasi-periodic and chaotic oscillations.

Thermoelastic geometrically nonlinear vibrations of straight and curved beams are analysed using p-version of final element method in [6]. In this paper the role of temperature for periodic and non-periodic motions is illustrated. The effect of parameters such as temperature variation, the thickness and the ratio of curvature on the beam on the nonlinear dynamics is studied.

Many practical issues of current interest in the field of thermoelasticity are faced with significant computational difficulties when treated by traditional analytical methods. This fact leads many scientists to propose numerical methods, and several attempts are made to find finite-element solutions to various thermal deformation problems. The finite element method (FEM)-based thermoelastic beam models allow the study of deformation of beams, including shear and longitudinal displacements and the inertia of the cross-section rotation. However, for steady states, transient dynamics, bifurcation or chaos analysis such models are not effective. Thus, in [7–9] thermoelastic low order of beam models are introduced. The suggested models of reduced nonlinear beams consider only the most important mechanical and thermal effects. Based on one mode reduction the authors presented a bifurcation scenario versus varied temperature or transition to chaotic oscillations and then compared the results with selected cases of the full model. The results for circular plate model for elevated temperature, based on extended Mindlin plate theory, have been presented in [10, 11]. The complete model considered transversal and longitudinal displacements due to nonlinear displacement field, shear deformation, inertia terms due to rotation of the cross-section and thermal and mechanical loads. After assumption of a constant elevated temperature and neglecting longitudinal inertia terms, the exact thermomechanical model of the circular plate defined by partial differential equations (PDEs) has been reduced to one degree of freedom system. Based on this model, a temperature effect on the first resonance zone as well on the bifurcation scenario, leading to buckling and chaotic oscillations have been determined.

The complex research of local and global dynamics of the plate for different thermal and mechanical loads has been carried out in [12, 13].

In the present work a thermoelastic model of a circular plate is analysed. This paper aims to study nonlinear oscillations of a heated plate subjected to dynamic loading. A reduced model of thermoelastic vibration is created and an analytical solution of the response of the plate in the frequency domain is developed. The influence of the temperature changes is studied.

2 Physical Model

We consider a circular plate with radius R and thickness h shown in Fig. 1. We also suppose fully axisymmetric problem. The motion of the plate is considered in a cylindrical coordinate system with axis r ($0 \leq r \leq R$), θ ($0 \leq \theta \leq 2\pi$) and z ($-h/2 \leq z \leq h/2$). Displacements of the element are represented by $u(r, t)$ coordinate which is the in-plane displacement, $w(r, t)$ the transverse displacement, $\psi(r, t)$ the cross-section rotation angle.

The considered in this paper thermoelastic model of a circular plate is based on Mindlin plate theory. The model is extended taking into account the geometrically nonlinear deformation of the plate (large displacements).

The strain and curvature-displacements relationships associated with the mid-plane which consider large displacements and shear are represented as:

$$\varepsilon_r^0 = \frac{\partial u}{\partial r} + \frac{1}{2}\left(\frac{\partial w}{\partial r}\right)^2, \varepsilon_t^0 = \frac{u}{r}, \Gamma_{rz}^0 = \psi + \frac{\partial w}{\partial r}, \tag{1}$$

$$\kappa_{\mathrm{r}}^0 = \frac{\partial \psi}{\partial r}, \quad \kappa_{\mathrm{t}}^0 = \frac{\psi}{r}$$

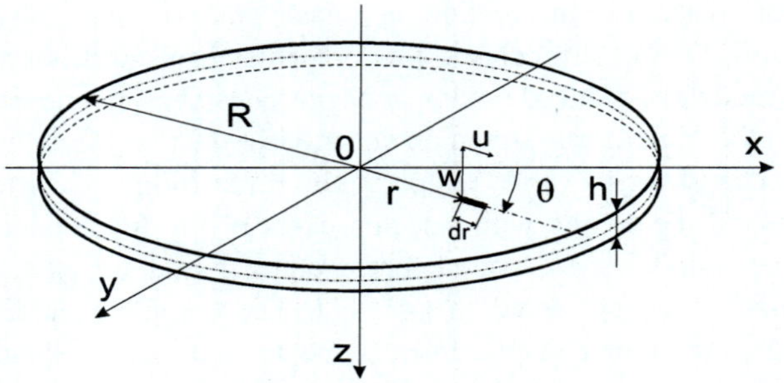

Fig. 1 A model of a circular plate with indicated coordinates and dimensions

and the strain vector is the following:

$$\boldsymbol{\varepsilon} = \left\{ \varepsilon_r^0 + zk_r^0,\, \varepsilon_t^0 + zk_r^0,\, f(z)\frac{1}{2}\Gamma_{rz}^0 \right\}^T, \tag{2}$$

where ε_r, ε_t, ε_{rz} are strains in radial and tangent directions and in *r-z* plane, κ_{r}, κ_{t} indicate curvatures of the mid-surface in radial and tangent directions, and $f(z)$ is a function of the distribution of the shear strain along the plate thickness. The superscript 0 associated with the mid-surface is omitted.

Assuming that the material of the plate is linear elastic and isotropic the relations between the component of the stress vector $\mathrm{S} = \{\sigma_{\mathbf{r}}, \sigma_t, \sigma_{\mathbf{rz}}\}$ and the vector of strain $\varepsilon = \{\varepsilon_{\mathbf{r}}, \varepsilon_t, \varepsilon_{\mathbf{rz}}\}$ take the form:

$$\begin{aligned} \sigma_r &= \tfrac{E}{1-\nu^2}\left[\varepsilon_r + \nu\varepsilon_t\right] - \tfrac{E}{1-\nu^2}\alpha_T\left(T - T_0\right) = \tfrac{E}{1-\nu^2}\left[\varepsilon_r + \nu\varepsilon_t - (1+\nu)\,\alpha_T\left(T - T_0\right)\right] \\ \sigma_t &= \tfrac{E}{1-\nu^2}\left[\varepsilon_t + \nu\varepsilon_r\right] - \tfrac{E}{1-\nu^2}\alpha_T\left(T - T_0\right) = \tfrac{E}{1-\nu^2}\left[\varepsilon_t + \nu\varepsilon_r - (1+\nu)\,\alpha_T\left(T - T_0\right)\right] \end{aligned} \tag{3}$$

$$\tau_{rz} = kG\Gamma_{rz}$$

where E is the Young modulus, G is the shear modulus, ν is Poisson ratio, α_T is the coefficient of thermal expansion, and k is the shear correction factor. In Eq. (3) *T*(*r*, *z*, *t*) is the current temperature and T_0 is the initial temperature.

The bending moments M_r and M_t, the shear force Q_r and the in-plane stress resultant per unit length N_r and N_t, presented in Fig. 2, are expressed as follows:

$$\begin{aligned} &M_r = D\left[\kappa_r + \nu\kappa_t\right] - A\alpha_T\kappa^T,\ M_t = D\left[\kappa_t + \nu\kappa_r\right] - A\alpha_T\kappa^T \\ &N_r = Ah\left[\varepsilon_r^0 + \nu\varepsilon_t^0\right] - A\alpha_T\gamma^T,\ N_t = Ah\left[\varepsilon_t^0 + \nu\varepsilon_r^0\right] - A\alpha_T\gamma^T,\ Q_r = kGh\left(\tfrac{\partial w}{\partial r} + \psi\right) \end{aligned} \tag{4}$$

The remaining coefficients and functions are described as:

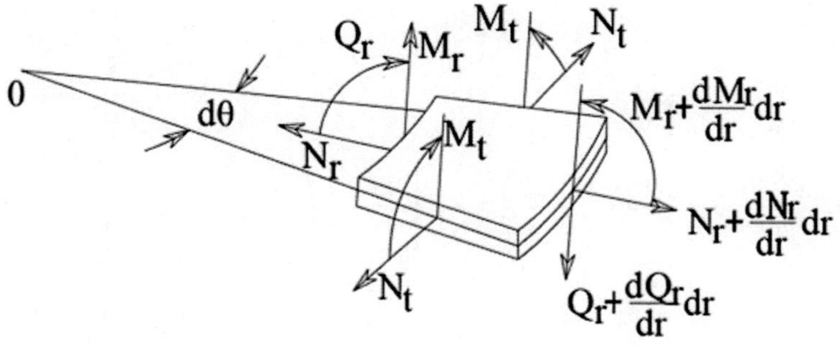

Fig. 2 In-plane and transversal stress resultants acting on an infinitesimal component of a plate

$$A = \frac{E}{1-\nu^2}, \quad D = Ah^3/12$$

$$\kappa^T(r) = \int_{-h/2}^{h/2} \Delta T(r,z)\, z dz, \quad \gamma^T(r) = \int_{-h/2}^{h/2} \Delta T(r,z)\, dz$$

The equations representing the thermoelastic vibration of the plate subjected to heating and mechanical load with intensity $p(r,t)$ are:

$$\frac{\partial N_r}{\partial r} + \frac{N_r - N_t}{r} - \rho h \frac{\partial^2 u}{\partial t^2} = 0 \tag{5}$$

$$\frac{\partial M_r}{\partial r} + \frac{M_r - M_t}{r} - Q_r - c_2 \frac{\partial \psi}{\partial t} - \frac{\rho h^3}{12} \frac{\partial^2 \psi}{\partial t^2} = 0$$

$$\frac{\partial Q_r}{\partial r} + \frac{Q_r}{r} + N_r \left(\frac{\partial^2 w}{\partial r^2} + \frac{1}{r} \frac{\partial w}{\partial r} \right) + \frac{\partial N_r}{\partial r} \frac{\partial w}{\partial r} - c_1 \frac{\partial w}{\partial t} - \rho h \frac{\partial^2 w}{\partial t^2} = -p(r,t)$$

where c_1 and c_2 indicate the damping coefficients and ρ is the material density.

In this work it is accepted that the plate gets an elevated temperature instantly and the heat propagation problem is not considered. Assuming that temperature is distributed uniformly along the whole plate κ^T and γ^T from Eq. (4) become: $\kappa^T = 0$ and $\gamma^T = \Delta T$ where ΔT is the difference between the initial temperature and the current temperature.

Following these assumptions and accepting that the inertia term in mid-plane can be neglected the equations of the plate vibration become:

$$Ah \left[\frac{\partial^2 u}{\partial r^2} + \frac{\partial w}{\partial r} \frac{\partial^2 w}{\partial r^2} + \frac{1}{r} \frac{\partial u}{\partial r} - \frac{u}{r^2} + \frac{1}{2r} (1-\nu) \left(\frac{\partial w}{\partial r} \right)^2 \right] = 0$$

$$D\left[\frac{\partial^2\psi}{\partial r^2}+\frac{1}{r}\frac{\partial\psi}{\partial r}-\frac{\psi}{r^2}\right]-k^2Gh\left(\frac{\partial w}{\partial r}+\psi\right)-c_2\frac{\partial\psi}{\partial t}-\frac{\rho h^3}{12}\frac{\partial^2\psi}{\partial t^2}=0 \quad (6)$$

$$k^2Gh\left(\frac{\partial^2 w}{\partial r^2}+\frac{1}{r}\frac{\partial w}{\partial r}+\frac{\partial\psi}{\partial r}+\frac{\psi}{r}\right)+Ah\left[\frac{\partial u}{\partial r}+\frac{1}{2}\left(\frac{\partial w}{\partial r}\right)^2+\frac{\nu}{r}u-\frac{1+\nu}{h}\alpha_T\Delta\mathrm{T}\right]$$

$$\left(\frac{\partial^2 w}{\partial r^2}+\frac{1}{r}\frac{\partial w}{\partial r}\right)+$$

$$Ah\left[\frac{\partial^2 u}{\partial r^2}+\frac{\nu}{r}\frac{\partial u}{\partial r}-\frac{\nu}{r^2}u+\frac{\partial w}{\partial r}\frac{\partial^2 w}{\partial r^2}\right]\cdot\left(\frac{\partial w}{\partial r}\right)+c_1\frac{\partial w}{\partial t}-\rho h\frac{\partial^2 w}{\partial t^2}=-p\,(r,t)$$

The boundary conditions for a clamped, in-plane fixed plate are:

$$u\,(0,t)=u\,(R,t)=w\,(R,t)=0 \quad \psi\,(R,t)=0$$

3 Analytical Solutions by Harmonic Balance Method

The formulated thermoelastic problems are quite complicated from a computational point of view, therefore, simplified techniques which allow reducing the governing partial differential equations into a reduced number of modal coordinates are applied. Such an approach is very effective for nonlinear models as it allows deep bifurcation analysis.

The model of the plate represented by partial differential equations is reduced to ordinary differential equations by the Galerkin method based on the modes projection. In contrary to the paper [14] in this analysis we consider one mode reduction assuming clamped boundary conditions of the plate and take into account excitation distributed according to the first mode shape. According to Galerkin method the solution is obtained by series:

$$w\,(r,t)=\sum_{n=1}^{N}w_n(r)q_n(t) \quad \psi\,(r,t)=\sum_{n=1}^{N}\psi_n(r)q_n(t) \quad (7)$$

where $w_n(r)$ and $\psi_n(r)$ are space functions representing vibration modes which should satisfy the geometrical boundary conditions and $q_n(t)$ is the time function generalized coordinate.

Substituting Eq. (7) into PDEs and multiplying by the selected mode function and then integrating through the plate radius, we get ODE of motion. For the first mode reduction ($N = 1$) we obtain just one nonlinear differential equation:

$$\ddot{q}_1 + 2\xi_1\omega_1\dot{q}_1 + \omega_1^2 q_1 + F_{NL1}q_1^3 + F_{NT1}\Delta T q_1 = F_{NP1}\sin\omega t \tag{8}$$

Equation (8) is written in dimensionless form with respect to the space coordinates $\overline{u} = u/R$, $\overline{w} = w/R$ but time is given in seconds. The 'dot' denotes time derivative and q_1 first dimensionless generalized coordinate. F_{NL1} is a coefficient occurring due to nonlinear geometrical terms, F_{NT1} is a coefficient related to temperature variation, F_{NP1} is a coefficient depending on the amplitude of mechanical loading and the modal damping is introduced by coefficient ξ_1.

The obtained reduced nonlinear one degree of freedom model with cubic nonlinearity and temperature influence is studied analytically by the extended harmonic balance method.

For a simplicity, the following substitutions are introduced:

$$\gamma = \mathrm{F}_{NL1},\ \ \mathrm{P} = \mathrm{F}_{NP1},\ \ \lambda = \mathrm{F}_{NT1},\ \ \mathrm{q} = \mathrm{x}$$

and then we get the equation below:

$$\ddot{x} + 2\xi\omega_1\dot{x} + \omega_1^2 x + \gamma x^3 + \lambda\Delta T x = P\sin\omega t \tag{9}$$

The solution is sought as:

$$x = A_1(t)\sin\omega t + A_2(t)\cos\omega t \tag{10}$$

where $A_1(t)$, $A_2(t)$ are unknown amplitudes assumed as slow functions of time (so called slow flow).

The velocity and the acceleration of x take the form:

$$\dot{x} = \dot{A}_1\sin\omega t + \omega A_1\cos\omega t + \dot{A}_2\cos\omega t - \omega A_2\sin\omega t \tag{11}$$

$$\ddot{x} = 2\omega\dot{A}_1\cos\omega t - 2\omega\dot{A}_2\sin\omega t - A_2\omega^2\cos\omega t - A_1\omega^2\sin\omega t + \tag{12}$$

$$\ddot{A}_1\sin\omega t + \ddot{A}_2\cos\omega t$$

and the cubic term is expressed as:

$$\gamma x^3 = \gamma\left(A_1(t)\sin\omega t + A_2(t)\cos\omega t\right) = \tfrac{3}{4}\gamma\sin\omega t\left(A_1^3 + A_1A_2^2\right) + \tfrac{3}{4}\gamma\cos\omega t\left(A_2^3 + A_1^2A_2\right) +$$
$$\tfrac{1}{4}\gamma\sin 3\omega t\left(-A_1^3 + 3A_1A_2^2\right) + \tfrac{1}{4}\gamma\cos 3\omega t\left(A_2^3 - 3A_1^2A_2\right) \tag{13}$$

Equations (10)–(13) are substituted into Eq. (9). Then the small terms $\ddot{A_1}$, $\ddot{A_2}$, the amplitude derivatives in power higher than one and harmonics sin$3\omega t$, cos$3\omega t$ are omitted and we obtain the modulation equation for amplitude components A_1 and A_2:

$$\dot{A}_1(t) = -\frac{1}{16(\xi^2\omega_1{}^2+\omega^2)}\begin{pmatrix} -8P\xi\omega_1 + 8\lambda\Delta T\xi\omega_1 A_1(t) + 8\xi\omega_1^3 A_1(t)+ \\ 8\xi\omega_1\omega^2 A_1(t) + 6\gamma\xi\omega_1 A_1^3(t) + 8\lambda\Delta T\omega A_2(t) \\ + 8\omega_1^2\omega A_2(t) - 16\xi^2\omega_1^2\omega A_2(t) - 8\omega^3 A_2(t) \\ + 6\gamma\omega A_1^2(t)A_2(t) + 6\gamma\xi\omega_1 A_1(t)A_2^2(t) + 6\gamma\omega A_2^3(t) \end{pmatrix}$$
$$\dot{A}_2(t) = -\frac{1}{16(\xi^2\omega_1{}^2+\omega^2)}\begin{pmatrix} 8P\omega - 8\lambda\Delta T\omega A_1(t) - 8\omega_1^2\omega A_1(t) + 16\xi^2\omega_1^2\omega A_1(t)+ \\ 8\omega^3 A_1(t) - 6\gamma\omega A_1^3(t) + 8\lambda\Delta T\xi\omega_1 A_2(t) + 8\xi\omega_1^3 A_2(t)+ \\ 8\xi\omega_1\omega^2 A_2(t) + 6\gamma\xi\omega_1 A_1^2(t)A_2(t)- \\ 6\gamma\omega A_1(t)A_2^2(t) + 6\gamma\xi\omega_1 A_2^3(t) \end{pmatrix} \tag{14}$$

For a steady state, the amplitudes are constant and therefore, their derivatives are equal to zero thus, we get two nonlinear algebraic equations:

$$\begin{aligned} &-\omega^2 A_1 + \omega_1^2 A_1 - 2\xi\omega_1\omega A_2 + \tfrac{3}{4}\gamma A_1^3 + \tfrac{3}{4}\gamma A_1 A_2^2 + \lambda\Delta T A_1 = P \\ &-\omega^2 A_2 + \omega_1^2 A_2 + 2\xi\omega_1\omega A_1 + \tfrac{3}{4}\gamma A_2^3 + \tfrac{3}{4}\gamma A_1^2 A_2 + \lambda\Delta T A_2 = 0 \end{aligned} \tag{15}$$

From the above set of equations, after some algebraic manipulations, and introducing resultant amplitude $A = \sqrt{A_1^2 + A_2^2}$ we obtain a single algebraic equation for the amplitude:

$$-16P^2 + 9A^6\gamma^2 + A^4\left(24\gamma\lambda\Delta T + 24\gamma\omega_1^2 - 24\gamma\omega^2\right) + \tag{16}$$

$$A^2\left(16\lambda^2\Delta T^2+32\lambda\Delta T\omega_1^2+16\omega_1^2-32\lambda\Delta T\omega^2-32\omega_1^2\omega^2+64\xi^2\omega_1^2\omega^2+16\omega^4\right)=0$$

Then, substituting $z = A^2$ a third order algebraic equation is obtained, which can be solved analytically. Depending on the parameters value we can get one or three real roots and then we find vibrations amplitude. For the sake of brevity, the formulas for the roots are not presented in this paper. On the basis of Eq. (16) we determine the resonance curve of the plates reduced model and also we may study an influence of structural parameters on the system response.

4 Numerical Examples of Thermoelastic Vibrations

The response of the plate at elevated temperatures subjected to harmonic loading can be analysed by FEM. Using this approach we can obtain time history diagrams and to study the influence of loading parameters and the temperature on response of the plate in time. However, if we want to study the steady state nonlinear response of the plate in the parameters domain, or create bifurcation diagrams and to analyse other specific nonlinear features the application of FE model is not efficient due to limitations of the software and time consuming simulations. Therefore, the reduced low order model is applied.

In the present study we have used the harmonic balance method to obtain the dependence between the amplitudes, the frequency of vibration and other parameters. Harmonic balance method (HBM) was applied to the reduced model taking into account the first vibration mode. The coefficients of Eq. (9) takes values:

$$\begin{aligned}
\omega_1 &= 3928.205 \\
\xi &= 0.01 \\
\gamma &= 0.37344\text{x}10^{12} \\
\mathrm{P} &= 1.0754246\text{x}10^{4} \\
\lambda &= -0.40946\text{x}10^{6}
\end{aligned}$$

Frequency of excitation ω is varied around the first natural frequency ω_1, and in some cases amplitude of excitation P can be changed as well. The influence of elevated temperature is tested by varying ΔT parameter.

The frequency-response Eq. (16) allow us to obtain the resonance curves for different values of loading amplitude and temperature.

The resonance curves at different temperatures and fixed amplitude of the loading obtained analytically by HBM are shown in Fig. 3. The analytical solution has been computed for three different values of the temperature—$\Delta T = -20$, $\Delta T = 0$ and $\Delta T = 20$. In order to verify the application of HBM the same value of parameters have been applied to create the resonance curves by direct simulation of Eq. (9) by the fourth order Runge-Kutta method and then applying the continuation technique. The results are shown in Fig. 4. As can be seen the results obtained by the two different methods are almost identical. This comparison proofs the correctness of the analytical method developed in Sect. 3.

The geometrical nonlinearity leads to strong stiffening behaviour of the resonant curves. The increasing of the frequencies of the excitation leads to increasing of the amplitudes of vibration until its peak value. After that instability solutions can appear and the amplitudes can jump to lower values. The variation of the temperature leads to shift of the curves. The lower temperature ($\Delta T = -20$) decreases amplitudes and moves the resonance curve to higher frequencies (blue curve). The high temperature—$\Delta T = 20$, on the other hand, increases the amplitudes and changes the resonance curve in the direction of lower frequencies (red curve). The resonance curve for $\Delta T = 0$ is shown in black colour.

The small peaks observed in Fig. 4 are not observed in the results obtained by HBM. More small peaks are observed in the resonance curve of the same problem studied by using three modes reduction model in [14]. Such peaks cannot

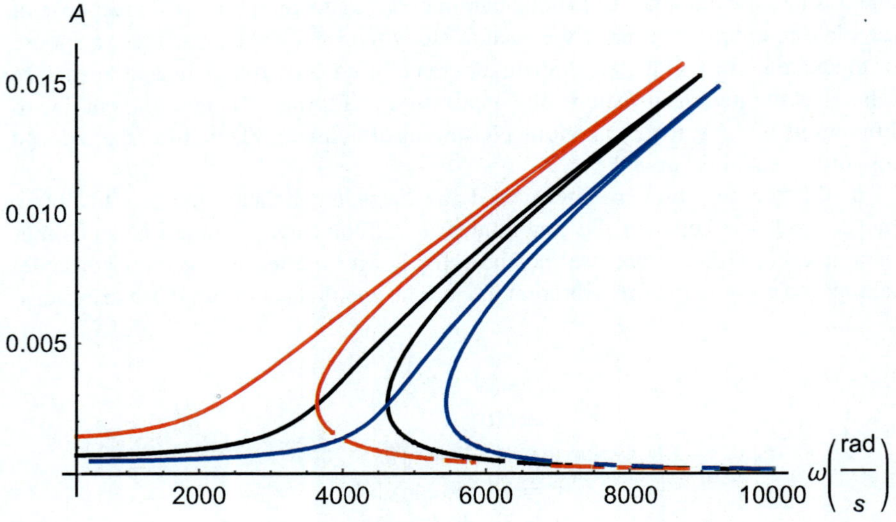

Fig. 3 Resonance curves by HBM for $\Delta T = 20$ (red), $\Delta T = 0$ (black) and $\Delta T = -20$ (blue)

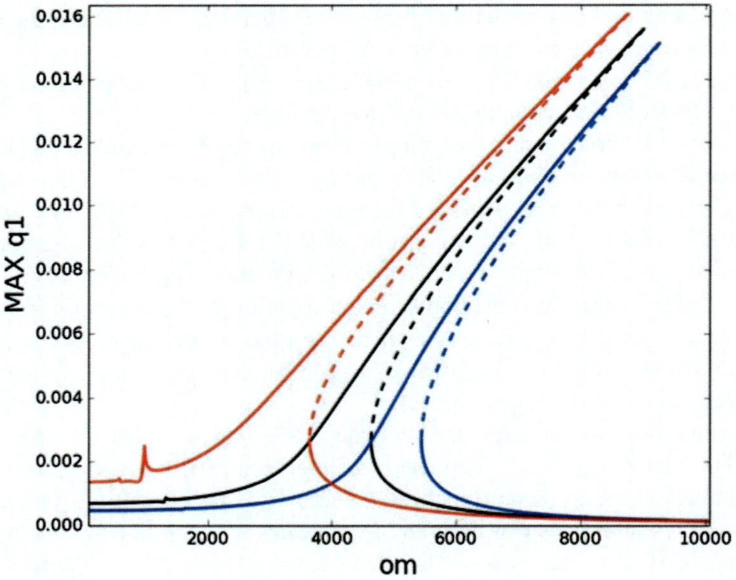

Fig. 4 Resonance curves from direct numerical simulation for $\Delta T = 20$ (red), $\Delta T = 0$ (black) and $\Delta T = -20$ (blue)

be obtained when only one mode is used in the solution and the single frequency response around resonance zone is sought. The general behaviour of the response of the plate in the frequency domain, however, obtained by the one mode reduction is the same as the ones obtained by the three mode reduction model and corresponds very well for FE analysis not presented in this paper.

5 Conclusions

The reduced one degree of freedom model of geometrically nonlinear thermoelastic Mindlin circular plate is developed in this paper. The reduced nonlinear model is studied by two different methods, by approximate analytical harmonic balance method and by direct simulation of the original ODE. It is demonstrated that analytical approach gives very accurate solutions with great agreement with numerical simulations.

Analysis shows that elevated temperature can change essentially the response of the plate and could provoke the plate to complex response, to buckling and bifurcation. These phenomena are more clearly seen when the excitation frequency is close to the first natural frequency. Based on the reduced model a dependence of amplitude and frequency of excitation is obtained by solving the nonlinear algebraic equation which represent a steady state. This allows obtaining easily the resonance curves and to study the influence of the loading parameters and elevated temperature on the behaviour of the plate. By using this model the detail bifurcation analysis of the thermo-elastic vibration of the plate will be performed in the future.

Acknowledgments This work has been accomplished with the financial support of the Bulgarian Ministry of Education and Science, Grant No. D01-221/03.12.2018 for NCDSC—part of the Bulgarian National Roadmap on RIs. The second author has been financed within the framework of the project "Lublin University of Technology-Regional Excellence Initiative", funded by the Polish Ministry of Science and Higher Education (Contract No. 030/RID/2018/19).

References

1. Biot, M.A.: Thermoelasticity and irreversible thermodynamics. J. Appl. Phys. **27**(3), 240–253 (1956)
2. Nowacki, W.: Thermoelasticity. Addison-Wesley, Reading, MA (1962)
3. Boley, B.A., Weiner, J.H.: Theory of Thermal Stresses. Wiley, New York (1960)
4. Manoach, E., Samborski, S., Mitura, A., Warminski, J.: Vibration based damage detection in composite beams under temperature variations using Poincaré maps. Int. J. Mech. Sci. **62**(1), 120–132 (2012). https://doi.org/10.1016/j.ijmecsci.2012.06.006
5. Amabili, M., Carra, S.: Thermal effects on geometrically nonlinear vibrations of rectangular plates with fixed edges. J. Sound Vib. **321**(2), 936–954 (2009)
6. Ribeiro, P., Manoach, E.: The effect of temperature on the large amplitude vibrations of curved beams. J. Sound Vib. **285**(4–5), 1093–1107 (2005). https://doi.org/10.1016/j.jsv.2004.09.010
7. Warminska, A., Manoach, E., Warminski, J., Samborski, S.: Regular and chaotic oscillations of a Timoshenko beam subjected to mechanical and thermal loadings. Contin. Mech. Thermodyn. **27**(4–5), 719–737 (2015). https://doi.org/10.1007/s00161-014-0381-6

8. Warminska, A., Manoach, E., Warminski, J.: Nonlinear dynamics of a reduced multimodal timoshenko beam subjected to thermal and mechanical loadings. Meccanica. **49**(8), 1775–1793 (2014). https://doi.org/10.1007/s11012-014-9891-3
9. Manoach, E., Warminski, J., Warminska, A.: Large amplitude vibrations of heated Timoshenko beams with delamination. Proc. Inst. Mech. Eng. C. **230**(1), 88–101 (2015). https://doi.org/10.1177/0954406215570702
10. Warminska, A., Manoach, E., Warminski, J.: Dynamics of a circular Mindlin plate under mechanical loading and elevated temperature. MATEC Web Conf. **83**, 05013 (2016). https://doi.org/10.1051/matecconf/20168305013
11. Warminska, A., Manoach, E., Warminski, J.: A reduced model of a thermoelastic nonlinear circular plate. MATEC Web Conf. **148**, 06001 (2018). https://doi.org/10.1051/matecconf/201814806001
12. Saetta, E., Rega, G.: Third order thermomechanically coupled laminated plate: 2D nonlinear modeling, minimal reduction, and transient/post-buckled dynamics under different thermal excitations. Compos. Struct. **174**, 420–441 (2017)
13. Settimi, V., Saetta, E., Rega, G.: Local and global nonlinear dynamics of thermomechanically coupled composite plates in passive thermal regime. Nonlinear Dyn. **93**, 167 (2018). https://doi.org/10.1007/s11071-017-3648-1
14. Manoach, E., Warminska, A., Warminski, J., Doneva, S.: A reduced multimodal thermoelastic model of a circular Mindlin plate. Int. J. Mech. Sci. (2019). https://doi.org/10.1016/j.ijmecsci.2019.02.010

A Rulkov Neuronal Model with Caputo Fractional Variable-Order Differences of Convolution Type

Oana Brandibur, Eva Kaslik, Dorota Mozyrska, and Małgorzata Wyrwas

Abstract In this paper, a theoretical and numerical investigation is undertaken for a fractional-order version of the Rulkov neuronal model, involving Caputo fractional variable-order differences of convolution type. As the first step, using linearization techniques and the Z-transform method, sufficient conditions are explored which guarantee the stability or instability of the unique equilibrium point of the system. Numerical simulations are further carried out to illustrate the theoretical findings, emphasizing the differences between the current model and simpler versions involving fractional-order difference with constant fractional orders, as well as the classical integer-order Rulkov model.

Keywords Fractional-order difference equation · Variable-order fractional operator · Fractional-order Rulkov model · Neuronal model · Fractional-order system · Instability · Bursting.

1 Introduction

In this work, we discuss stability and instability properties in the framework of a generalized discrete-time Rulkov neuronal model [9, 10], achieved by introducing a Caputo-type fractional variable-order difference operator to model the evolution of the membrane potential. This generalized model is constructed based on experimental neuronal research [1, 4], which emphasizes that fractional-order operators can be successfully used in the mathematical modelling of neuronal dynamics. In fact,

---

O. Brandibur · E. Kaslik
West University of Timisoara, Faculty of Mathematics and Computer Science, Timişoara, Romania
e-mail: oana.brandibur@e-uvt.ro; eva.kaslik@e-uvt.ro

D. Mozyrska (✉) · M. Wyrwas
Bialystok University of Technology, Faculty of Computer Science, Białystok, Poland
e-mail: d.mozyrska@pb.edu.pl; m.wyrwas@pb.edu.pl

J. Awrejcewicz (ed.), *Perspectives in Dynamical Systems II: Mathematical and Numerical Approaches*, Springer Proceedings in Mathematics & Statistics 363,
https://doi.org/10.1007/978-3-030-77310-6_20

fractional-order membrane potential dynamics are known to introduce capacitive memory effects [11], proving their utility in reproducing more realistically the electrical activity of neurons.

In the qualitative theory of fractional order systems, the investigation of stability properties plays a leading role. When constant fractional orders are taken into account, necessary and sufficient conditions have been recently obtained for the stability of both linear continuous-time fractional systems [2, 3] and linear discrete-time fractional systems [6, 7]. In the case of linear fractional-order systems with variable fractional orders of subunitary values, stability properties have been explored in [5, 8].

In this paper, we employ the Caputo-type fractional variable-order difference operator of convolution type, originally defined in [8]. This choice is justified by the fact that the Caputo-type operator permits the existence of nonzero equilibrium points to the equations, while fractional-order equations with the Grünwald-Letnikov operator can only have zero as stationary point.

2 Preliminaries

The first aim of this section is to define the variable-order fractional difference operator of Caputo type in a constructive manner.

Definition 1 Considering the order function $\nu : \mathbb{Z} \to \mathbb{R}_+$, the *oblivion function* is defined as

$$a^{[\nu(l)]}(k) = \begin{cases} 0 & \text{for } k < 0 \\ 1 & \text{for } k = 0 \\ (-1)^k \frac{\nu(l)[\nu(l)-1]\cdots[\nu(l)-k+1]}{k!} & \text{for } k > 0 \end{cases} \quad , \ \forall\, k, l \in \mathbb{Z}. \tag{1}$$

It is easy to see that formula (1) in Definition 1 is equivalent to the following recurrence relation with respect to $k \in \mathbb{Z}_+$:

$$\begin{aligned} &a^{[\nu(l)]}(0) = 1\,, \\ &a^{[\nu(l)]}(k) = a^{[\nu(l)]}(k-1)\left[1 - \frac{\nu(l)+1}{k}\right] \text{ for } k \geq 1\,. \end{aligned} \tag{2}$$

In what follows, we denote by Δ the forward difference operator defined as

$$(\Delta y)(k) := y(k+1) - y(k) \quad , \ \forall\, y : \mathbb{Z} \to \mathbb{R},\ k \in \mathbb{Z}$$

and by $\Delta^n := \Delta \circ \cdots \circ \Delta$ its n-fold application, which can be expressed as

$$(\Delta^n y)(k) := \sum_{i=0}^{n} (-1)^{n-i} \binom{n}{i} y(k+i) \quad , \forall\, y : \mathbb{Z} \to \mathbb{R},\ k \in \mathbb{Z}.$$

Additionally, we define $\left(\Delta^0 y\right)(k) := y(k)$.

Definition 2 Considering the order function $\nu : \mathbb{Z} \to \mathbb{R}_+$, the *fractional variable-order sum of convolution type* is given by

$$\left(\Delta^{-\nu(\cdot)} y\right)(k) := \left(a^{[-\nu(\cdot)]} * y\right)(k) = \sum_{i=0}^{k} a^{[-\nu(i)]}(i) y(k-i) \quad , \forall\, y : \mathbb{Z} \to \mathbb{R}.$$

where $k \in \mathbb{Z}_+$ and "$*$" denotes the convolution operator.

Definition 3 Considering the order function $\nu : \mathbb{Z} \to (n-1, n]$, $n \in \mathbb{Z}_+^*$, the *Caputo fractional variable-order difference operator of convolution type* is defined as

$$\left(\Delta^{\nu(\cdot)} y\right)(k) = \left(\Delta^{-(n-\nu(\cdot))} \left(\Delta^n y\right)\right)(k) \quad , \forall\, y : \mathbb{Z} \to \mathbb{R},\ k \in \mathbb{Z}. \tag{3}$$

Remark 1 Some properties of the variable-order fractional difference operator of Caputo type are listed below

a. If $n = 1$ then $\left(\Delta^{\nu(\cdot)} y\right)(k) = \left(\Delta^{-(1-\nu(\cdot))} (\Delta y)\right)(k)$.
b. If $\nu(k) = n \in \mathbb{Z}_+$, for any $k \in \mathbb{Z}$, it follows that $\left(\Delta^{\nu(\cdot)} y\right)(k) = (\Delta^n y)(k)$.
c. If $n = 1$, the $\mathcal{Z}$-transform is given by

$$\mathcal{Z}\left[\Delta^{\nu(\cdot)} y\right](z) = ((z-1)Y(z) - zy(0))\, A_\nu(z), \tag{4}$$

where $Y(z) = \mathcal{Z}[y](z)$ and $A_\nu(z) = \mathcal{Z}\left[a^{[\nu(\cdot)-1]}\right](z) = \sum_{i=0}^{\infty} (-1)^i \binom{\nu(i)-1}{i} z^{-i}$.

3 Stability and Instability Results for Two-Dimensional Systems of Variable-Order Fractional Difference Equations

Let us consider the n-dimensional fractional-order system

$$\left(\Delta^{\nu}(\cdot) x\right)(k) = f(k, x(k)) \quad , \forall\, k \geq 1, \tag{5}$$

where $\nu(\cdot) = (\nu_1(\cdot), \nu_2(\cdot), \ldots, \nu_n(\cdot))$, $\nu_p : \mathbb{Z} \to (0, 1]$ are order functions and $f : \mathbb{Z}_+ \times \mathbb{R}^n \to \mathbb{R}^n$ is continuous on the whole domain of definition and Lipschitz-continuous with respect to the second variable, such that $f(k, 0) = 0$ for any

$k \in \mathbb{Z}_+$. Let $\varphi(k, x_0)$ denote the unique solution of (5) which satisfies the initial condition $x(0) = x_0$.

Definition 4

a. The trivial solution of (5) is called *stable* if for any $\varepsilon > 0$ there exists $\delta = \delta(\varepsilon) > 0$ such that for every $x_0 \in \mathbb{R}^n$ satisfying $\|x_0\| < \delta$ we have $\|\varphi(k, x_0)\| \leq \varepsilon$ for any $k \geq 0$.
b. The trivial solution of (5) is called *asymptotically stable* if it is stable and there exists $\rho > 0$ such that $\lim_{k\to\infty} \varphi(k, x_0) = 0$ whenever $\|x_0\| < \rho$.
c. Let $q \in (0, 1]$. The trivial solution of (5) is called $\mathcal{O}(k^{-q})$*-asymptotically stable* if it is stable and there exists $\rho > 0$ such that for any $\|x_0\| < \rho$ one has $\|\varphi(k, x_0)\| = \mathcal{O}(k^{-q})$ as $k \to \infty$.

In the following, we consider a two-dimensional linear fractional-order system with variable-order Caputo difference operators:

$$\begin{cases} \left(\Delta^{\nu_1(\cdot)}x\right)(k) = a_{11}x(k) + a_{12}y(k) \\ \left(\Delta^{\nu_2(\cdot)}y\right)(k) = a_{21}x(k) + a_{22}y(k) \end{cases} \tag{6}$$

where $A = (a_{ij})$ is a real two-dimensional matrix and $\nu_1, \nu_2 : \mathbb{Z} \to (0, 1]$ are the order functions.

Using the $\mathcal{Z}$-transform in system (6), based on (4) we deduce:

$$\left(\begin{bmatrix} (z-1)A_{\nu_1}(z) & 0 \\ 0 & (z-1)A_{\nu_2}(z) \end{bmatrix} - A\right)\begin{bmatrix} X(z) \\ Y(z) \end{bmatrix} = \begin{bmatrix} x(0)zA_{\nu_1}(z) \\ y(0)zA_{\nu_2}(z) \end{bmatrix}$$

where $\mathcal{Z}[x] = X$ and $\mathcal{Z}[y] = Y$ are the $\mathcal{Z}$-transforms of x and y, respectively and $A_{\nu_p}(z) = \mathcal{Z}\left[a^{[\nu_p(\cdot)-1]}\right](z)$, $p \in \{1, 2\}$, are the $\mathcal{Z}$-transforms corresponding to the oblivion functions.

The following characteristic equation is obtained:

$$\det\left(\operatorname{diag}\left((z-1)A_{\nu_1}(z), (z-1)A_{\nu_2}(z)\right) - A\right) = 0.$$

Hence, the characteristic function of the system (6) is:

$$\Delta(z; A, \nu_1(\cdot), \nu_2(\cdot)) = \left[(z-1)A_{\nu_1}(z) - a_{11}\right]\left[(z-1)A_{\nu_2}(z) - a_{22}\right] - a_{12}a_{21}.$$

Using similar techniques as in [2, 3], the following result can be proved for the characterization of the stability and instability properties of system (6):

Theorem 1

1. *If all the roots of the characteristic function* $\Delta(z; A, \nu_1(\cdot), \nu_2(\cdot))$ *are inside the unit circle (i.e.* $|z| < 1$*), system (6) is* $\mathcal{O}(n^{-q})$*-globally asymptotically stable, where* $q = \liminf_{k\in\mathbb{Z}_+} \min\{\nu_1(k), \nu_2(k)\}$.

2. *If* $\det(A) \neq 0$ *and* $\Delta(z; A, \nu_1(\cdot), \nu_2(\cdot))$ *has a root outside the closed unit circle, system (6) is unstable.*

Using basic mathematical tools, the following sufficient conditions for the instability of system (6) can be obtained, which do not depend on the variable fractional orders $\nu_1(\cdot)$ and $\nu_2(\cdot)$. These order-independent instability results are useful in applications where the exact order functions are not known precisely.

Theorem 2 (Fractional-Order Independent Instability Results) *If one of the following two conditions hold:*

1. $\det(A) < 0$;
2. $a_{11} > 0$ *and* $a_{11}a_{22} \geq \det(A) > 0$.

then system (6) is unstable regardless of the variable orders $\nu_1(\cdot)$ *and* $\nu_2(\cdot)$.

4 A Variable-Order Fractional Rulkov-Type Neuronal Model

As an application to the theoretical results presented above, we analyze the following discrete-time fractional-order Rulkov-type model, describing the spiking behaviour of a biological neuron:

$$\begin{cases} \left(\Delta^{\nu_1(\cdot)}x\right)(k) = \dfrac{\alpha}{1+x(k)^2} - x(k) + y(k) \\ \left(\Delta^{\nu_2(\cdot)}y\right)(k) = -\mu(x(k) - \sigma) \end{cases} \tag{7}$$

where x represents the membrane potential, y is a gating variable, with $0 < \mu \ll 1$, σ acts as an external current applied to the neuron and $\alpha > 0$ is a nonlinearity parameter. We consider $\nu_1 : \mathbb{Z} \to (0, 1]$ and for simplicity, we will further assume that the order function ν_2 is constant, i.e. $\nu_2(k) = 1$, for any $k \in \mathbb{Z}_+$. The reason for this assumption is that in the equation of the gating variable the use of a fractional-order difference cannot be justified biologically.

System (7) has a unique fixed point $(x^*, y^*) = \left(\sigma, \sigma - \dfrac{\alpha}{1+\sigma^2}\right)$. The Jacobian matrix of system (7) at (x^*, y^*) is:

$$A = \begin{pmatrix} -1 - \dfrac{2\alpha\sigma}{(1+\sigma^2)^2} & 1 \\ -\mu & 0 \end{pmatrix}$$

Taking into considerations the notations from the previous section, we have:

$$a_{11} = a_{11}(\alpha, \sigma) = -1 - \frac{2\alpha\sigma}{(1+\sigma^2)^2} < 0, \quad a_{22} = 0, \quad 0 < \det(A) = \mu \ll 1.$$

It is easy to notice that Theorem 2 cannot be applied in this case. With the aim of determining the stability and instability region with respect to the parameters α and σ, one has to investigate the roots of the characteristic equation

$$(z-1)^2 A_{\nu_1}(z) - a_{11}(z-1) + \mu = 0. \tag{8}$$

Looking for roots of the form $z = e^{i\omega}$, $\omega \in (0, 2\pi)$ allows us to determine the boundary of the stability and instability regions, respectively. Indeed, $z = e^{i\omega}$ is a root of the characteristic equation (8) if and only if

$$a_{11} = (e^{i\omega} - 1) A_{\nu_1}(e^{i\omega}) + \mu (e^{i\omega} - 1)^{-1}.$$

Taking the imaginary part in the previous equation, it follows that $\omega \in (0, 2\pi)$ is a root of the equation

$$\Im \left[(e^{i\omega} - 1) A_{\nu_1}(e^{i\omega}) + \mu (e^{i\omega} - 1)^{-1} \right] = 0 \tag{9}$$

For a given order function ν_1 and a fixed value of the parameter μ, Eq. (9) is solved numerically. For every root $\omega^* \in (0, 2\pi)$, the curve from the (α, σ)-plane defined implicitly as

$$a_{11}(\alpha, \sigma) = \Re \left[(e^{i\omega^*} - 1) A_{\nu_1}(e^{i\omega^*}) + \mu (e^{i\omega^*} - 1)^{-1} \right]$$

is a part of the boundary of the stability and instability regions.

It is important to emphasize that since our aim is to observe neuronal bursting behavior in the fractional variable-order Rulkov models, we search for values of the parameters α and σ for which the unique equilibrium of system (7) is guaranteed to be unstable.

For all numerical simulations, $\mu = 0.001$ has been chosen, considering the following order functions:

(a) $\nu_1(k) = 1$—integer order classical Rulkov model;
(b) $\nu_1(k) = 0.8$—constant fractional order;
(c) $\nu_1(k) = 0.8 - \frac{0.2}{0.01k+1}$—rational order function;
(d) $\nu_1(k) = 0.8 - 0.2\exp(-0.1k)$—exponential order function;
(e) $\nu_1(k) = 0.8 + 0.1\cos(0.01k\pi)$—periodic order function involving a cosine;
(f) $\nu_1(k) = 0.8 + 0.1\,\mathrm{SquareWave}(k/200)$—square wave order function;

The graphs of the non-constant order functions are shown in Fig. 1. The integer order case (a), as well as the constant order case (b) are considered for comparison purposes. The rational and exponential order functions (c)–(d) converge to 0.8 as $k \to \infty$, with a faster convergence rate for the exponential function (d). The order functions (e) and (f) are periodic, oscillating around the value 0.8.

Figure 2 shows the stability and instability regions in the (α, σ)-plane for system (7) for the considered order functions ν_1 given above.

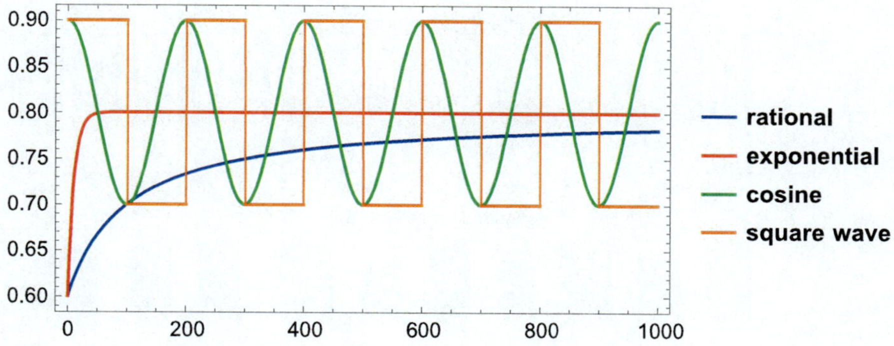

Fig. 1 Order functions (c)–(f) used in numerical simulations

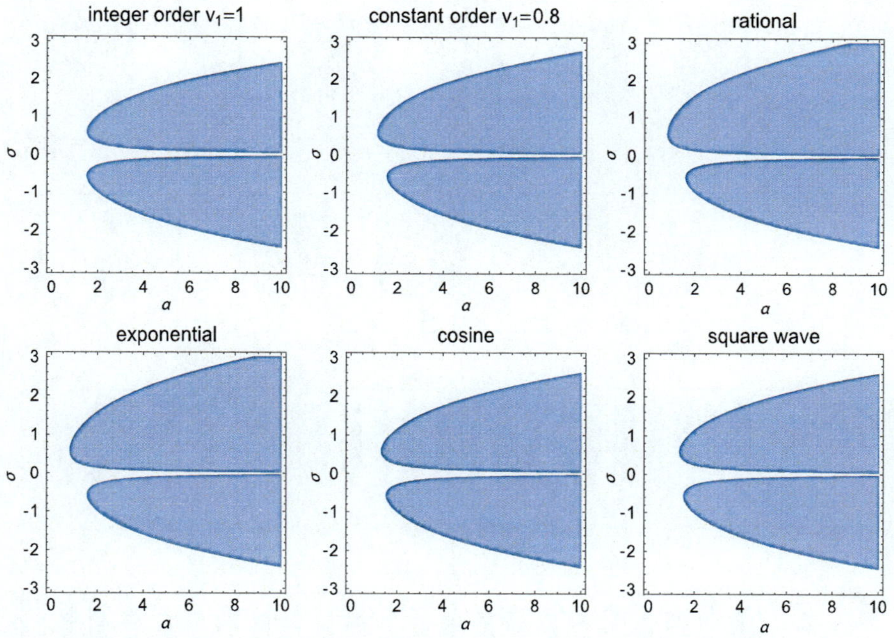

Fig. 2 Stability (white) and instability (blue) regions in the (α, σ)-plane for system (7) for order functions (a)–(f)

It is interesting to notice that the largest instability regions are obtained for the rational and exponential functions (c) and (d), respectively. Different types of bursting behavior is observed in Fig. 3 for $\alpha = 4.5$, $\sigma = -1$ belonging to the instability regions plotted in Fig. 2, in each of the cases (a)–(f) for the choice of the order function ν_1.

Compared to the integer-order Rulkov model (a), the fractional-order model with constant order $\nu_1 = 0.8$ (b) presents longer bursts followed by longer

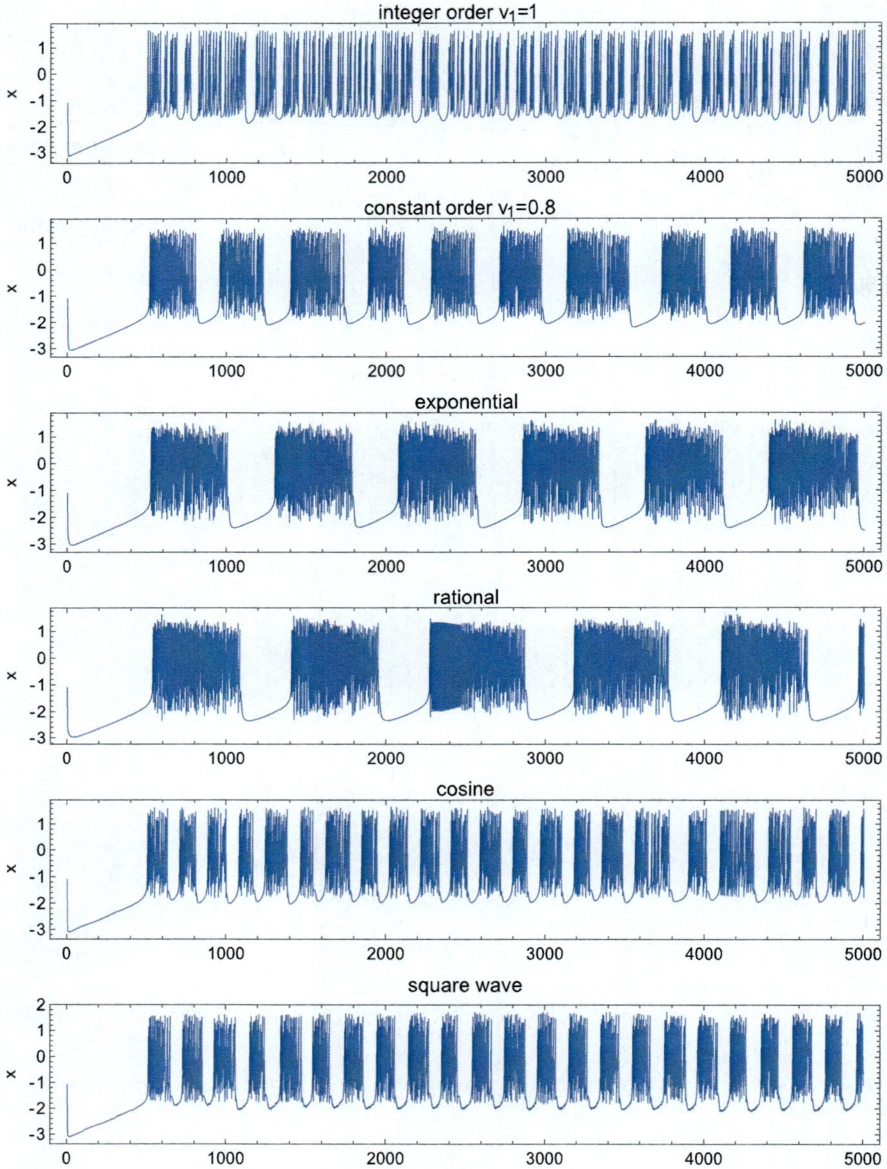

Fig. 3 Bursting behavior in the Rulkov model (7) with $\mu = 0.001$, $\alpha = 4.5$, $\sigma = -1$ and the fractional order functions (a)–(f)

recovery periods. Burst length and recovery period length are further increased when considering rational and exponential variable orders (c) and (d). Periodic variable orders (e) and (f) produce shorter and more frequent bursts.

5 Conclusions

Sufficient conditions are presented to characterise stability and instability properties of two-dimensional systems of variable-order fractional difference equations, which are further applied to a variable-order fractional Rulkov neuronal model. Numerical simulations show that chaotic bursting can be modulated by the order function associated to the Caputo variable-order fractional difference operator used to model the evolution of membrane potential.

Acknowledgments This article is based upon work from COST Action CA15225, a network supported by COST (European Cooperation in Science and Technology). The work was supported by Polish funds of National Science Center, granted on the basis of decision DEC-2016/23/B/ST7/03686.

References

1. Anastasio, T.J.: The fractional-order dynamics of brainstem vestibulo-oculomotor neurons. Biol. Cybern. **72**(1), 69–79 (1994). https://doi.org/10.1007/BF00206239
2. Brandibur, O., Kaslik, E.: Stability properties of a two-dimensional system involving one Caputo derivative and applications to the investigation of a fractional-order Morris-Lecar neuronal model. Nonlinear Dyn. **90**(4), 2371–2386 (2017). https://doi.org/10.1007/s11071-017-3809-2
3. Brandibur, O., Kaslik, E.: Stability of two-component incommensurate fractional-order systems and applications to the investigation of a Fitzhugh-Nagumo neuronal model. Math. Methods Appl. Sci. **41**(17), 7182–7194 (2018). https://doi.org/10.1002/mma.4768
4. Lundstrom, B.N., Higgs, M.H., Spain, W.J., Fairhall, A.L.: Fractional differentiation by neocortical pyramidal neurons. Nat. Neurosci. **11**(11), 1335–1342 (2008). https://doi.org/10.1038/nn.2212
5. Mozyrska, D., Ostalczyk, P.: Variable-fractional-order Grünwald-Letnikov backward difference selected properties. In: Proceedings of the 39th International Conference on Telecommunications and Signal Processing (2016). https://doi.org/10.1109/TSP.2016.7760959
6. Mozyrska, D., Wyrwas, M.: Explicit criteria for stability of fractional h-difference two-dimensional systems. Int. J. Dyn. Control **5**(1), 4–9 (2017). https://doi.org/10.1007/s40435-016-0239-9
7. Mozyrska, D., Wyrwas, M.: Stability by linear approximation and the relation between the stability of difference and differential fractional systems. Math. Methods Appl. Sci. **40**(11), 4080–4091 (2017). https://doi.org/10.1002/mma.4287
8. Mozyrska, D., Wyrwas, M.: Stability of linear systems with Caputo fractional-, variable-order difference operator of convolution type. In: 2018 41st International Conference on Telecommunications and Signal Processing (TSP). IEEE, Athens (2018). https://doi.org/10.1109/TSP.2018.8441360
9. Rulkov, N.F.: Regularization of synchronized chaotic bursts. Phys. Rev. Lett. **86**(1), 183–186 (2001). https://doi.org/10.1103/PhysRevLett.86.183
10. Rulkov, N.F.: Modeling of spiking-bursting neural behavior using two-dimensional map. Phys. Rev. E **65**(4), 041922 (2002). https://doi.org/10.1103/PhysRevE.65.041922
11. Weinberg, S.H.: Membrane capacitive memory alters spiking in neurons described by the fractional-order hodgkin-huxley model. PloS One **10**(5), e0126,629 (2015). https://doi.org/10.1371/journal.pone.0126629

Electrostatically Actuated Initially Curved Micro Beams: Analytical and Finite Element Modelling

Nadezhda Mozhgova, Alexey Lukin, Ivan Popov, and Dmitriy Indeitsev

Abstract Nowadays, industrial production microelectromechanical systems (MEMS) is developing rapidly, so they are widely used in various spheres of human activity: medicine, energy, various systems navigation in the automotive and petroleum industries, etc. Regardless of the purpose of the MEMS, sensitive elements commonly undergo an initial curvature imperfection, due to the microfabri cation process. Initial curvature imperfection significantly affects the mechanical behavior of microplates, beams, etc. For example, initially curved microbeams loaded by concentrated forces may exhibit bistability (the existence of two different stable equilibria under the same loading). The transition between two stable states in these structures is commonly referred to as a snap-through buckling. In the present article, the basic sensitive element of MEMS—an initially curved beam—was taken into consideration. Equilibria forms branching for various initial curvature and geometry parameters of sensitive element was investigated utilizing model order reduction technique (MOR) and numerical continuation methods. Finite element modeling of the above-mentioned problems of electroelasticity was carried out in the ANSYS software system and conclusions were drawn on the degree of applicability of FEM and ROM-FEM methods under various conditions.

Keywords MEMS · Reduced-order model · Bistable beam

N. Mozhgova (✉)
Peter the Great St. Petersburg Polytechnic University, Saint Petersburg, Russia
e-mail: mozhgova.nv@edu.spbstu.ru

A. Lukin · I. Popov · D. Indeitsev
Peter the Great St. Petersburg Polytechnic University, Saint Petersburg, Russia

Institute for Problems in Mechanical Engineering of the Russian Academy of Sciences (IPME RAS), Saint Petersburg, Russia

J. Awrejcewicz (ed.), *Perspectives in Dynamical Systems II: Mathematical and Numerical Approaches*, Springer Proceedings in Mathematics & Statistics 363,
https://doi.org/10.1007/978-3-030-77310-6_21

1 Introduction

The relevance of the paper is due to the intensive development of the industry of nano- and microsystems in application to the problems of modern high-tech instrumentation. The current level of development of nano- and microsystem technologies creates opportunities for the production and implementation in various fields of human activity of a wide range of nano- and microelectromechanical systems (NEMS and MEMS), often based on physical principles, unrealizable at the macro-scale level [1].

Microbeams are widely used in MEMS as a basic sensitive element in a various sensors [2]. Over the past decade, electrostatically actuated initially straight double-clamped micro beam became a kind of benchmark problem, which was intensively used for the evaluation of various analytical, numerical and experimental approaches. One of the distinguishing features of such a micro device is that it is loaded by an electrostatic force, which is a nonlinear function of the beam's deflections [3].

In contrast to straight beams, initially curved electrostatically actuated double-clamped beams combine both geometric mechanical nonlinearity and generic electrostatic softening nonlinearity. The behavior of curved bistable beams is well understood: in these structures, able to stay in two different configurations at the same loading, the transition between the two states is through the snap-through buckling mechanism [4]. If such a structure is loaded above some critical value, it may buckle so that its deflection suddenly increases.

In [5] authors are considered the asymmetric buckling of a shallow initially curved stress-free micro beam subjected to distributed nonlinear deflection-dependent electrostatic force is studied. And in [14] authors are demonstrate a flow velocity measurement technique based on snap-through detection of an electrostatically actuated, bistable micromechanical beam.

The purpose of our study is investigating the equilibria forms branching for various initial curvature and geometry parameters of sensitive element. We will look for direct numerical solutions using model order reduction technique (MOR) and numerical continuation method. After that we carry out finite element modeling of the above-mentioned problems of electroelasticity and analyze the degree of applicability of FEM and ROM-FEM methods under various conditions.

The rest of the paper is organized as follows. The nonlinear equation describing the nonlinear beams model with midplane stretching is derived in Sect. 2. The direct numerical solutions and their analysis are presented in Sect. 3. A discussion of the various finite element methods, some of the results obtained with their help and compare with analytics you can see in Sect. 4. Finally, the main conclusions are summarized in Sect. 5.

2 Problem Statement

We consider a flexible initially curved double clamped prismatic micro beam of length L having a rectangular cross-section of width b and thickness d as shown in Fig. 1. The beam is made of homogeneous isotropic linearly elastic material with Young's modulus E. The initial shape of the beam is described by the function $z_0(x) = h \bullet w_0(x)$, where h is the initial elevation of the beam's central point above it's ends, and $w_0(x)$ is a non-dimensional function such that $\max_{x\in[0,L]} |w_0(x)| = 1$. The beam is subjected to a distributed electrostatic force provided by an electrode located at a distance g_0 (the gap) from the beam and extended beyond its ends. Also, the potentials difference is applied on the ends of the beam. We assume that d $\ll$ L, and that the deflections are moderately large compared to beam's thickness, thus the midplane stretching is induced. In this case, the beam tends to stretch and undergo large deformation in response to large forcing. This induces tensile axial stress, thereby changing the stiffness of the beam in a nonlinear way that resembles a cubic effect. The induced axial stress couples the in-plane and out-of-plane motions of the beam. Midplane stretching affects microstructures, beams, plates, and diaphragms of clamped or near fixed edge conditions and it is by far the most significant source of geometric nonlinearity in MEMS.

Thus, our system is governed by the following equilibrium equations [3, 6]:

$$EI\left(z'''' - z_0''''\right) = \left(N + \frac{EA}{2L}\int_0^L \left(z'^2 - z_0'^2\right)dx\right)z'' + F, \tag{1}$$

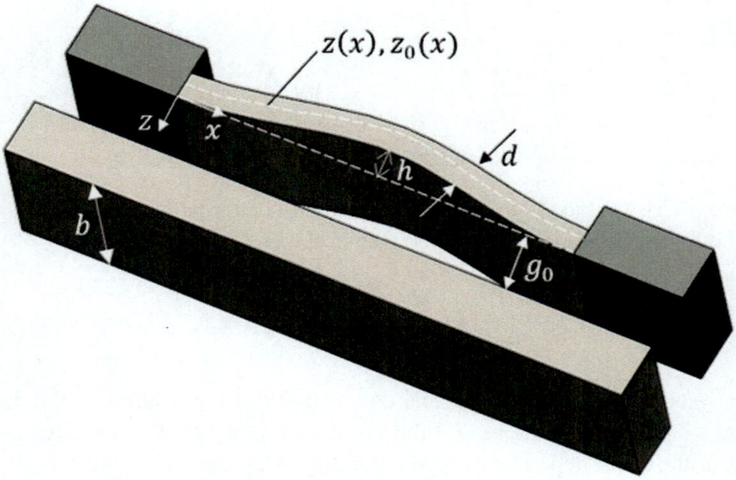

Fig. 1 Schematic of an electrostatically actuated microbeam with immovable edges and initial curvature

where $I = \frac{bd^3}{12}$—cross-section moment of inertia, $z(x)$—deflections function, N—axial force, $A = bd$—cross-section area, F—electrostatic force.

Electrostatic force in most models is described by the following expression:

$$F = -\frac{\lambda}{(g_0 + z)^2}, \tag{2}$$

where $\lambda = \frac{\varepsilon b[V_{DC}]^2}{2}$—dimensionless parameter of the electric field, ε—the relative permittivity of the medium in the space between the electrode and the beam, b—beam width, V_{DC}—DC voltage.

But if we consider work [7], it becomes clear that such an expression for the force is not quite accurate. Authors analyzed the approximations of the electrostatic field in the modeling of electrostatically controlled micro-and nanoelectromechanical systems. Using a uniformly suitable approximation [8] to estimate the strength of the electric field, the authors concluded that there are terms that act everywhere, not just at the boundary. Following this reasoning, the authors proposed a "corner-corrected theory" which does not include corrections acting only in the boundary layer, but includes those acting along the entire length of the sensing element:

$$F = -\frac{\lambda\left(1 + \delta^2 z'^2\right)}{(g_0 + z)^2} \tag{3}$$

where $\delta = \frac{g_0}{L}$—dimensionless parameter of edge effect of electric field.

So in this work we will use this expression for the electrostatic force. Thus Eq. (1) converted to:

$$EI\left(z'''' - z_0''''\right) = \left(N + \frac{EA}{2L}\int_0^L \left(z'^2 - z_0'^2\right) dx\right) z'' - \frac{\varepsilon b[V_{DC}]^2\left(1 + \delta^2 z'^2\right)}{2(g_0 + z)^2}. \tag{4}$$

Axial force N is produced by Joule effect, which occurs from current through the beam. The Fourier equation for temperature of the beam is as follows [9]:

$$-\frac{d}{dx}\left(k(T)\frac{dT}{dx}\right) = J^2 \rho_e(T), \tag{5}$$

where $k(T)$—is a coefficient of thermal conductivity of the beam material depends nonlinearly on the temperature, J—the current density, $\rho_e(T)$—resistivity of the beam material (linear dependence on temperature is assumed).

The current density can be expressed in terms of the voltage applied to the ends of the beam V_{th}:

$$J = \frac{V_{th}}{\rho_e(T)L}. \tag{6}$$

Then the Fourier law is rewritten in the following form:

$$-\frac{d}{dx}\left(k(T)\frac{dT}{dx}\right) = \frac{{V_{th}}^2}{\rho_e(T)L^2}. \tag{7}$$

From this equation, knowing V_{th}, the temperature at each point of the beam can found.

The axial force depends on the temperature as follows:

$$N = -\frac{EA}{l}\int_0^l \alpha(T)\,(T\,[x] - T_0)\,dx, \tag{8}$$

where $\alpha(T)$—a coefficient of thermal expansion, T_0—initial (room) temperature of the beam. The sign «−» is because axial force N is compressive, but in equation term of axial force is included as stretchable (with sign «+»). It must be taking into account when interpretating the results.

For electrostatic problems, it is convenient to normalize the deflection of the beam with respect to g_0. Hence, the following nondimensional variables (denoted by hats) are introduced:

$$\hat{z} = \frac{z}{g_0}, \quad \hat{x} = \frac{x}{L}. \tag{9}$$

Substituting nondimensional variables in equation of motion and dropping the hats from the dimensionless variables for convenience, the following nondimensional equation is derived:

$$\frac{\partial^4 z}{\partial x^4} - \frac{\partial^4 z_0}{\partial x^4} = \left[\alpha_1 \int_0^1 \left(z'^2 - {z_0'}^2\right) dx + N_{non}\right]\frac{\partial^2 z}{\partial x^2} - \frac{\alpha_2\left(V_{DC}^2 + \delta^2 z'^2\right)}{(1+z)^2} \tag{10}$$

The parameters appearing in equation above are defined as:

$$\alpha_1 = 6\left(\frac{g_0}{d}\right)^2, \quad \alpha_2 = \frac{6\varepsilon L^4}{Eb^3 d^3}, \quad N_{non} = \frac{12NL^2}{Ebd^3} \tag{11}$$

To derive a reduced-order model, we apply the Galerkin method, i.e. seek an approximate solution of the above system in the form of

$$z(x) = \phi_0(x) + \sum_{i=1}^{n} C_i \phi_i, \tag{12}$$

where $\phi_0(x) = 0$, because the boundary conditions of the beam are homogenous.

We choose ϕ_i, $i = 1, 2, \ldots$ to be the beam's linear orthonormal modeshapes by equation:

$$\phi_i^{IV} = \omega_{non,i}^2 \phi_i, \tag{13}$$

where $\omega_{non,i}^2$ is a natural frequency square.

Also, boundary conditions were considering:

$$\int_0^1 \phi_j \left(1 - \sum_{l=1}^{n} C_l \phi_l\right)^2 \left(\sum_{i=1}^{n} C_i \omega_{non,i}^2 \phi_i\right) dx$$

$$-\alpha_1 \int_0^1 \phi_j \left(1 - \sum_{l=1}^{n} C_l \phi_l\right)^2 \left\{\sum_{i=1}^{n} C_i \phi_i'' \int_0^1 \left(\sum_{k=1}^{n} C_k \phi_k'\right)^2\right\} dx + \alpha_2 V_{DC}^2 \int_0^1 \phi_j dx = 0$$

where ϕ_i and C_i is a modeshapes and coefficients for them, respectively.

Thus, we have integral-differential equation with respect to the coefficients C_i, where i is the number of modeshapes considered in the decomposition.

3 Direct Numerical Solutions

All the solutions given in this section were obtained with the help of the MATLAB software package, namely in MatCont. MatCont is a MATLAB-based software package developed under the supervision of W. Govaerts and Yu. A. Kuznetsov [10] for interactive numerical study of dynamical systems. This package has a number of capabilities (1) MatCont accesses MATLAB's ODE integrators and helps us integrate systems of ODEs without having to perform the actual MATLAB function calls. (2) It allows us to compute equilibrium solutions or fixed points to the system of ODEs and continue those equilibria with respect to a parameter of interest. And (3) it is able to detect Hopf bifurcations, branch points, saddle-node bifurcations (or fold bifurcations), etc. [11].

3.1 Initially Straight Beam

To begin with, a bifurcation diagram was constructed for the initial straight beam, i.e. $h = 0$. This was done for comparison with the known results, since this problem is widely covered in the scientific literature.

Geometric parameters of the beam and a few parameter of the model are presented in Table 1.

As you can see in Fig. 2, the character of bifurcation diagram is the same as in other papers, for example in [12].

Table 1 Parameters of the model

Beam's length	L	1000 μm
Beam's width	b	20 μm
Beam's thickness	d	2 μm
Gap	g_0	10 μm
Number of modeshapes	N_{static}	2

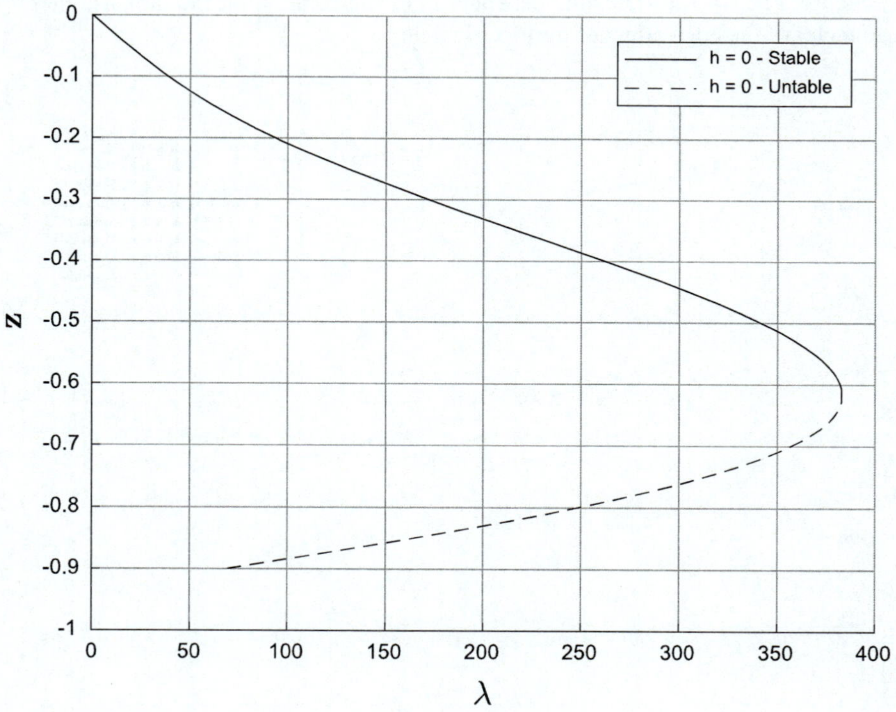

Fig. 2 Bifurcation diagram ($h = 0, \alpha_1 = 150, N_{non} = 0, \delta = 0$)

3.2 Initially Curved Beam and Choice of Geometric Parameters of the Beam

Next step in our work is analyzing dependence character of bifurcation diagram on the height of the initial deflection.

The curves shown in Fig. 3 built with the parameter value $\boldsymbol{\alpha_1} = \mathbf{150}$. The graph clearly shows that not all values of the amplitude the initial loss of the sensing element will exhibit the property of bistability (existence of two stable equilibrium positions at the same value of loading-parameter $\lambda = \frac{\varepsilon b[V_{DC}]^2}{2}$).

Then, at a fixed value of the amplitude of the initial loss, at which bistability is manifested, the parameter $\boldsymbol{\alpha_1}$ was varied.

The graph (Fig. 4) shows a significant dependence of the type of bifurcation charts from this parameter. For values $\alpha_1 < \alpha_1{}^*$ and given the amplitude of the initial deflection, the system does not show bistability.

Thus, to obtain a bistable system, it is necessary to carefully select the geometric parameters.

Figure 5 below illustrates the dependence of the solution on the dimensionless parameter of the edge effect of the electric field.

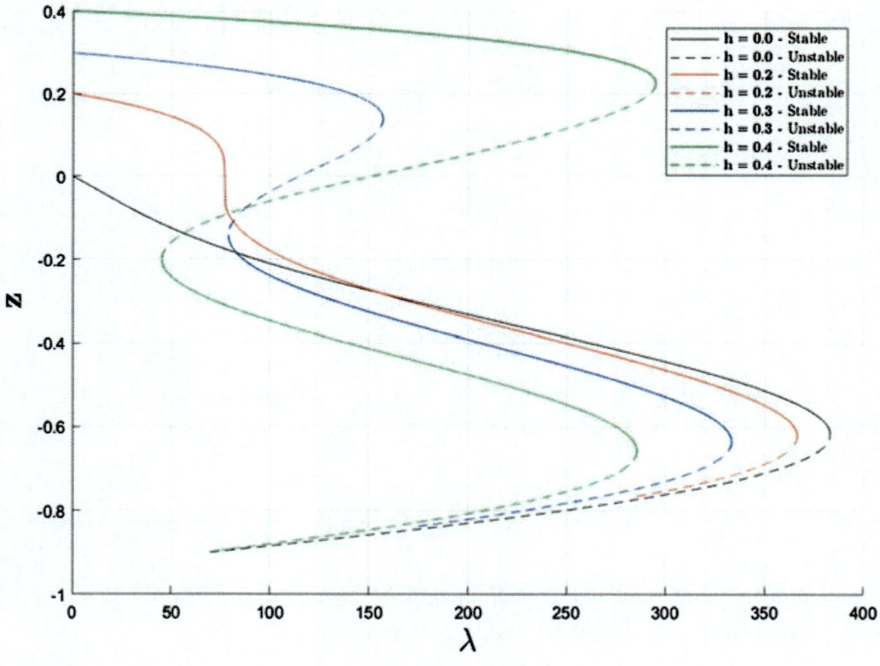

Fig. 3 Bifurcation diagram for different initial deflection ($\boldsymbol{\alpha_1} = \mathbf{150}, N_{non}, \delta = \mathbf{0}$)

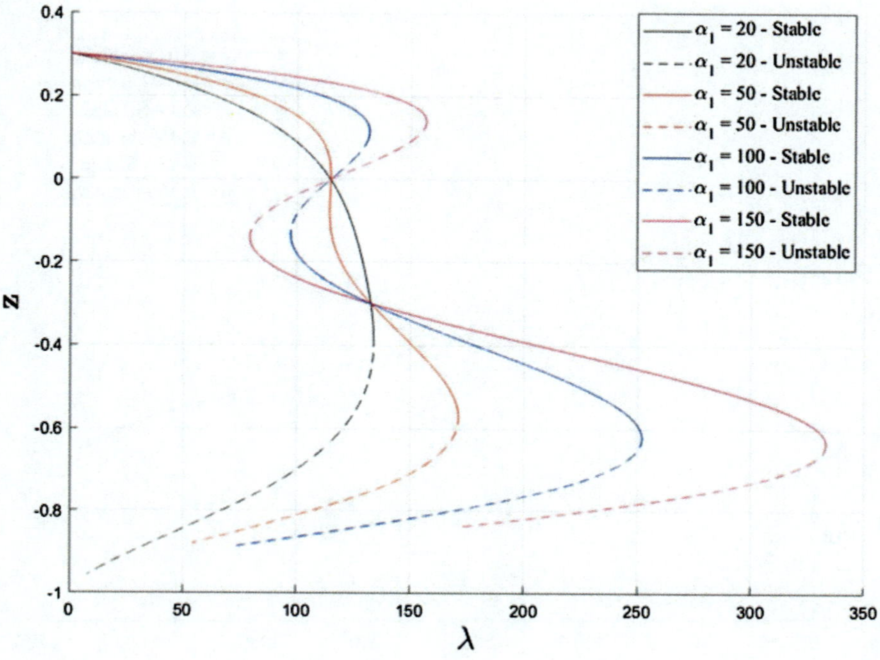

Fig. 4 Dependence on the parameter α_1 at large amplitude of initial deflection

This figure demonstrates that the influence of edge effect of electric field on this system is low, so further in this work it will be neglected.

The last step in this part is an investigation the behavior of system on dependence of axial force. All previous graphs were described cases when axial force is equal to zero. Now in Fig. 6 is shown the diagram of equilibria states in three-dimensional space of parameters—λ, z, N_{non}.

It may be noted that the impact of axial force on system's behavior is significant. With the increase of axial force, the critical value of electrostatic force's parameter, at which the stability of the system is lost, is increase too. And in states, which are close to unstable branch, the behavior of the systems with increase axial force is different unlike other stable states—they are turn to unstable equilibrium states. With decrease of axial force, the behavior of system is similar.

The following graph in Fig. 7 shows the same curve only in 2D.

In this figure the dependence of the initial elevation amplitude on the axial force parameter is clearly visible. With increase the axial force, the initial elevation amplitude is increases too, since the beam is stretched.

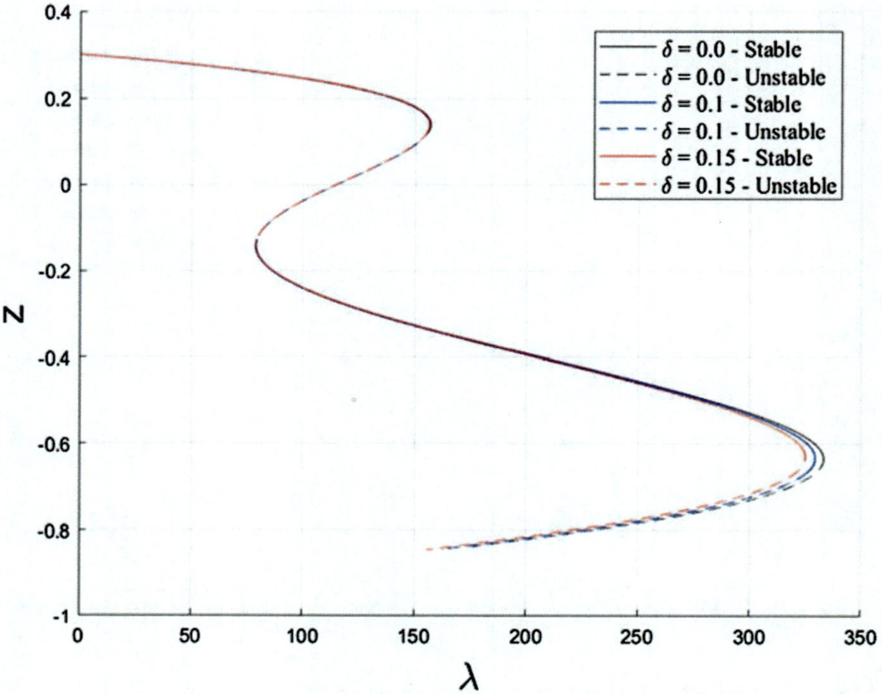

Fig. 5 Dependence on the parameter $\boldsymbol{\delta}$ at big amplitude of initial deflection

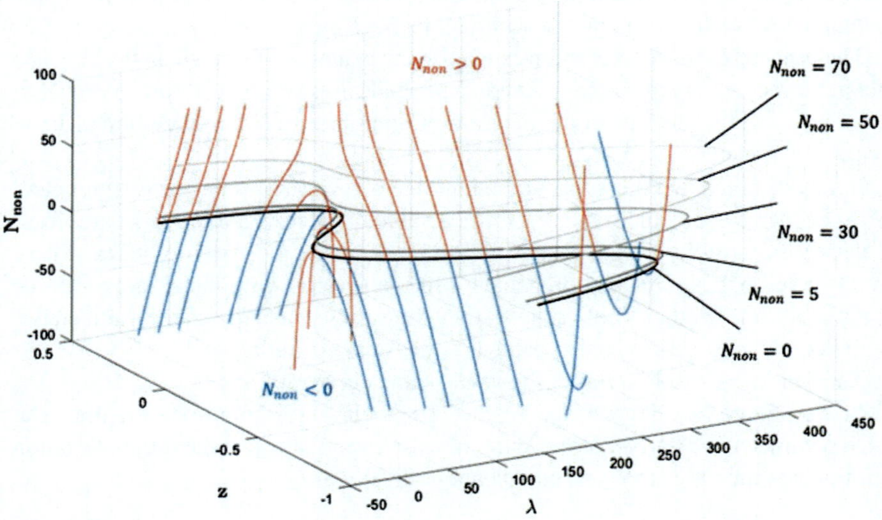

Fig. 6 Bifurcation diagram in three-dimensional space of parameters $\boldsymbol{\lambda}, z, N_{non}$

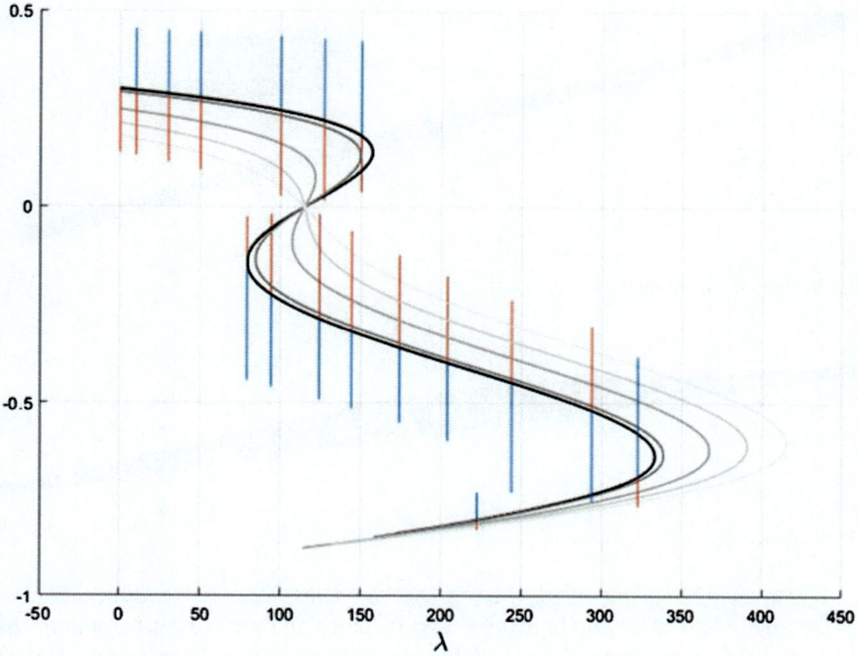

Fig. 7 Bifurcation diagram in two-dimensional space of parameters λ,z for different N_{non}

4 Finite Element Methods

Finite element (FE) modeling was carried out in the system of finite element analysis ANSYS. In this software package exists a few methods to solving coupled-field problems:

- Direct Coupled-Field Analysis
- Load Transfer Methods
  - Load Transfer Coupled Physics Analysis
  - Unidirectional Load Transfer

And ANSYS also offers the following additional coupled-field methods:

- Coupled Physics Circuit Simulation
- Reduced Order Modeling

The direct method usually involves just one analysis that uses a coupled-field element type containing all necessary degrees of freedom. Coupling is handled by calculating element matrices or element load vectors that contain all necessary terms.

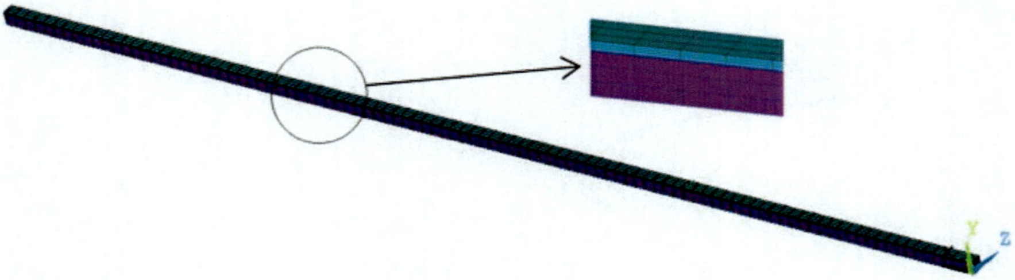

Fig. 8 Finite-element model

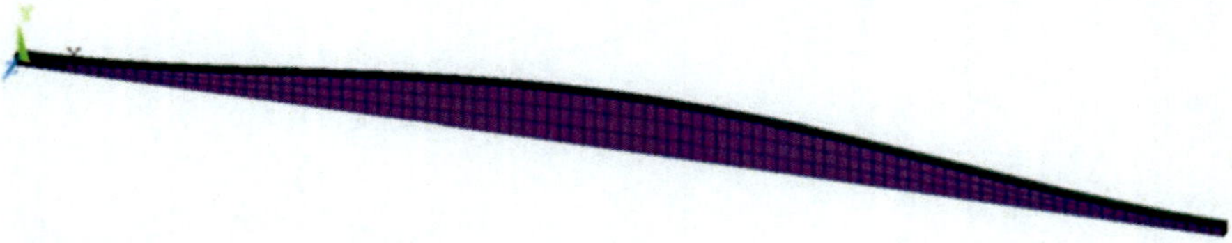

Fig. 9 Fundamental mode shape

The load transfer methods involve two or more analyses, each belonging to a different field. You couple the two fields by applying results from one analysis as loads in another analysis. There are different types of load transfer analyses.

Reduced Order Modeling describes a solution method for efficiently solving coupled-field problems involving flexible structures. The reduced order modeling (ROM) method is based on a modal representation of the structural response. The deformed structural domain is described by a factored sum of the mode shapes (eigenvectors). The resulting ROM is essentially an analytical expression for the response of a system to any arbitrary excitation. This methodology has been implemented for coupled electrostatic-structural analysis and is applicable to micro-electromechanical systems (MEMS) [13].

In this work we presented Direct Coupled-Field Analysis and comparing with results described in previous section. Figure 8 demonstrates the FE model of the system.

Geometrical parameters are given in Sect. 3. For mechanical structure we used SOLID185 element, for air gap—SOLID226,1001 element. Figure 9 demonstrates fundamental mode shape of the beam.

To begin with, we compare «analytical» result with finite-element in the absence of a prestressed state.

In this graph (Fig. 10), there is a fairly good coincidence of the results obtained by two different methods, especially at small deflections.

In Fig. 11, the coincidence of the results is also small, from which we can conclude that in general, the numerical scheme works correctly. At this stage, the study of this problem is not fully performed. As a further research we plan to study the ROM method in ANSYS, compare it with the ROM method in MATLAB for

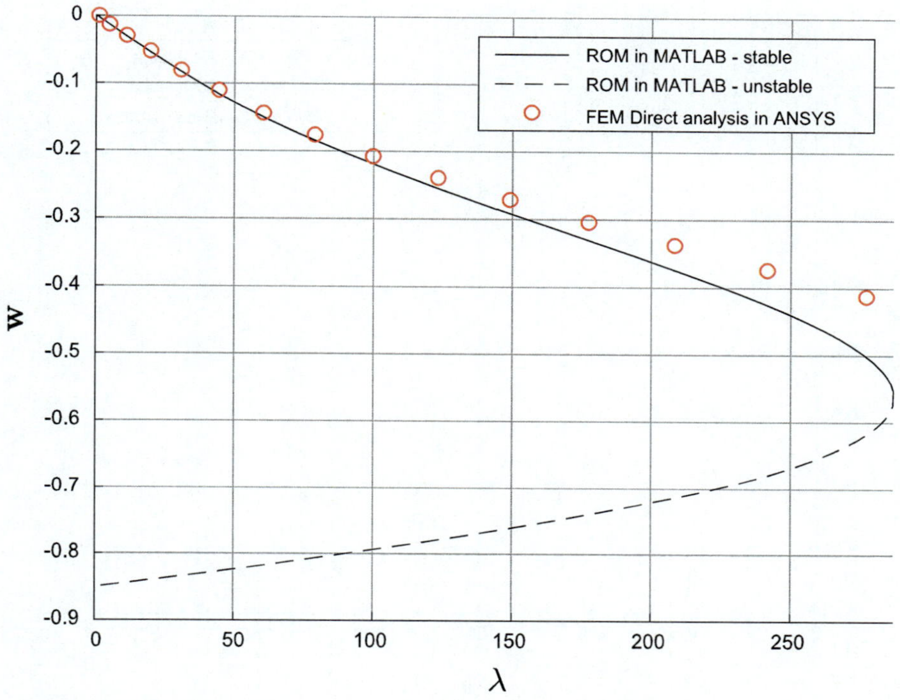

Fig. 10 Diagram without prestress ($\boldsymbol{h = 0}$, $\boldsymbol{\alpha_1 = 150}$, $\boldsymbol{P_{coeff} = 0}$, $\boldsymbol{\delta = 0}$)

both the initially straight beam and the initially imperfect one. It is also interesting to see the same dependencies in the case of the initial imperfection assignment not by the first proper form, but by the second.

5 Conclusions

In this work, buckling of an initially curved microbeam loaded by a nonlinear, configuration dependent, electrostatic force was analyzed. Direct numerical solutions were obtained by Galerkin decomposition with linear undamped eigenmodes of an associated straight beam as base functions. Bifurcation deflection-voltage diagrams were obtained for different values of the initial loss amplitude and geometric parameters. The parameters of the system at which it exhibits the bistability property were determined. In the ANSYS finite element analysis system, a Direct Coupled-Field Analysis for solving coupled-field problems for an initially straight beam was applied. The results are compared with those obtained by the ROM method in MATLAB. It was concluded that this method can be used with satisfaction accuracy only in the case of small deflection values.

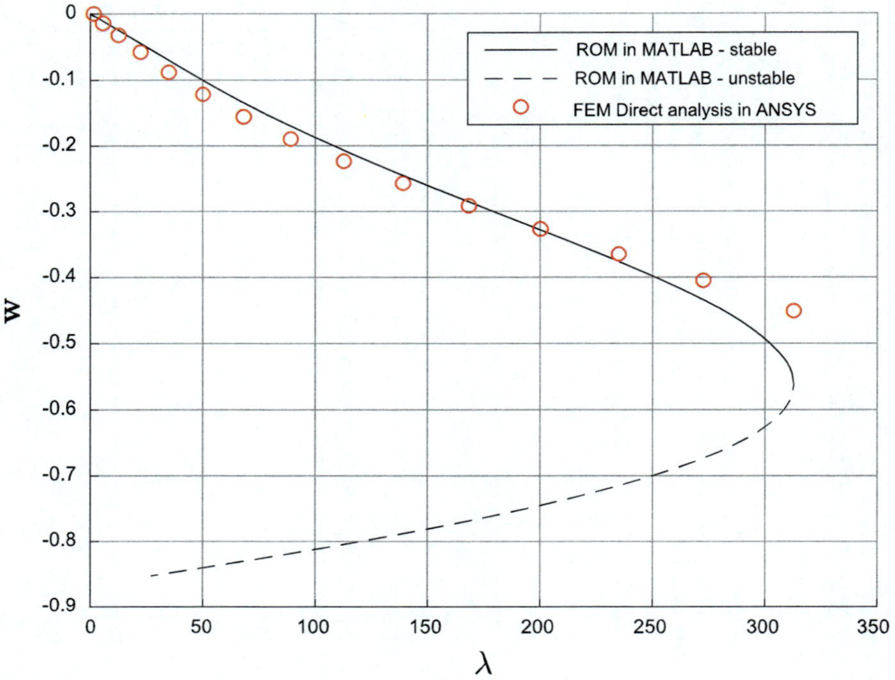

Fig. 11 Diagram with prestress ($\boldsymbol{h = 0, \alpha_1 = 150, P_{coeff} = -0.4, \delta = 0}$)

Acknowledgments The work was supported by RFBR grant 20-01-00537.

References

1. Bouchaala, A., Nayfeh, A.H., Younis, M.I.: The effect of an added mass on the frequency shifts of a clamped-clamped microbeam for bio-mass detection. In: ASME 2016 International Design Engineering Technical Conferences and Computers and Information in Engineering Conference. American Society of Mechanical Engineers Digital Collection (2016). https://doi.org/10.1115/DETC2016-59667
2. Kouravand, S.: Design and modeling of some sensing and actuating mechanisms for MEMS applications. Appl. Math. Model. **35**(10), 5173–5181 (2011). https://doi.org/10.1016/j.apm.2011.04.015
3. Medina, L., Gilat, R., Krylov, S.: Symmetry breaking in an initially curved micro beam loaded by a distributed electrostatic force. Int. J. Solids Struct. **49**(13), 1864–1876 (2012). https://doi.org/10.1016/j.ijsolstr.2012.03.040
4. Shojaeian, M., Beni, Y.T., Ataei, H.: Size-dependent snap-through and pull-in instabilities of initially curved pre-stressed electrostatic nano-bridges. J. Phys. D Appl. Phys. **49**(29), 295303 (2016). https://doi.org/10.1088/0022-3727/49/29/295303
5. Hajjaj, A.Z., Alcheikh, N., Younis, M.I.: The static and dynamic behavior of MEMS arch resonators near veering and the impact of initial shapes. Int. J. Non-Linear Mech. **95**, 277–286 (2017). https://doi.org/10.1016/j.ijnonlinmec.2017.07.002

6. Younis, M.I.: MEMS Linear and Nonlinear Statics and Dynamics, vol. 20. Springer Science & Business Media, New York (2011). https://doi.org/10.1007/978-1-4419-6020-7
7. Pelesko, J.A., Driscoll, T.A.: The effect of the small-aspect-ratio approximation on canonical electrostatic MEMS models. J. Eng. Math. **53**(3–4), 239–252 (2005). https://doi.org/10.1007/s10665-005-9013-2
8. Найфэ, А. Введение в методы возмущений (1984)
9. Hajjaj, A.Z., et al.: Electrothermally tunable arch resonator. J. Microelectromech. Syst. **26**(4), 837–845 (2017). https://doi.org/10.1109/JMEMS.2017.2676006
10. Dhooge, A., Govaerts, W., Kuznetsov, Y.A.: MATCONT: a MATLAB package for numerical bifurcation analysis of ODEs. ACM Trans. Math. Softw. **29**(2), 141–164 (2003). https://doi.org/10.1145/779359.779362
11. Holmes, W.R., Mata, M.A., Edelstein-Keshet, L.: Local perturbation analysis in MatCont. User's Guide & Tutorials, Oct (2014)
12. Medina, L., et al.: Experimental investigation of the snap-through buckling of electrostatically actuated initially curved pre-stressed micro beams. Sens. Actuators A Phys. **220**, 323–332 (2014). https://doi.org/10.1016/j.sna.2014.10.016
13. ANSYS: ANSYS 7.1—Coupled-Field Analysis Guide. ANSYS, Inc., Canonsburg, PA (2002)
14. Kessler, Y., et al.: Flow sensor based on the snap-through detection of a curved micromechanical beam. J. Microelectromech. Syst. **27**(6), 945–947 (2018). https://doi.org/10.1109/JMEMS.2018.2868776

Numerical and Analytical Investigation of Chatter Suppression by Parametric Excitation

Fadi Dohnal, Wolfgang Alois Hörtnagel, and Mariusz Zamojski

Abstract A concept for increasing process stability during milling is presented utilizing the time-periodic modulation of the tool support. A simple time-delayed system describing the effect of regenerative chatter is enhanced by a time-periodic variation of the support. Such a system leads to entirely new dynamics. Numerical results of stability charts are discussed in terms of spindle speed and cut depth and show classic chatter lobes that are modified by the parametric excitation. This kind of parametric excitation is more general than the one occurring for varying spindle speed because its frequency is independent of the cutting frequency of the tool and therefore independent of the spindle speed and number of teeth. First analytical approximations on the stability of the modified lobes are benchmarked against numerical predictions. This study is a preparation for experimental tests.

Keywords Chatter · Stability · Time-periodic system

1 Introduction

Machine tool vibrations affect the wear, tool life and surface quality [9] leading to an increase of production cost and time. A simple model for chatter is the regenerative effect that is summarised in [6, 7]. Chatter occurs typically within instability lobes in the spindle speed diagram. Tools and methods for influencing (shift and distortion) these lobes are discussed in detail in the pioneering work [2]. Several countermeasures can be derived like tuning the support and tool stiffness, the cutting feed, the spindle speed, the geometry of the cutter profile and the number of teeth. The so-called process damping helps also and is always present in real

F. Dohnal (✉) · W. A. Hörtnagel · M. Zamojski
UMIT – Private University for Health Sciences, Medical Informatics and Technology, Division for Mechatronics Lienz, Lienz, Austria
e-mail: fadi.dohnal@umit.at
https://www.umit.at/melz

J. Awrejcewicz (ed.), *Perspectives in Dynamical Systems II: Mathematical and Numerical Approaches*, Springer Proceedings in Mathematics & Statistics 363,
https://doi.org/10.1007/978-3-030-77310-6_22

machines. Another possibility is attaching a passive linear or nonlinear vibration absorber to the cutting tool as introduced recently in [5]. A semi-active mean was proposed in [1] in which the bearing stiffness of the spindle was modulated time-periodically showing an increase in damping. Active means in this context employ piezoelectric actuators mounted on the workpiece directly, see for example [3].

All these measures have benefits and drawbacks and a successful implementation depends strongly on parameters like machine throughput and complexity of the cut which directly translates into cost and time. The present contribution addresses the regenerative effect in metal cutting as defined in [7] but extends the system with a time-harmonic modulation of the support stiffness of the cutting tool. The reason for this is motivated by the observation of parametric anti-resonances in [4].

2 Regenerative Effect

The simplest mechanical model for regenerative chatter is shown in Fig. 1. The equations of motion of this single degree-of-freedom model of a cutting tool in turning machinery is given by

$$m\ddot{x} + c\dot{x} + kx = F_x \tag{1}$$

where m is the mass, c the damping coefficient and k the stiffness coefficient of the cutting tool. The workpiece is assumed to be rigid. F_x is the component of the cutting force in the cutting direction. It can be expressed by the empirical power law [2]

$$F_x(t) = K_x\, w\, h^r(t) \tag{2}$$

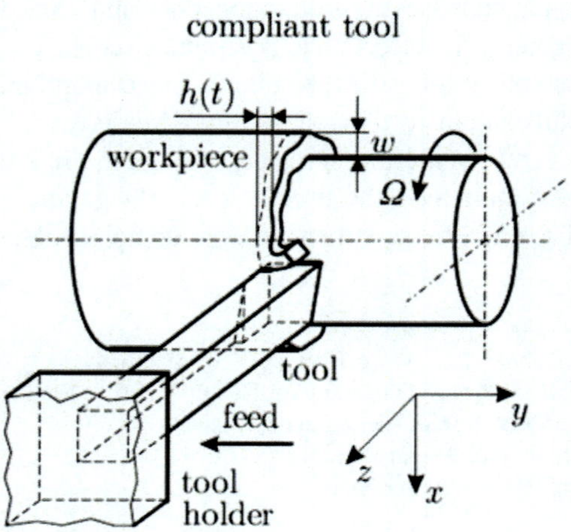

Fig. 1 Mechanical system according to [6]

where w is the chip depth, K_x is the cutting force coefficient, r the cutting force exponent and $h(t)$ the time-dependent chip thickness

$$h(t) = v_f \tau + x(t-\tau) - x(t) \tag{3}$$

where v_f is the velocity of the feeding tool and τ is the period of rotation $(2\pi/\Omega)$. Choosing $r = 3/4$ and expanding the cutting force F_x and the tool displacement x into its Taylor series around $h_0 = v_f \tau$ yields the linearized, delayed equations of motion in x with constant coefficients (see [5, 7] for more details)

$$m\ddot{x} + c\dot{x} + kx = k_1 \left(x(t-\tau) - x(t)\right) \tag{4}$$

or

$$\ddot{x} + 2\zeta\omega_n\dot{x} + \omega_n^2 x = \frac{k_1}{m} \left(x(t-\tau) - x(t)\right) \tag{5}$$

where $\zeta = c/(2\sqrt{km})$, $\omega_n = \sqrt{k/m}$ and $k_1 = 3/4\, K_x\, w\, h_0^{-1/4}$. Analytical stability conditions can be derived for predicting the onset of unstable vibrations (instability lobes) as described in more detail in [7]. Inserting the ansatz $x = A\exp(\lambda t)$ yields a complex-valued characteristic equation. At the stability limit curves we have a purely imaginary eigenvalue of the form $\lambda = i\omega$ which gives

$$-\omega^2 + \omega_n^2 + \frac{k_1}{m}\big(1 - \cos(\omega\tau)\big) = 0, \qquad 2\zeta\omega_n\omega + \frac{k_1}{m}\sin(\omega\tau) = 0 \tag{6}$$

These equations can be transformed to [5, 7]

$$k_{1,cr} = \frac{m}{2}\frac{(\omega-\omega_n)^2 + (2\zeta\omega_n\omega)^2}{\omega^2 - \omega_n^2}, \qquad n_{cr} = \frac{30\omega}{j\pi - \arctan\left(\dfrac{\omega^2-\omega_n^2}{2\zeta\omega_n\omega}\right)} \qquad j = 1, 2, \ldots \tag{7}$$

defining the values of the cutting force parameter $k_{1,cr}$ and the tool speed n_{cr} at the stability boundary.

System parameters for an example system are chosen from [7] and are listed in Table 1. The direct numerical integration of the equations of motion in Eq. (4) in the parameter space n-k_1 is shown in Fig. 2. Green dots indicate a stable and red dots an unstable system response. The analytically predicted stability limits in Eq. (7) fit perfectly.

Table 1 System parameters for example system taken from [7]

Symbol	Value
m	347 kg
k	$97 \cdot 10^6$ N/m
c	9173 Ns/m

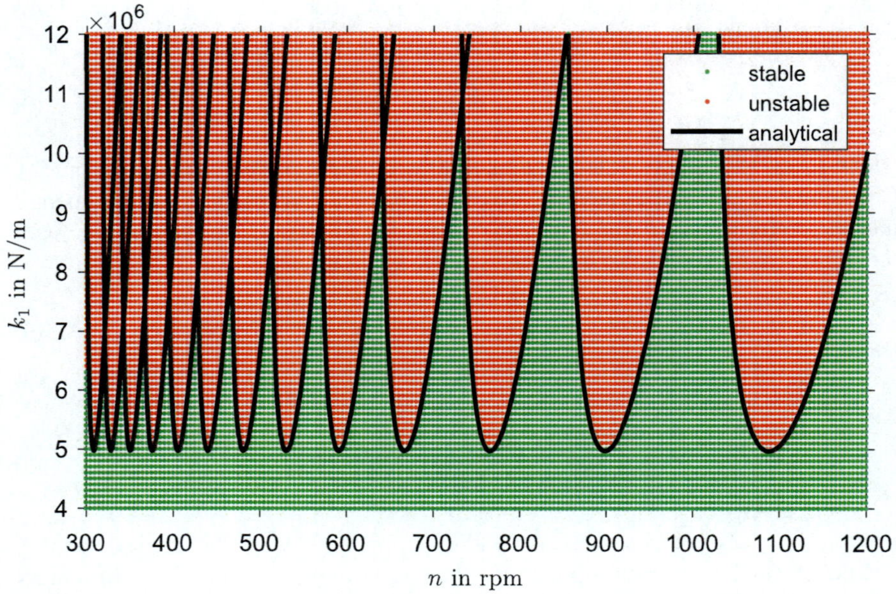

Fig. 2 Stability chart for chatter tool with constant support characteristic: direct numerical simulation in comparison with analytical prediction in Eq. (7)

3 Regenerative Effect at Time-Harmonic Modulation of the Tool Stiffness

We add in the system in Eq. (4) a time-periodic modulation of the tool support stiffness

$$m\ddot{x}(t) + c\dot{x}(t) + k\big(1 + \varepsilon \sin(\Omega_{PE}t)\big)\, x(t) = k_1\big(x(t-\tau) - x(t)\big) \tag{8}$$

This is a generalisation of the commonly studied delayed differential equation with parametric excitation because we assume that $\Omega_{PE} \neq 60/\tau$, i.e. the frequency of the parametric excitation is *not* a multiple of the spindle speed. A parametric excitation introduces in general a modulation of the system response that leads to side-bands in the frequency spectra, see e.g. [4]. The focus of our investigation lies on the stability boundary in the parameter space, more specifically the distortion of the stability boundary in Fig. 2 by the newly introduced parameters ε and Ω_{PE} in Eq. (8). The stability boundaries at $\varepsilon = 5\%$ and $\varepsilon = 10\%$ at an arbitrarily chosen parametric excitation frequency $\Omega_{PE} = 40$ rad/s are shown in Fig. 4. The stability boundaries show a shift towards higher values of k_1 (larger cutting depth) in the vicinity of 570 rpm. The chosen value of Ω_{PE} is indicated by the vertical solid line n_{PE} while the beneficial region lies at n_{opt}.

Two exemplary time histories are shown in Fig. 3 together with the corresponding frequency content in the steady-state region. This comparison confirms the frequency modulation induced by parametric excitation at $\omega \mp \Omega_{PE}$. Increasing the strength of the time-periodicity by increasing ε also leads to sidebands at $\omega \mp j\Omega_{PE}$ for $j = 1, 2, \ldots$. The analytical stability boundary is approximated by applying the method of harmonic balance [8]. The observation above allows for the following ansatz

$$x(t) = c_0 e^{i\omega t} + c_{1p} e^{i(\omega+\Omega_{PE})t} + c_{1m} e^{i(\omega-\Omega_{PE})t} + \text{complex conjugate} + \mathcal{O}(\varepsilon^2) \quad (9)$$

For achieving convergence we assume that c_0 is of order 1 and $c_{1p,1m}$ of order ε. Inserting Eq. (9) into Eq. (4) and collecting coefficients of the exponential functions $\mp i\omega t, \mp i(\omega \mp \Omega_{PE})t$ yields

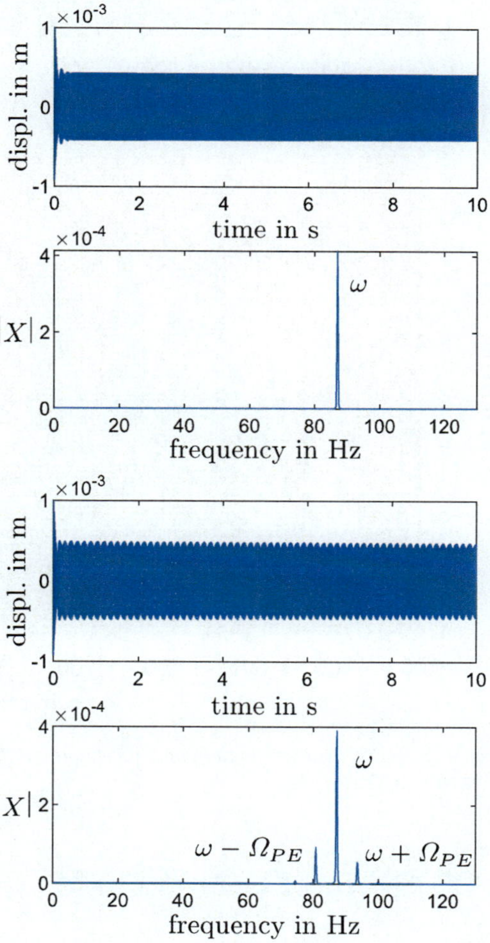

Fig. 3 Time histories of chatter tool at 600 rpm at stability boundary: (left plots) constant tool stiffness ($\varepsilon = 0\%$) at $k_1 = 5.24 \cdot 10^6$ N/m, (right plots) time-periodic tool support ($\varepsilon = 5\%$) at $k_1 = 5.75 \cdot 10^6$ N/m and $\Omega_{PE} = 40$ rad/s

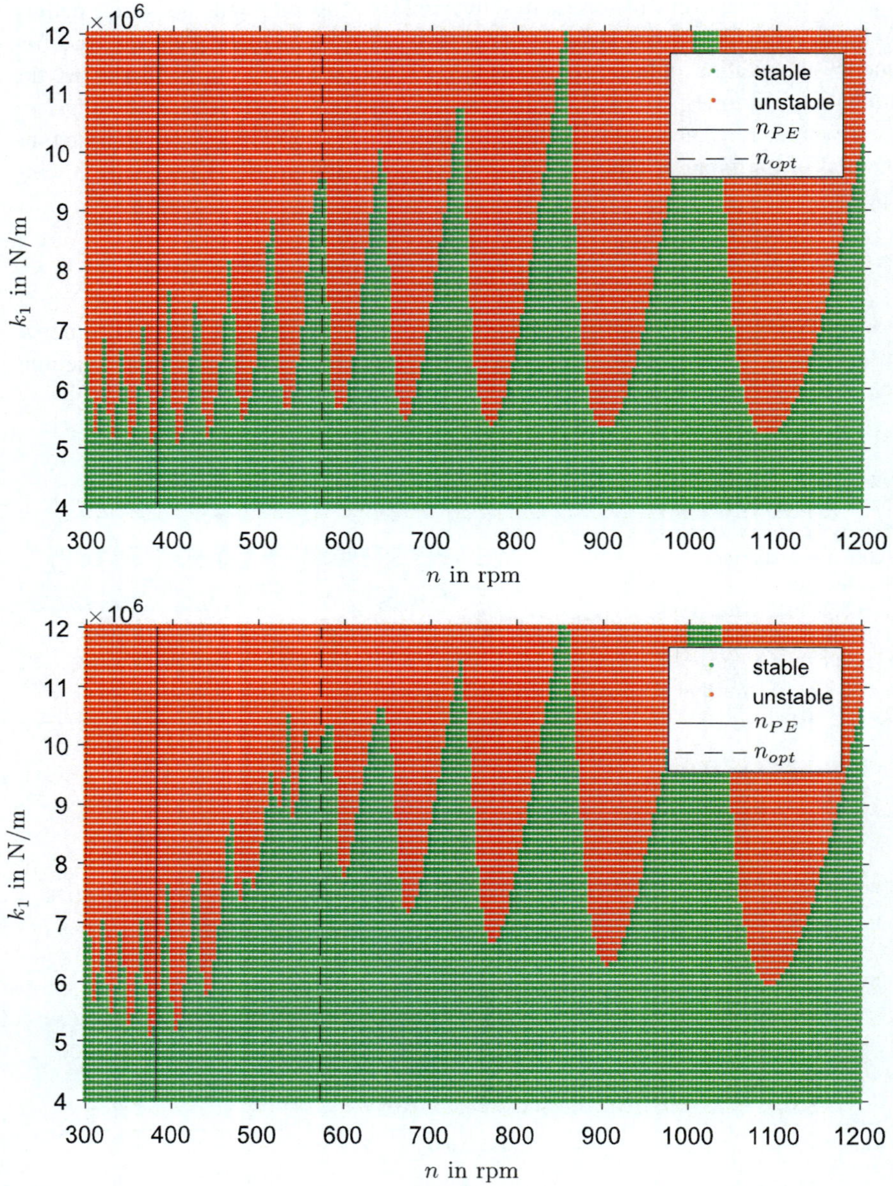

Fig. 4 Stability chart for chatter tool with support stiffness modulated at 40 rad/s: (top) $\varepsilon = 5\%$, (bottom) $\varepsilon = 10\%$

$$\begin{bmatrix} g(\omega) & \varepsilon k/2 & \varepsilon k/2 \\ \varepsilon k/2 & g(\omega+\Omega_{PE}) & 0 \\ \varepsilon k/2 & 0 & g(\omega-\Omega_{PE}) \end{bmatrix} \begin{bmatrix} c_0 \\ c_{1p} \\ c_{1m} \end{bmatrix} = \mathbf{0} \tag{10}$$

with the abbreviation

$$g(\omega) = k - m\omega^2 + i\omega c + k_1\big(1 - e^{i\omega\tau}\big) \tag{11}$$

An equivalent set of equations is obtained for the complex conjugate coefficients $\overline{c}_i$ which is omitted here. The corresponding characteristic equation for a non-trivial solution in Eq. (10) reads

$$4g(\omega)\, g(\omega+\Omega_{PE})\, g(\omega-\Omega_{PE}) - \varepsilon^2 k^2 \Big(g(\omega+\Omega_{PE}) + g(\omega-\Omega_{PE})\Big) = 0 \tag{12}$$

The numerical evaluation of this stability boundary matches well with the pointwise numerical time integration of the system in Fig. 4. However, the expressions are cumbersome and further simplifications are needed.

For the numerical values in Table 1 and Fig. 4, the parametric excitation frequency is close to the natural frequency of the system which justifies a Taylor expansion of the form

$$\bar{g}\left(1 \pm \frac{\Omega_{PE}}{\omega}\right) = \bar{g}(1) \pm \Big(-2m\omega + i\big(c - k_1 e^{i\omega\tau}\big)\Big)\Omega_{PE} + \mathcal{O}\left(\frac{\Omega_{PE}}{\omega}\right) \tag{13}$$

which approximates Eq. (12) to

$$g(\omega)^2 - \Big(-2m\omega + i\big(c - k_1 e^{i\omega\tau}\big)\Big)^2 \Omega_{PE}^2 - \varepsilon^2 k^2/2 \approx 0 \tag{14}$$

This equation can be rearranged to a quadratic polynomial in k_1

$$a_2(\omega\tau, \Omega_{PE}^2)\, k_1^2 + a_1(m, c, k, \omega, \tau, \Omega_{PE}^2)\, k_1 + a_0(m, c, k, \omega, \Omega_{PE}^2) - \varepsilon^2 k^2/2 \approx 0 \tag{15}$$

with coefficient functions $a_i(\cdot)$. Evaluation of Eq. (15) for the system parameters chosen in Table 1 and $\Omega_{PE} = 40$ rad/s and 570 rpm gives a lower limit of $k_{1,cr} \approx 9 \cdot 10^6$ N/m for both values of ε. This fits well to the stability boundary curves shown in Fig. 4 close to the speed n_{opt}. Finally, the approximate relation in Eq. (14) can be evaluated for finding the necessary parametric excitation frequency Ω_{PE} at given system and operation parameters.

4 Conclusions

The mitigation of regenerative chatter using a time-periodic support of the tool was investigated. The model equations and analytical stability limit curves of the classical regenerative chatter model are revisited and extended to a delayed and parametrically excited equation of motion. The frequency of parametric excitation in chatter vibrations is usually assumed to occur at a multiple of the spindle speed, depending on the number of tool teeth. In the present work we deliberately introduce a parametric excitation frequency ω_{PE} which is independent of the tool speed. This first study shows that such a time-modulation is capable of distorting the stability limit curves and creating large regions of larger cutting depths for certain speed intervals. Further investigations are needed for improving the quality of the analytical prediction and for experimental validation of the benefit of the proposed concept.

References

1. Abele, E., Dohnal, F., Feulner, M., Sielaff, T., Daume, C.: Numerical investigation of chatter suppression via parametric anti-resonance in a motorized spindle unit during milling. Prod. Eng. **12**, 309–317 (2018)
2. Altintas, Y.: Manufacturing Automation: Metal Cutting Mechancis, Machine Tool Vibrations, and CNC Design. Cambridge University Press, Cambridge (2012)
3. Cao, H., Zhang, X., Chen, X.: The concept and progress of intelligent spindles: a review. Int. J. Mach. Tools Manuf. **112**, 21–52 (2017)
4. Dohnal, F.: A contribution to the mitigation of transient vibrations, Parametric anti-resonance: theory, experiment and interpretation. Habilitation Thesis, Technical University Darmstadt (2012)
5. Habib, G., Kerschen, G., Stepan, G.: Chatter mitigation using the nonlnear tuned vibration absorber. In: KU Leuven, D.W. (ed.) Proceedings of ISMA 2016 - International Conference on Noise and Vibration Engineering and USD2016 - International Conference on Uncertainty in Structural Dynamics, Leuven, pp. 3671–3685 (2016)
6. Insperger, T., Lehotzky, D., Stepan, G.: Regenerative delay, parametric forcing and machine tool chatter: a review. In: Proceedings of 12th IFAC Workshop on Time Delay Systems, Ann Arbor, pp. 322–327 (2015)
7. Stepan, G.: Modelling nonlinear regenerative effects in metal cutting. Philos. Trans. R. Soc. Lond. A Math. Phys. Eng. Sci. **359**, 739–757 (2001)
8. Thomsen, J.J.: Vibration and Stability: Advanced Theory, Analysis and Toolss, 2nd edn. Springer, Berlin (2003)
9. Tlusty, J., Spacek, L.: Self-excited vibrations on machine tools (in Czech). Nakl CSAV, Prague (1954)

Nonlinear Study of a Pneumatic Artificial Muscle (PAM) Under Superharmonic Resonance Condition Using Method of Multiple Scales

Bhaben Kalita and Santosha K. Dwivedy

Abstract In this work, the nonlinear behaviour exhibit in the Pneumatic Artificial Muscle (PAM) has been studied. For the analysis, a single degree of freedom system is considered where the nonlinear Pneumatic Artificial Muscle (PAM) is attached with an external spring to provide additional support the system. The nonlinear equation of motion is solved with the help of the method of multiple scales to find out the reduced equations for superharmonic resonance condition. The dynamic stability and bifurcation of the system have been studied from the reduced equations. The frequency responses have been plotted to understand the effect of the different parameters on the system amplitude. Basin of attraction also have been plotted to verify the frequency plots. Finally, with the help of this work, the designers and researchers working in this field will get an idea to know about the safe range of various system parameters to operate for different applications of PAMs.

Keywords Pneumatic artificial muscle · Method of multiple scales · Superharmonic resonance condition

1 Introduction

Pneumatic Artificial Muscle (PAM) is an actuator which converts the pneumatic force obtains from the air pressure to a pulling force. The PAMs have significant advantages over the traditional pneumatic actuator like lower weight, high force, easy to install, absence of mechanical wear, soft and flexible in nature, low cost and safe human interaction. Therefore, the artificial muscles are widely used in the field of medical and robotics because of their ability to produce linear forces and displacement with the help of a simple mechanism [1–3]. However, due to the presence of compressibility of air and natural properties of viscoelastic

B. Kalita (✉) · S. K. Dwivedy
Mechanical Engineering Department, Indian Institute of Technology Guwahati, Guwahati, India
e-mail: k.bhaben@iitg.ac.in; dwivedy@iitg.ac.in

J. Awrejcewicz (ed.), *Perspectives in Dynamical Systems II: Mathematical and Numerical Approaches*, Springer Proceedings in Mathematics & Statistics 363,
https://doi.org/10.1007/978-3-030-77310-6_23

material, a high nonlinear characteristic is found in the PAM which makes them difficult to model and control. These qualities and drawbacks make the PAM as an attractive topic for many researchers and industry to study the behavior in various environments.

The PAM was first introduced by the physician, Joseph L. McKibben in 1950s which was used in artificial hand for the polio patients [1]. This McKibben muscle actuator was consists of a rubber tube (bladder) which was covered by a braided mesh shell and two ends were closed. The one end of the muscle have been connected to the air inlet and other connected with the load. Later, in 1980s, a more powerful PAM was developed by the Bridgestone Company for various medical applications as well as service and industrial robotics [4]. These type of PAM caused the hysteresis behavior due to the friction produced by the braided mesh shell for variation of air pressure. So to avoid this drawbacks, a number of modified PAMs have been introduced for various applications in the different field of advanced robotics and rehabilitation [3]. Nowadays, company like Bridgestone Co. and Festo AG modified the traditional McKibben artificial muscle which is widely used because of its structural advantages. These type of PAMs contain a flexible hose with non-elastic fibers organized in rhomboidal structure which results in a 3D grid pattern. The grid pattern will deformed when the air pressure is applied to the PAM and results a pulling force in axial direction [4, 5].

Various models have been proposed in the literature to understand the nonlinear behavior exhibits in the PAM. Chou and Hannaford [1] along with Tondu and Lopez [6] described the models based on the virtual work principle of an infinitely thin inner tube and continuously cylindrical shape. An experimental model is derived by Li et al. [7] which establish a relation between the operating air pressures, muscle force along with the contraction of the PAM. Furthermore, the other major parameters of the PAM like putting force, material properties, length and diameter are also added more nonlinearity on the dynamic behavior of the muscle. Along with this, the inter relation between the parameters are vary from one PAM to another PAM due to these nonlinear parameters. This is explained by Kalita and Dwivedy [8] with numerical model of the muscle dynamics when the natural frequency of the system is nearly equal to the external excitation frequency of the system. In another work [9], they studied the dynamics of the PAM when the natural frequency of the system is nearly twice the external excitation frequency of the system. The authors studied the effect of various parameters by adding a cubic nonlinearity to the muscle force. PAMs generally operated with help of the antagonistic arrangement for various applications to obtain the required actuation [10–12]. The dynamic model of the antagonistic arrangement of the PAMs has been explained by Tóthová and Pitel [11] with the help of an advanced geometric muscle model. In another work, Balara and Tóthová [12] described the static and dynamic properties of the nonlinear characteristics of various parameters of the PAM.

The dynamic model of the PAM is very difficult for practical control and use, because of the presence of high nonlinearity. In this study, the new type of PAM developed by Kalita and Dwivedy [13] is used for the analysis purpose where the system is supposed to be a single degree of freedom system. An external spring is attached to the system in an antagonistic arrangement to provide support to the system with an external load is applied. A second order nonlinear governing equation can be obtained which is solved by using the method of multiple scales. The obtained reduced equations are used to find out the response and stability of the system by plotting the time and frequency response curves. The safe range of the various system parameters to operate can also found out with the help these response plots which can be used by the researchers and designers to know about the dynamics of the artificial muscles. The mathematical modeling along with approximate analytical solutions have been discussed in the following section.

2 Mathematical Modeling

Figure 1a shows the experimental setup for the antagonistic arrangement of the PAM and the spring to achieve the actuation of the muscle for lifting a load to a particular position. The schematic diagram of such arrangement has been depict in Fig. 1b to understand the actuation of the PAM.

In Fig. 2a, the system has been modeled as a single degree of freedom system and external force $F \sin \omega t$ is applied to the system. This type of system has been used in many medical as well as in industrial robotic applications. Figure 2b shows

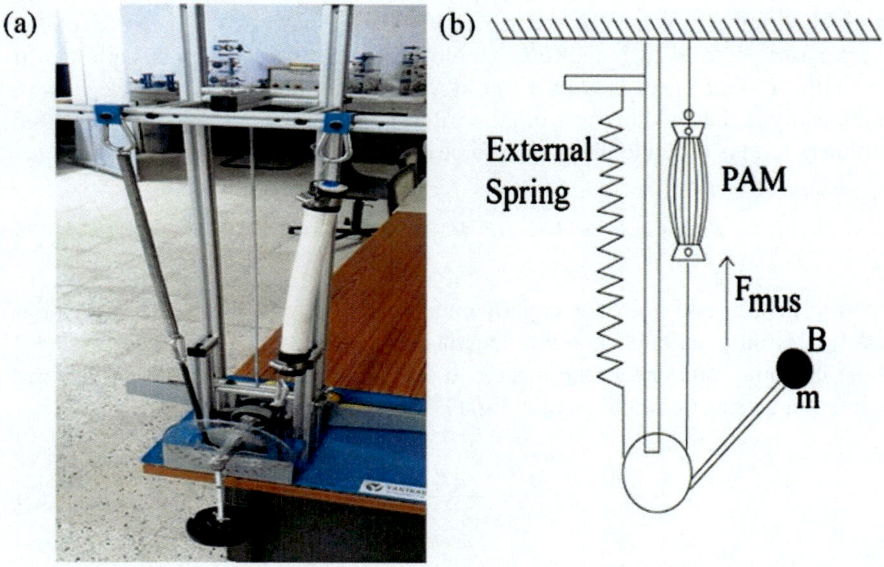

Fig. 1 (**a**) Experimental setup with PAM and spring in antagonistic connection. (**b**) Schematic diagram of the setup

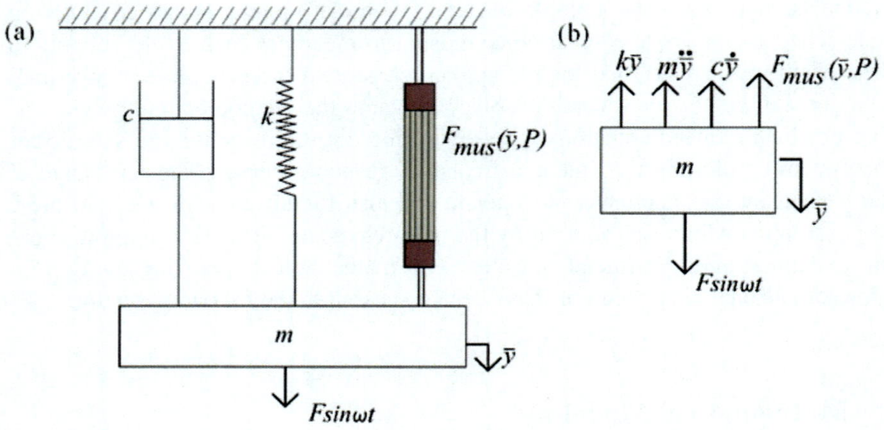

Fig. 2 (**a**) Equivalent spring-mass-damper system, (**b**) Free-body diagram

the free body diagram to represent the various forces acting on the whole system and the governing equation can be found out using Newton's law as follows.

$$m\ddot{\bar{y}} + c\dot{\bar{y}} + k\bar{y} + F_{mus} = F \sin \omega t \tag{1}$$

where the dot indicates the differentiation with respect to time t and $\bar{y}$ is the displacement to the static equilibrium position of the system. The stiffness of the external spring is given by k and the coefficient of the viscous damping of the damper is c. Here, F_{mus} is the force employed by the PAM which is supposed to be similar to that mentioned by Li et al. [7]. Since, there is a presence of high nonlinearity in the PAM because of the various nonlinear parameters, an extra cubic nonlinear term is added to the muscle force equation as follows.

$$F_{mus}(\bar{y}, P) = \left(c_1 + c_2 P + c_3 P^2\right)\left(\frac{\bar{y}}{l_{\max}}\right) + \eta \bar{y}^3 \tag{2}$$

where c_1, c_2, c_3 and η are the experimental constants for a particular application and $l_{\max}$ is the maximum possible length that can be attained by the muscle. P is the operating pressure in the muscle to actuate. One may obtain the following expression by substituting Eq. (2) in Eq. (1).

$$\ddot{\bar{y}} + \frac{c}{m}\dot{\bar{y}} + \left[\frac{k}{m} + \frac{\left(c_1 + c_2 P + c_3 P^2\right)}{m l_{\max}}\right]\bar{y} + \frac{\eta}{m}\bar{y}^3 = \frac{F}{m}\sin \omega t \tag{3}$$

Now, a non-dimensional time $\tau = \omega_0 t$ is considered for the system where ω_0 is the fundamental natural frequency which can be written as below.

$$\omega_0 = \sqrt{\frac{k}{m} + \frac{\left[c_1 + c_2 P + c_3 P^2\right]}{m l_{\max}}} \tag{4}$$

A nondimentional displacement is considered $\overline{y} = ry$ where r is the scaling factor. The different other non-dimensional parameters are considered as given below.

$$\Omega = \frac{\overline{\Omega}}{\omega_0}, \mu = \frac{c}{2\varepsilon m \omega_0}, \alpha = \frac{r^2 \eta}{\varepsilon m \omega_0^2}, f = \frac{F}{m r \omega_0^2} \tag{5}$$

Now, Eq. (3) can be simplified to the temporal equation of motion as follows.

$$\ddot{y} + 2\varepsilon\mu\dot{y} + y + \varepsilon\alpha y^3 = f \sin \Omega\tau \tag{6}$$

The dot represents the differentiation with respect to nondimentional time τ. The book keeping parameter ε is less than 1 and μ is the non-dimensional damping parameter. It may be noted that the nondimentional parameter f is not dependent on the operating pressure P. Hence, the amplitude of the external force is a constant term which is not a function of air pressure supplied to the system. This is very useful in the field of robotics and medical applications. The temporal equation (6) contains various nonlinear terms and it is very difficult have a closed form solution. Therefore, one can obtain the approximate analytical solution with the help perturbation technique like the method of multiple scales [14, 15].

In the method of multiple scales, the displacement can be expressed in terms of different time scales (T_0, T_1) with a book keeping parameter ε as follows.

$$y(\tau; \varepsilon) = y_0 (T_0, T_1) + \varepsilon y_1 (T_0, T_1) + O\left(\varepsilon^2\right) \tag{7}$$

By following the similar procedure as mentioned by Nayfeh and Mook [15], one may obtain the first order modulation and phase equations for superharmonic resonance condition as follows.

$$a' = \varepsilon\left(-\mu a - \alpha\Lambda^3 \sin\gamma\right) \tag{8}$$

$$a\gamma' = 3a\varepsilon\sigma - \varepsilon\left(\alpha\Lambda^3 \cos\gamma + \frac{3}{8}\alpha a^3 + 3\alpha a\Lambda^2\right) \tag{9}$$

where, $\Lambda = \frac{f}{2(1-\Omega^2)}$. After finding out the particular solution of y_0 and y_1, substituting them in Eq. (7) the total time response of the system $y(\tau)$ for superharmonic resonance condition is given below.

$$
\begin{aligned}
y &= a\cos(3\Omega T_0-\gamma)+2\Lambda\cos(\Omega T_0)+\tfrac{4\varepsilon\Omega\mu\Lambda}{(1-\Omega^2)}\sin(\Omega T_0)\\
&-\alpha\varepsilon\Big[-\tfrac{a^3}{32}\cos(9\Omega T_0-3\gamma)+\tfrac{3a^2\Lambda}{(1-\Omega^2)}\cos(\Omega T_0)\\
&+\tfrac{3a\Lambda^2}{(1-(1-2\Omega)^2)}\cos(\Omega T_0-\gamma)+\tfrac{3a\Lambda^2}{(1-(1+2\Omega)^2)}\cos(5\Omega T_0-\gamma)\\
&+\tfrac{3a^2\Lambda}{2(1-(2+\Omega)^2)}\cos(7\Omega T_0-2\gamma)+\tfrac{3a^2\Lambda}{2(1-(2-\Omega)^2)}\cos(5\Omega T_0-2\gamma)\\
&+\tfrac{6\Lambda^3}{(1-\Omega^2)}\cos(\Omega T_0)\Big]
\end{aligned}
\tag{10}
$$

Now, for steady state response (a_0, γ_0), the frequency response equation of the system can be written as follows.

$$
\alpha^2\Lambda^6=\left(\mu^2+\left(3\sigma-\frac{3}{8}\alpha a^2-3\alpha\Lambda^2\right)^2\right)a^2 \tag{11}
$$

From Eq. (11), one may noticed that there is no trivial state response exist in the case of superharmonic resonance condition. But the nontrivial state response for the system can be achieved by solving Eqs. (8) and (9) simultaneously. Hence, substituting $a = a_0 + a_1$ and $\gamma = \gamma_0 + \gamma_1$ where a_0 and γ_0 are the equilibrium points in Eqs. (8) and (9) the stability of the steady state response can be found by determining the eigenvalues of the Jacobian matrix (J). The Jacobian matrix (J) is as follows.

$$
J=\varepsilon\begin{pmatrix} -\mu & -3a_0\sigma+\frac{3}{8}\alpha a_0^3+3\alpha a_0\Lambda^2\\ \frac{3\sigma}{a_0}-\frac{9}{8}\alpha a_0-\frac{3\alpha\Lambda^2}{a_0} & -\mu\end{pmatrix} \tag{12}
$$

The system will be stable for the superharmonic resonance condition if all the real parts of the eigenvalues of the Jacobian matrix (J) in (12) are negative.

3 Numerical Results and Discussion

For the numerical analysis, the various parameters present in the PAM is considered to be as Table 1 which is similar to that mentioned in Li et al. [7] and Kalita and Dwivedy [8, 9]. Due to the existence of various nonlinear terms in the temporal equation (6), the system will depict a typical nonlinear behavior. Here, the external excitation frequency of the system is considered to be equal to the one third of the natural frequency of the system i.e., superharmonic resonance condition. The influence of the different parameters like operating pressure P, damping μ, external force F, stiffness of the spring k, nonlinearity α along with the muscle parameter c_3 have been studied with the help frequency plots. From these plots one can understand the effect of different parameters on the dynamics behavior of the system. In all the frequency response plots, the blue solid line indicates the stable solutions whereas the red solid line depicts the unstable solution.

Figure 3 shows the frequency response of the system by taking the system parameters value as mentioned in Table 1. In this case, there is no trivial solution of the system. So, the PAM will always depend for actuation with the amplitude to the nontrivial response of the system as shown in Fig. 3. It may be noted from the nontrivial response that upto point P the system has single stable state and after point P it has bi-stable state. So, one may attain either of the stable state which is depending on the initial conditions. One may notice that with decrease in the frequency, the system will involve with a jump-up phenomenon after the excitation go beyond the critical point P ($\sigma = 4.118$), which is a saddle-node bifurcation point. Here, the unstable solution at point P will always have an affinity to jump up to the point Q which have a stable solution as in Fig. 3. Hence, to achieve the required displacement of the PAM, the system parameters should be chosen properly by

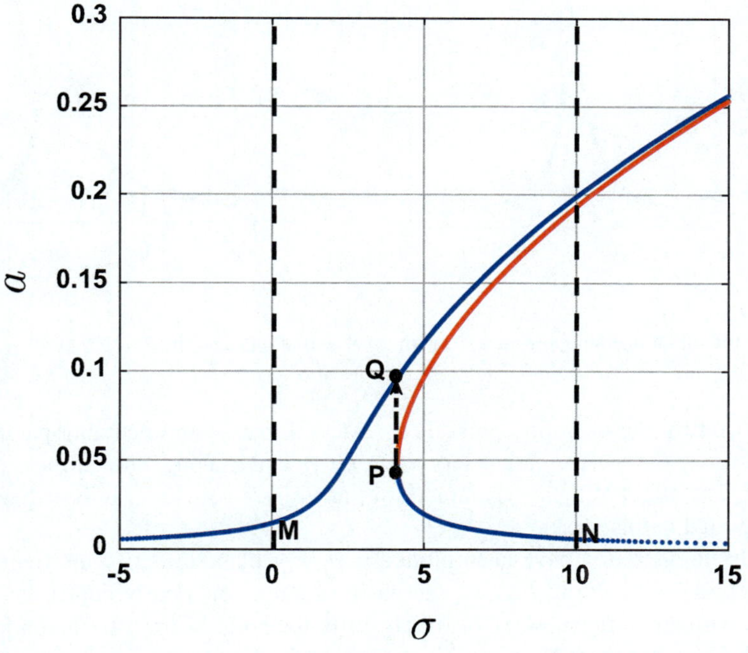

Fig. 3 Frequency response with system parameters value as mentioned in Table 1

Table 1 System parameters values for simulation

Parameter	Numerical value	Parameter	Numerical value	Parameter	Numerical value
l_{max}	74 mm	μ	0.01	F	8 kN
c_1	−234.25 N	P	500 kPa	r	1
c_2	1.96 N/kPa	m	6 N	α	1500
c_3	0.3 N/kPa^2	k	12 N/mm	ε	0.1

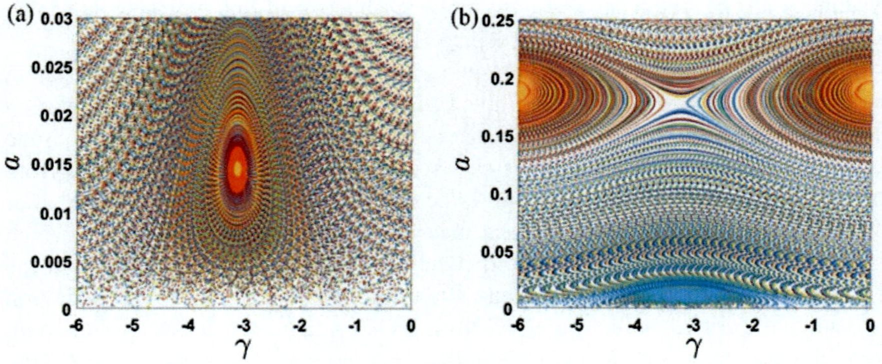

Fig. 4 Basin of attraction: (**a**) $\sigma = 0.01$ and (**b**) $\sigma = 10$ from Fig. 3

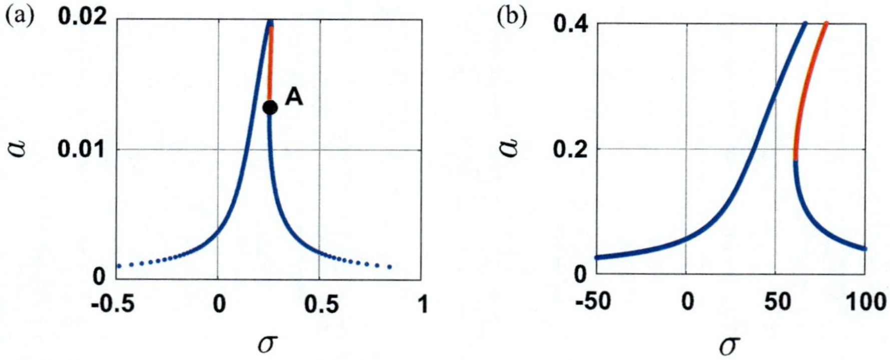

Fig. 5 Variation of operating pressure (P) with (**a**) $P = 1000$ kPa and (**b**) $P = 250$ kPa

actively control the operating pressure P, external load F and detuning parameter σ or passively the muscle parameters μ, c_1, c_2, c_3 and α along with stiffness of the spring k. The effect of muscle parameters c_1 and c_2 are very less, so these have not been reported in this work.

The basin attraction have been plotted in $a \sim \gamma$ plane to verify the frequency responses in Fig. 4. From Fig. 4a, the basin of attraction clearly depicts a stable solution with amplitude $a = 0.014$ which verify the point M ($\sigma = 0.01$) marked in Fig. 3. For another point N at $\sigma = 10$, the basin of attraction in Fig. 4b depicts that the system exhibit two stable and one unstable solution in the nontrivial state.

Figure 5 depicts the frequency response plots with two unlike values of operating pressure of the muscle (P). It can be observed in the system that the maximum response amplitude decreases with increase in P ($P = 1000$ kPa) and Hopf bifurcation can be observed at detuning parameter $\sigma = 0.252$ and $\sigma = 0.261$ in Fig. 5a. But decrease in the value P i.e., $P = 250$ kPa (Fig. 5b), the maximum response amplitude is increased and the range of detuning parameter σ becomes very large as compared to Fig. 3. The time response and phase portraits have been

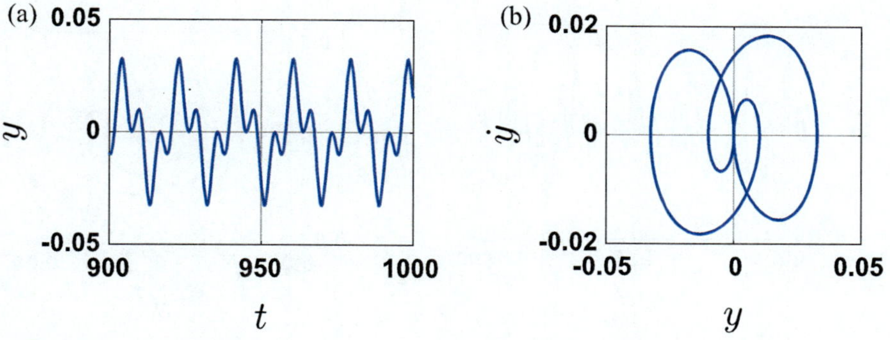

Fig. 6 (**a**) Time response and (**b**) phase portrait corresponding to the point 'A' in Fig. 5a

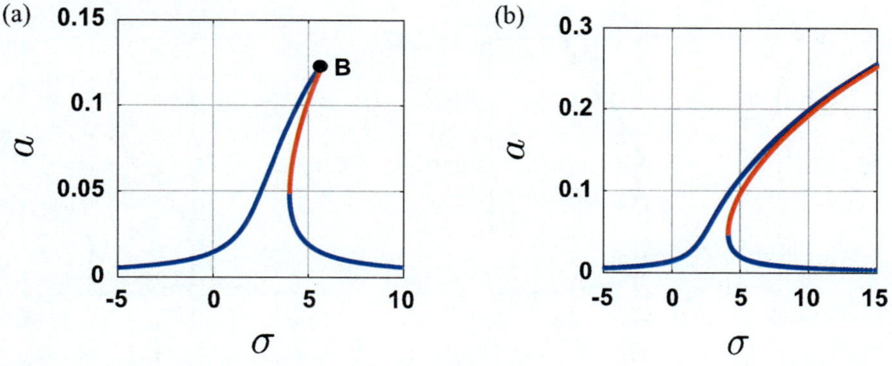

Fig. 7 Variation of damping (μ) with (**a**) $\mu = 0.1$ and (**b**) $\mu = 0.001$

plotted in Fig. 6 at point A ($\sigma = 0.252$) in Fig. 5a to understand the behavior of the Hopf bifurcation points in Fig. 5a.

From Fig. 7a, with increase in the damping parameter ($\mu = 0.1$), the maximum response amplitude has been decreased in the system and the saddle node bifurcation point can be observed at B where the maximum response amplitude of the system is $a = 0.123$. Beyond this point B the system has only one stable nontrivial solution. In Fig. 7b, with decrease in the damping parameter ($\mu = 0.001$) the maximum response amplitude slightly increases as compared to Fig. 3. The time response and phase portraits have been plotted in Fig. 8 to observe the nature of the saddle node bifurcation point B at Fig. 7a.

In Fig. 9a, with increase in the value of the external load ($F = 16$ kN) the response amplitude will increase along with the range of the detuning parameter also large as compared to Fig. 3. But with decrease in the value of the external load ($F = 4$ kN), the maximum response of the amplitude $a = 0.154$ will occur at the saddle node bifurcation point $\sigma = 5.14$.

The effect of stiffness of the external spring k is very less compared to the other parameter of the system as shown in Fig. 10. With increase in the value of stiffness

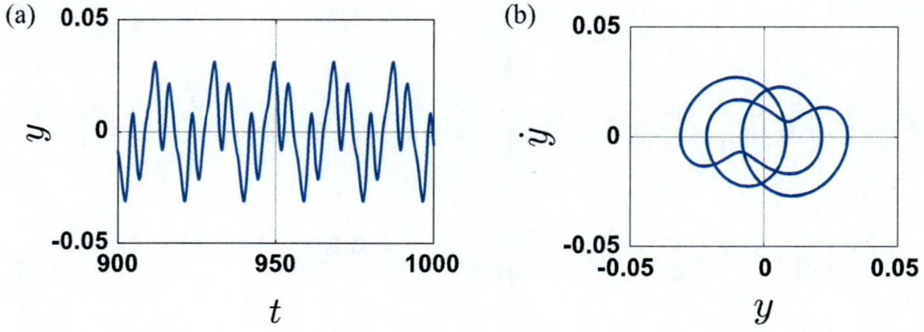

Fig. 8 (**a**) Time response and (**b**) phase portrait corresponding to the point 'B' in Fig. 7a

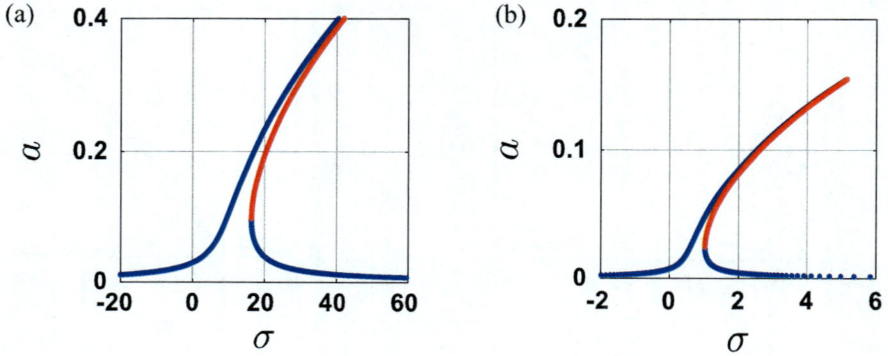

Fig. 9 Variation of external load (F) with (**a**) $F = 16$ kN and (**b**) $F = 4$ kN

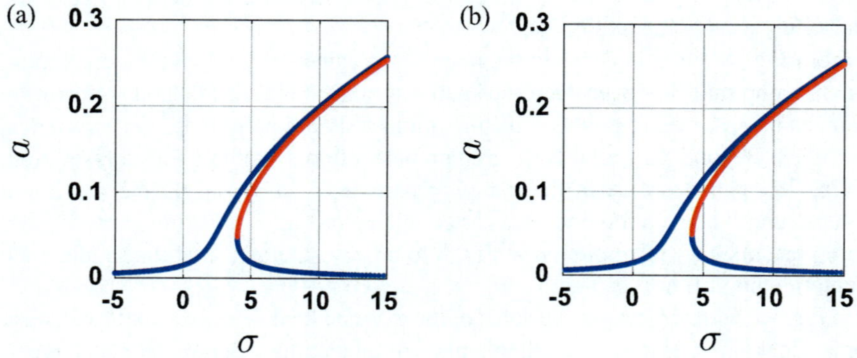

Fig. 10 Variation of stiffness of the external spring (k) with (**a**) $k = 24$ N/mm^2 and (**b**) $k = 6$ N/mm^2

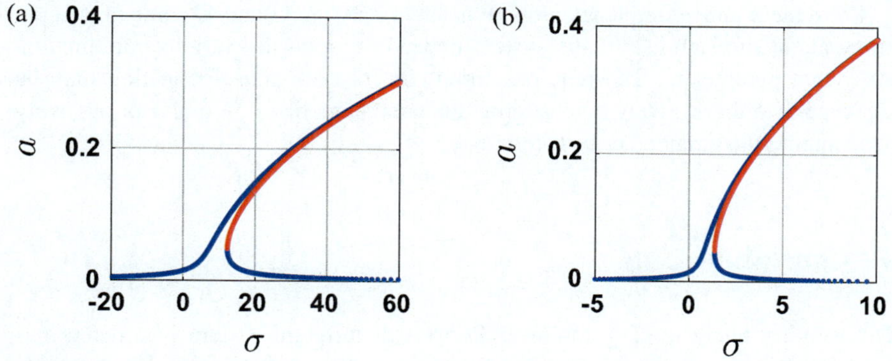

Fig. 11 Variation of nonlinearity (α) with (**a**) $\alpha = 4500$ and (**b**) $\alpha = 500$

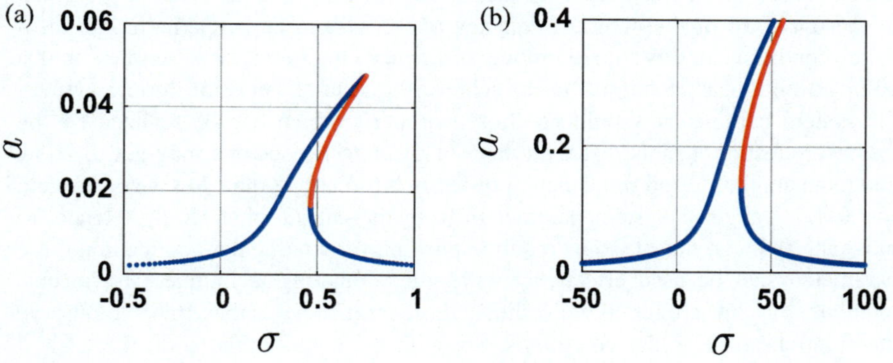

Fig. 12 Variation of muscle parameter (c_3) with (**a**) $c_3 = 0.9$ N/kPa^2 and (**b**) $c_3 = 0.1$ N/kPa^2

($k = 24$ N/mm^2), the response amplitude increases in the system by around 0.4% and decrease in the value of stiffness ($k = 6$ N/mm^2) the response amplitude decreases by around 0.2% as compared to Fig. 3.

From Figs. 3 and 11a, it can be noticed that with increase in the value nonlinearity α the response amplitude is decreased. Similarly, with decrease in the value of α, the response amplitude of the system increased as shown in Fig. 11b. Hence, conclusion can be made by triple the nonlinear parameter α ($\alpha = 4500$), around 62% decrease in the response amplitude and by reducing the nonlinear parameter α to its one third value i.e., $\alpha = 500$, around 92% increase in the response amplitude.

In Fig. 12a, with the increase in the value of muscle parameter c_3 ($c_3 = 0.9$ N/kPa^2) in the system the maximum response amplitude $a = 0.047$ will be at saddle node bifurcation point at $\sigma = 0.742$. But with decrease in the value of c_3 ($c_3 = 0.1$ N/kPa^2), the maximum response amplitude will be occur at high value of detuning parameter σ as compared to Fig. 3.

From these above frequency plots (Figs. 3, 5, 7, 9, 10, 11 and 12), one may get to know about the stability of the system depending upon the various combination of system parameters. The required amplitude for particular application may be achieved by either actively by changing the parameters like P, F and σ or passively by changing the parameters μ, k, α and c_3.

4 Conclusion

The stability analysis of a nonlinear PAM with different system parameters has been investigated in this work. The governing system equation has been derived and solved to approximate analytical solution with the help the method of multiple scales. The external excitation frequency in this case is considered to be equal to the one third of the natural frequency of the system i.e., superharmonic resonance condition. In this superharmonic resonance conditions the system exhibits a complex nonlinear response. So, to achieve the required position by the system, the system parameters should be chosen properly which can be realized by the frequency response plots. With the help of these responses one may get to know about the amplitude and frequency of different bifurcation points to save the system for a wide range of system parameters from the catastrophic failure. Hence, to move the artificial muscle to its required position for a particular application, these parameters can be used effectively with soft computing techniques and inverse methods. Basin of attraction will validate these frequency response plots to observe the solutions from initial conditions. So, with the help of this work, the desired system response can be achieved with the proper control of different passive or active system parameters under the superharmonic resonance conditions.

References

1. Chou, C.P., Hannaford, B.: Measurement and modeling of McKibben pneumatic artificial muscles. IEEE Trans. Robot. Autom. **12**(1), 90–102 (1996)
2. Klute, G.K., Czerniecki, J.M., Hannaford, B.: McKibben artificial muscles: pneumatic actuators with biomechanical intelligence. In: Int. Conf. Advanced Intelligent Mechatronics, pp. 221–226. IEEE/ASME, Atlanta, GA (1999)
3. Daerden, F., Lefeber, D.: Pneumatic artificial muscles: actuators for robotics and automation. Eur. J. Mech. Environ. Eng. **47**(1), 11–21 (2002)
4. Tondu, B., Ippolito, S., Guiochet, J., Daidie, A.: A seven-degrees-of-freedom robot-arm driven by pneumatic artificial muscles for humanoid robots. Int. J. Robot. Res. **24**(4), 257–274 (2005)
5. Wickramatunge, K.C., Leephakpreeda, T.: Empirical modeling of pneumatic artificial muscle. In: Proc. Int. Multi Conf. of Engineers and Computer Scientists, Hong Kong, p. 2 (2009)
6. Tondu, B., Lopez, P.: Modeling and control of McKibben artificial muscle robot actuators. IEEE Control. Syst. Mag. **20**(2), 15–38 (2000)
7. Li, H., Kawashima, K., Tadano, K., Ganguly, S., Nakano, S.: Achieving haptic perception in forceps' manipulator using pneumatic artificial muscle. IEEE/ASME Trans. Mech. **18**(1), 74–85 (2011)

8. Kalita, B., Dwivedy, S.K.: Nonlinear dynamics of a parametrically excited pneumatic artificial muscle (PAM) actuator with simultaneous resonance condition. Mech. Mach. Theory. **135**, 281–297 (2019)
9. Kalita, B., Dwivedy, S.K.: Dynamic analysis of pneumatic artificial muscle (PAM) actuator for rehabilitation with principal parametric resonance condition. Nonlinear Dyn. **97**(4), 2271–2289 (2019)
10. Kang, B.S., Kothera, C.S., Woods, B.K., Wereley, N.M.: Dynamic modeling of Mckibben pneumatic artificial muscles for antagonistic actuation. In: Int. Conf. Robotics and Automation, pp. 182–187. IEEE, Kobe (2009)
11. Tóthová, M., Pitel, J.: Dynamic model of pneumatic actuator based on advanced geometric muscle model. In: 9th Int. Conf. Computational Cybernetics, pp. 8–87. IEEE, Tihany (2013)
12. Balara, M., Tóthová, M.: Static and dynamic properties of the pneumatic actuator with artificial muscles. In: 10th Jubilee Int. Sym. Intelligent Systems and Informatics, pp. 577–581. IEEE, Subotica (2012)
13. Kalita, B., Dwivedy, S.K.: Dynamic analysis of a parametrically excited golden Muga silk embedded pneumatic artificial muscle. In: 14th Int. Conf. on Vibration Engineering and Technology of Machinery, Lisbon, Portugal, p. 02008 (2018)
14. Nafeh, A.H., Balachandran, B.: Applied Non-linear Dynamics. Wiley Interscience, New York (1994)
15. Nayfeh, A.H., Mook, D.T.: Nonlinear Oscillations. John Wiley & Sons, New York (2008)

Two-Mode Long-Wave Low-Frequency Approximations for Anti-Plane Shear Deformation of a High-Contrast Asymmetric Laminate

Mohammed Alkinidri, Julius Kaplunov, and Ludmila Prikazchikova

Abstract The anti-plane shear of a three-layered laminate of an asymmetric structure is considered. The chosen geometry of the laminate assumes coupling its symmetric and anti-symmetric modes, which is not a feature of a symmetric structure. A high contrast in mechanical properties of the inner and outer layers is assumed. A specific contrast setup supporting an asymptotically small lowest shear cut-off frequency is studied. For a laminate with traction-free faces two-mode long-wave low-frequency approximation of the full dispersion relations incorporating both the fundamental mode and the first harmonic is derived. The accuracy of the derived approximations is tested by numerical comparison with the exact solution. The 1D partial differential equation corresponding to the aforementioned two-mode shortened dispersion relation is also presented.

Keywords Asymptotic · Contrast · Laminate · Two-mode · Asymmetric · Wave

1 Introduction

Multi-layered sandwich type structures with high contrast properties have numerous applications in various high-tech domains, including automotive and aerospace industries, see e.g. [1, 2]. In particular, asymmetric three-layered structures are used nowadays for manufacturing modern prototypes of photovoltaic modules [3, 4]. Mechanical behaviour of such structures can not be always tackled within

M. Alkinidri · L. Prikazchikova
School of Computing and Mathematics, Keele University, Keele, UK
e-mail: j.kaplunov@keele.ac.uk

J. Kaplunov (✉)
School of Computing and Mathematics, Keele University, Keele, UK

Faculty of Industrial Engineering Novo mesto , Novo mesto, Slovenia

J. Awrejcewicz (ed.), *Perspectives in Dynamical Systems II: Mathematical and Numerical Approaches*, Springer Proceedings in Mathematics & Statistics 363,
https://doi.org/10.1007/978-3-030-77310-6_24

conventional engineering models even taking into account shear deformation and rotation inertia [5, 6]. This motivates developing of more elaborated advanced approaches. Among the latter, asymptotic analysis of dynamic phenomena, specific for high-contrast layered plates, seems to be of particular theoretical and practical interest and importance [7–13].

The recent multi-parametric analysis of a symmetric three-layered sandwich plate subject to in-plane bending deformation [7] demonstrates that the value of the lowest shear cut-off frequency tends to zero for several setups of high-contrasting material and geometrical parameters. As a result, an extra shear mode may be excited over the long-wave low-frequency range, along with the fundamental bending one. The related two-mode polynomial shortened forms of the full dispersion relation are derived in the cited paper [7]. The scenario supporting uniformly valid shortened dispersion equations are revealed.

Along with a vector problem in [7], a simpler scalar anti-plane shear problem is considered for the same symmetric three-layered sandwich plate in [8]. In this case, the same as above small cut-off frequency is characteristic of antisymmetric motion. The asymptotic expansions for the associated vibration mode are derived not only for the dispersion equations but also for the equations of motion. For the latter the asymptotic approach originally developed for analysing high-frequency near-cut-off behaviour, e.g. see [6, 14, 15] and references therein, is adapted for the low-frequency band.

In this paper we insert asymmetry in the formulation of [8] leading to two low-frequency cut-offs. One of them is a natural generalisation of the shear cut-off considered in [8], whereas the second one, as might be expected, corresponds to the fundamental mode and is equal to zero. In the degenerate case of a symmetric plate the fundamental mode characteristic of symmetric motion is decoupled from antisymmetric ones treated in [8].

In what follows, we deal with two-mode long-wave low-frequency approximations of the full dispersion relation, similarly to [7], but for a simpler scalar problem. This looks promising for further insight in dynamics of high-contrast layered structures, including asymptotic considerations of two-mode expansion of the equations of motions. At the same time, the algebra in this paper is obviously more involved than that in [8], since the plate motion now cannot be split into symmetric and antisymmetric components.

The paper is organised as follows. First we formulate the anti-plane problem for a three-layered asymmetric plate and derive the exact dispersion relation. Then, we choose a high-contrast setup corresponding to a plate with stiff outer layeres and soft inner layer, supporting a small shear cut-off. Next, we derive a two-mode uniform asymptotic approximation of the dispersion relation and illustrate numerically a good agreement between the exact and asymptotic results.

2 Statement of the Problem

Consider a three-layered asymmetric laminate with the isotropic layers of thickness h_1, h_2 and h_3, see Fig. 1. The Cartesian coordinate system is chosen in such a way that the axis x_1 goes through the mid-plane of the core layer. In what follows two outer layers have the same material parameters.

For the antiplane shear motion the only non-zero displacement is orthogonal to the x_1x_2 plane. Hence, the equations of motion for each layer can be written as

$$\frac{\partial \sigma_{13}^l}{\partial x_1} + \frac{\partial \sigma_{23}^l}{\partial x_2} - \rho_l \frac{\partial^2 u_l}{\partial t^2} = 0, \qquad l = 1, 2, 3, \tag{1}$$

with

$$\sigma_{i3}^l = \mu_l \frac{\partial u_l}{\partial x_i}, \quad i = 1, 2, \tag{2}$$

where σ_{i3}^l are shear stresses, $u_l = u_l(x_1, x_2)$ are out of plane displacements, t is time, μ_l are Lamé parameters, and ρ_l are mass densities. As we have already mentioned, $\mu_1 = \mu_3$ and $\rho_1 = \rho_3$.

The continuity and traction-free boundary conditions are given by

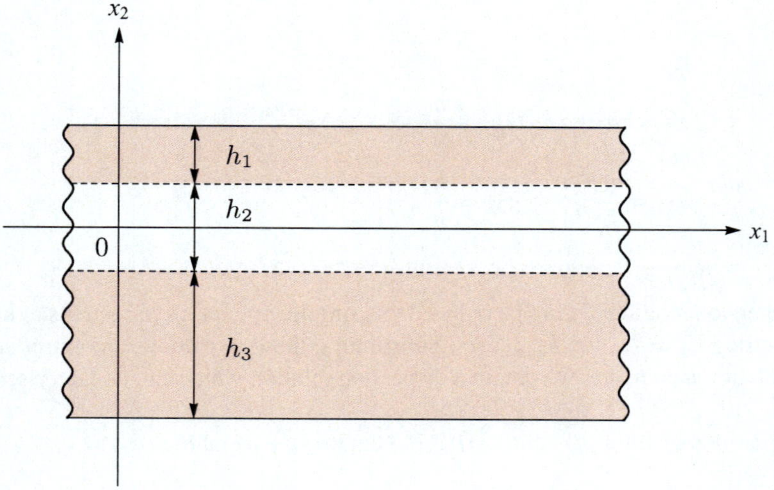

Fig. 1 A three-layered asymmetric plate

$$u_1 = u_2, \quad \sigma_{23}^1 = \sigma_{23}^2 \quad \text{at} \quad x_2 = \frac{h_2}{2},$$
$$u_2 = u_3, \quad \sigma_{23}^2 = \sigma_{23}^3 \quad \text{at} \quad x_2 = -\frac{h_2}{2}, \tag{3}$$

and

$$\sigma_{23}^1 = 0 \quad \text{at} \quad x_2 = \frac{h_2}{2} + h_1,$$
$$\sigma_{23}^3 = 0 \quad \text{at} \quad x_2 = -\frac{h_2}{2} - h_3, \tag{4}$$

respectively. We seek solution of the formulated problem (1)–(4) in the form of a travelling wave $e^{i(kx_1 - \omega t)}$, where k is the wave number and ω is frequency. The related dispersion relation becomes

$$\mu\alpha_1\alpha_2\big(\tanh(h_{12}\alpha_1) + \tanh(h_{32}\alpha_1)\big) + \mu^2{\alpha_2}^2\tanh(\alpha_2) + \\ + {\alpha_1}^2\tanh(h_{12}\alpha_1)\tanh(h_{32}\alpha_1)\tanh(\alpha_2) = 0 \tag{5}$$

where

$$\alpha_1 = \sqrt{K^2 - \frac{\mu}{\rho}\Omega^2}, \qquad \alpha_2 = \sqrt{K^2 - \Omega^2} \tag{6}$$

and

$$K = kh_2, \qquad \Omega = \frac{\omega h_2}{c_2}, \qquad \mu = \frac{\mu_2}{\mu_1}, \qquad \rho = \frac{\rho_2}{\rho_1},$$
$$h_{12} = \frac{h_1}{h_2}, \qquad h_{32} = \frac{h_3}{h_2}, \tag{7}$$

with $c_2 = \sqrt{\mu_2/\rho_2}$.

Dispersion relation (5) can be reduced to a simpler one for a symmetric sandwich plate setting $h_1 = h_3$ and $h_2 = 2\tilde{h}_2$. Substituting these into above and introducing new notation $h = h_1/\tilde{h}_2$ we obtain a dispersion relation which can be factorised as

$$\big(2\mu\alpha_2 + \alpha_1\tanh(\alpha_1 h)\tanh(\alpha_2)\big)\big(2\mu\alpha_2\tanh(\alpha_2) + \alpha_1\tanh(\alpha_1 h)\big) = 0. \tag{8}$$

The first and second aggregates in the left-hand side of (8) correspond to the dispersion relations for symmetric and antisymmetric waves, respectively.

We also present the formulae for displacements

$$
\begin{aligned}
u_1 &= 2\beta\mu\alpha_2 \cosh\left(\alpha_1\left(h_{12}+\frac{1}{2}-\xi\right)\right),\\
u_2 &= \beta\left((\mu\alpha_2+\alpha_1)\cosh\left(\alpha_1 h_{12}-\alpha_2\xi+\frac{\alpha_2}{2}\right)\right.\\
&\quad\left.+(\mu\alpha_2-\alpha_1)\cosh\left(\alpha_1 h_{12}+\alpha_2\xi-\frac{\alpha_2}{2}\right)\right),\\
u_3 &= \frac{\beta}{2\alpha_1}\left(-(\mu\alpha_2-\alpha_1)^2\cosh\left(\alpha_1\left(h_{12}-\frac{1}{2}-\xi\right)-\alpha_2\right)\right.\\
&\quad+(\mu\alpha_2+\alpha_1)^2\cosh\left(\alpha_1\left(h_{12}-\frac{1}{2}-\xi\right)+\alpha_2\right)\\
&\quad+(\mu^2\alpha_2^2-\alpha_1^2)\cosh\left(\alpha_1\left(h_{12}+\frac{1}{2}+\xi\right)-\alpha_2\right)\\
&\quad\left.-(\mu^2\alpha_2^2-\alpha_1^2)\cosh\left(\alpha_1\left(h_{12}+\frac{1}{2}+\xi\right)+\alpha_2\right)\right),
\end{aligned}
\tag{9}
$$

where $\xi = x_2/h_2$ and

$$
\beta = A\left((\mu\alpha_2-\alpha_1)\sinh\left(\alpha_1 h_{12}-\frac{\alpha_2}{2}\right)-(\mu\alpha_2+\alpha_1)\sinh\left(\alpha_1 h_{12}+\frac{\alpha_2}{2}\right)\right)^{-1},
$$

with A being an arbitrary constant.

3 Asymptotic Analysis

First, setting K = 0 in dispersion relation (5), we have for the cut-off frequencies

$$
\begin{aligned}
&\sqrt{\mu\rho}\left(\tan\left(h_{12}\sqrt{\frac{\mu}{\rho}}\Omega\right)+\tan\left(h_{32}\sqrt{\frac{\mu}{\rho}}\Omega\right)\right)+\mu\rho\tan(\Omega)\\
&\quad-\tan\left(h_{12}\sqrt{\frac{\mu}{\rho}}\Omega\right)\tan\left(h_{32}\sqrt{\frac{\mu}{\rho}}\Omega\right)\tan(\Omega)=0.
\end{aligned}
\tag{10}
$$

Consider the contrast in the material parameters of the outer and core layers given by

$$
\mu \ll 1, \quad \rho \sim \mu, \quad h_{12} \sim 1, \quad h_{32} \sim 1. \tag{11}
$$

These formulae specify an asymmetric laminate with stiff outer layers and a soft core. In this case, apart from usual zero cut-off ($\Omega = 0$) we have an extra small one approximated by

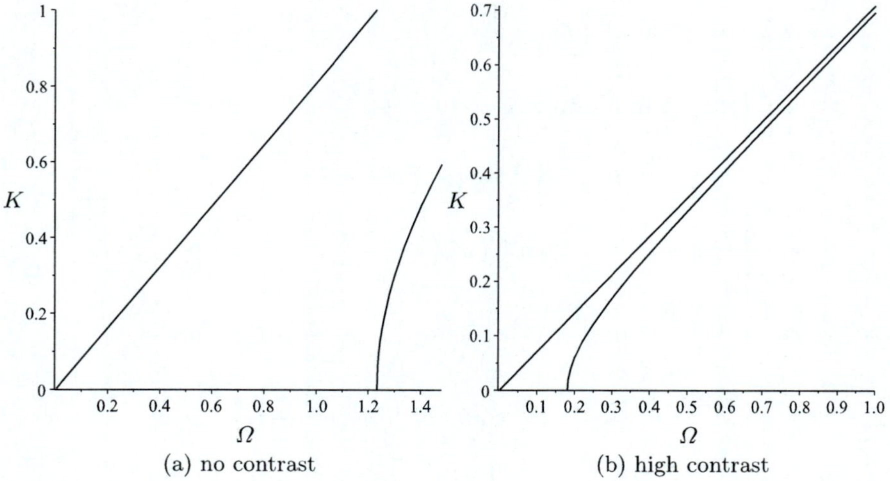

Fig. 2 Dispersion curves (5) for $h_{12} = 1.0$, $h_{32} = 1.5$ and (a) $\mu = 1.0$ and $\rho = 2.0$, (b) $\mu = 0.01$ and $\rho = 0.02$

$$\Omega \approx \sqrt{\frac{(h_{12} + h_{32})\rho}{h_{12}h_{32}}} \ll 1. \tag{12}$$

Hence, for the assumed contrast material properties we have two cut-offs over the low frequency band. This is not the case for a non-contrast setup which allows only a zero cut-off. This observation is illustrated numerically in Fig. 2, where dispersion curves (5) are plotted for both non-contrast and contrast setups.

Next, expanding all trigonometric functions in (10) in asymptotic Taylor series at $\Omega \ll 1$ and $K \ll 1$ and assuming relations (11) to be valid, we derive a polynomial dispersion relation, which can be written as

$$\begin{aligned} &\gamma_1 K^2 + \gamma_2 \Omega^2 + \gamma_3 K^4 + \gamma_4 K^2 \Omega^2 + \gamma_5 \Omega^4 + \gamma_6 K^6 \\ &\quad + \gamma_7 K^4 \Omega^2 + \gamma_8 K^2 \Omega^4 + \gamma_9 \Omega^6 + \cdots = 0, \end{aligned} \tag{13}$$

where

$$\gamma_1 = \mu\,(h_{12} + h_{32} + \mu)\,,$$

$$\gamma_2 = -\frac{\mu^2}{\rho}\,(h_{12} + h_{32} + \rho)\,,$$

$$\gamma_3 = h_{12}h_{32} - \frac{\mu}{3}\left(h_{12}^3 + h_{32}^3 + \mu\right),$$

$$\gamma_4 = \frac{2\mu}{3\rho}\left(h_{12}^3\mu + h_{32}^3\mu - 3h_{12}h_{32} + \mu\rho\right),$$

$$\gamma_5 = -\frac{\mu^2}{3\rho^2}\left(h_{12}^3\mu + h_{32}^3\mu - 3h_{12}h_{32} + \rho^2\right),$$

$$\gamma_6 = \frac{2\mu}{15}\left(h_{12}^5 + h_{32}^5 + \mu\right) - \frac{h_{12}h_{32}}{3}\left(h_{12}^2 + h_{32}^2 + 1\right),$$

$$\gamma_7 = -\frac{1}{15\rho}\left(6\mu^2\left(h_{12}^5 + h_{32}^5 + \rho\right) - 5h_{12}h_{32}\left(3h_{12}^2\mu + 3h_{32}^2\mu + 2\mu + \rho\right)\right),$$

$$\gamma_8 = \frac{\mu}{15\rho^2}\left(6\mu\left(h_{12}^5\mu + h_{32}^5\mu + \rho^2\right) - 5h_{12}h_{32}\left(3h_{12}^2\mu + 3h_{32}^2\mu + \mu + 2\rho\right)\right),$$

$$\gamma_9 = -\frac{\mu^2}{15\rho^3}\left(2\left(h_{12}^5\mu^2 + h_{32}^5\mu^2 + \rho^3\right) - 5h_{12}h_{32}\left(h_{12}^2\mu + h_{32}^2\mu + \rho\right)\right). \tag{14}$$

At leading order coefficients $\gamma_i \approx \gamma_i^0$ are given below

$$\begin{aligned}
\gamma_1^0 &= (h_{12} + h_{32})\,\mu,\\
\gamma_2^0 &= -\frac{h_{12} + h_{32}}{\rho_0}\mu,\\
\gamma_3^0 &= h_{12}h_{32},\\
\gamma_4^0 &= -\frac{2h_{12}h_{32}}{\rho_0},\\
\gamma_5^0 &= \frac{h_{12}h_{32}}{\rho_0^2},\\
\gamma_6^0 &= -\frac{h_{12}h_{32}}{3}\left(h_{12}^2 + h_{32}^2 + 1\right),\\
\gamma_7^0 &= \frac{h_{12}h_{32}}{3\rho_0}\left(3h_{12}^2 + 3h_{32}^2 + \rho_0 + 2\right),\\
\gamma_8^0 &= -\frac{h_{12}h_{32}}{3\rho_0^2}\left(3h_{12}^2 + 3h_{32}^2 + 2\rho_0 + 1\right),\\
\gamma_9^0 &= \frac{h_{12}h_{32}}{3\rho_0^3}\left(h_{12}^2 + h_{32}^2 + \rho_0\right),
\end{aligned} \tag{15}$$

where $\rho_0 = \rho/\mu$. From (15) we observe that $\gamma_1 \sim \gamma_2 \sim \mu$, and $\gamma_i \sim 1$, $i = 3, \ldots, 9$. As a result, the leading order shortened approximation, involving the fundamental mode with a zero cut-off along with the lowest harmonic with the cut-off of order $O(\sqrt{\mu})$ given by (12),takes the form

$$\gamma_1^0 K^2 + \gamma_2^0 \Omega^2 + \gamma_3^0 K^4 + \gamma_4^0 K^2\Omega^2 + \gamma_5^0 \Omega^4 + \cdots = 0.$$

The above equation can be factorised

$$\left(K^2\rho_0 - \Omega^2\right)\left(h_{12}h_{32}\left(K^2\rho_0 - \Omega^2\right) + h_{12}\mu\rho_0 + h_{32}\mu\rho_0\right) = 0. \tag{16}$$

Therefore, for the fundamental mode and first harmonic we have

$$\Omega^2 = \rho_0 K^2 \tag{17}$$

and

$$\Omega^2 = \frac{\rho_0}{h_{12}h_{32}}\left(h_{12}\mu + h_{32}\mu + h_{12}h_{32}K^2\right), \tag{18}$$

respectively. It is worth mentioning that approximation (17) for the fundamental mode is valid over the whole low-frequency band $K \ll 1$, consequently, it does not fail at the vicinity of the cut-off (12), leading to a uniform approximation, see also [7] concerned with a similar analysis.

It might be also expected that the shortened dispersion relation (16) would correspond to a 1D partial differential equation, which can be presented in the original variables as

$$\Box^2 v - \frac{\mu_2(h_1 + h_3)}{\mu_1 h_1 h_2 h_3}\Box v = 0, \tag{19}$$

where $v(x_1, t)$ is a characteristic displacement and the d'Alembert operator $\Box$ is defined as

$$\Box = \frac{\partial^2}{\partial x_1^2} - \frac{1}{c_1^2}\frac{\partial^2}{\partial t^2}$$

with $c_1 = \sqrt{\mu_1/\rho_1}$.

A numerical comparison is shown for the exact dispersion curves (5) and approximations (17) and (18) in Fig. 3.

As an example we also plot in Fig. 4 the variation of properly normalised plate displacements u_i/β (9) across the thickness calculated at cut-off frequency (12), which takes the value $\Omega \approx 0.18$ for the same problems parameters as in Fig. 3. For the fundamental mode in Fig. 4a we have from (5) $K \approx 0.13$, where as for the first harmonic $K = 0$.

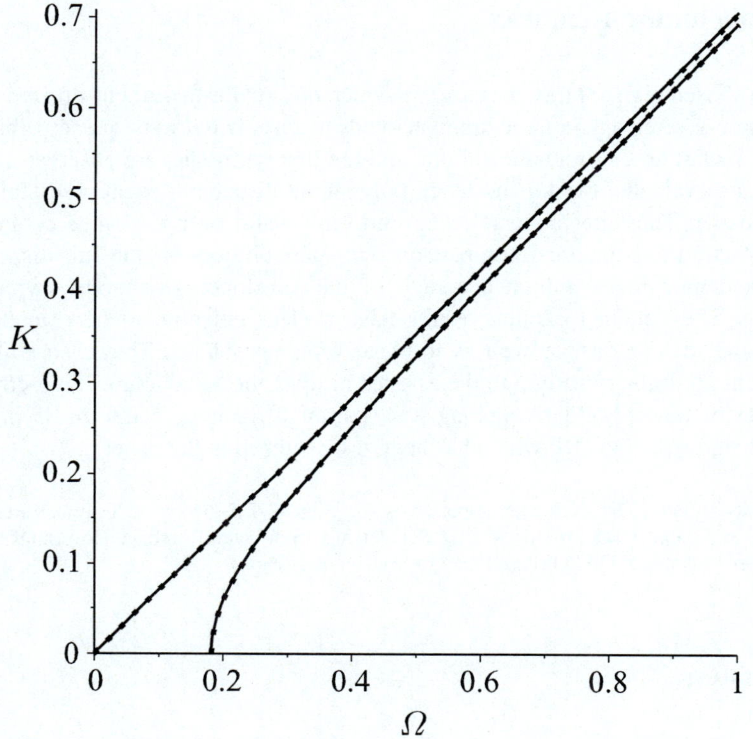

Fig. 3 Dispersion curves (5) (solid line) together with approximations (17) and (18) (dotted lines) for $h_{12} = 1.0$, $h_{32} = 1.5$, $\mu = 0.01$, and $\rho = 0.02$

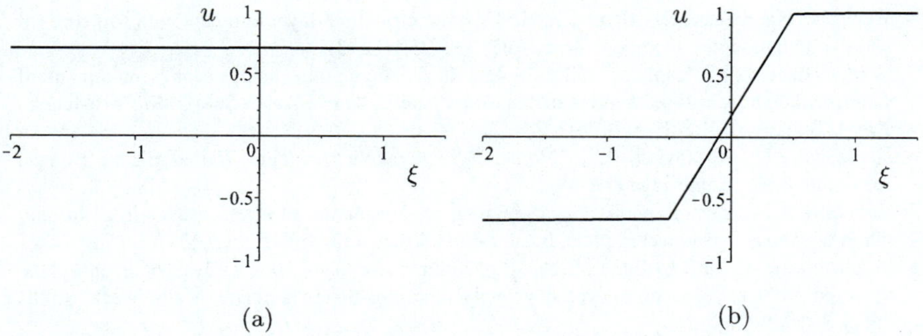

Fig. 4 Displacement variations at the cut-off frequency $\Omega \approx 0.18$ for $h_{12} = 1.0$, $h_{32} = 1.5$, $\mu = 0.01$, and $\rho = 0.02$ **(a)** fundamental mode, **(b)** first harmonic

4 Concluding Remarks

For the chosen high contrast scenario, in which the smallest shear cut-off frequency for a three-layered asymmetric laminate tends to zero, two low-frequency vibration modes, including the fundamental one and the first harmonic, are observed. These modes are evaluated from a shortened polynomial dispersion equation established in the paper. The latter appears to be uniformly valid over the range containing the first cut-off. Numerical comparison with the solutions of the full dispersion relation demonstrates a high accuracy of the developed two-mode asymptotic formula. The obtained explicit results have a clear potential to be extended to other types of contrast, as well as to plane vector problems. They also make an important preliminary insight to the essence of dynamic behaviour of high-contrast layered structures prior deriving long-wave partial differential models justifying and generalising equation (19) which has been just sketched in the paper.

Acknowledgments The work was supported by the grant J2-9224 from the Slovenian Research Agency. M.Alkinidri acknowledges PhD Scholarship from Saudi Arabian Government. The authors are also grateful to D.Prikazchikov for fruitful discussions.

References

1. Vinson, J.R.: The Behavior of Sandwich Structures of Isotropic and Composite Materials. CRC Press, Boca Raton (1999)
2. Sayyad, A.S., Ghugal, Y.M.: Bending, buckling and free vibration of laminated composite and sandwich beams: a critical review of literature. Compos. Struct. **171**, 486–504 (2017)
3. Rion, J., Leterrier, Y., Månson, J.A.E., Blairon, J.M.: Ultra-light asymmetric photovoltaic sandwich structures. Compos. A Appl. Sci. **40**(8), 1167–1173 (2009)
4. Weps, M., Naumenko, K., Altenbach, H.: Unsymmetric three-layer laminate with soft core for photovoltaic modules. Compos. Struct. **105**, 332–339 (2013)
5. Goldenveizer, A.L., Kaplunov, J.D., Nolde, E.V.: Asymptotic analysis and refinement of Timoshenko-Reisner-type theories of plates and shells. Trans. Acad. Sci. USSR. Mekhanika Tverd. Tela **25**(6), 126–139 (1990)
6. Kaplunov, J.D., Kossovitch, L.Y., Nolde, E.V.: Dynamics of Thin Walled Elastic Bodies. Academic Press, Cambridge (1998)
7. Kaplunov, J., Prikazchikov, D., Prikazchikova, L.: Dispersion of elastic waves in a strongly inhomogeneous three-layered plate. Int. J. Solids Struct. **113**, 169–179 (2017)
8. Prikazchikova, L., Ece Aydin, Y., Erbaş, B., Kaplunov, J.: Asymptotic analysis of an anti-plane dynamic problem for a three-layered strongly inhomogeneous laminate. Math. Mech. Solids **25**(1), 3–16 (2020)
9. Viverge, K., Boutin, C., Sallet, F.: Model of highly contrasted plates versus experiments on laminated glass. Int. J. Solids Struct. **102**, 238–258 (2016)
10. Boutin, C., Viverge, K.: Generalized plate model for highly contrasted laminates. Eur. J. Mech. A Solid. **55**, 149–166 (2016)
11. Aßmus, M., Naumenko, K., Altenbach, H.: Mechanical behaviour of photovoltaic composite structures: influence of geometric dimensions and material properties on the eigenfrequencies of mechanical vibrations. Compos. Commun. **6**, 59–62 (2017)

12. Berdichevsky, V.L.: An asymptotic theory of sandwich plates. Int. J. Eng. Sci. **48**(3), 383–404 (2010)
13. Andrianov, I.V., Awrejcewicz, J., Manevitch, L.I.: Asymptotical Mechanics of Thin-Walled Structures. Springer Science & Business Media, Springer, Berlin (2013)
14. Kaplunov, J.D., Markushevich, D.G.: Plane vibrations and radiation of an elastic layer lying on a liquid half-space. Wave Motion **17**(3), 199–211 (1993)
15. Kaplunov, J.D., Kossovich, L.Y., Rogerson, G.A.: Direct asymptotic integration of the equations of transversely isotropic elasticity for a plate near cut-off frequencies. Q. J. Mech. Appl. Math. **53**(2), 323–341 (2000)

A Study on the Coefficient of Restitution Effect on Single-Sided Vibro-Impact Nonlinear Energy Sink

Adnan S. Saeed and Mohammad A. Al-Shudeifat

Abstract Vibration mitigation is an essential factor in many engineering applications given the high risk of failure due to the frequent occurrence of earthquakes, blasts, collisions and fluid-structure interaction. Linear and nonlinear vibration absorbers have been continuously studied to be employed in such structures to decrease the vibration levels and therefore protect them from destruction. Up to date, the most effective and efficient passive vibration absorber is the single-sided vibro-impact (SSVI) nonlinear energy sink (NES) which consists of a small mass attached to the primary structure via linear stiffness and linear damping coupling elements in addition to a rigid barrier that enables it to engage in non-smooth inelastic impacts. It has been shown in the literature that an accurately optimized SSVI NES is capable of transferring and dissipating high percentages of the initial input energy into the primary structure. However, most of the investigations in the literature implement a coefficient of restitution of 0.7 corresponding to steel-to-steel impacts. Consequently, this paper investigates further improvements to the SSVI NES by studying the effect of changing the coefficient of restitution to increase the efficiency of targeted energy transfer (TET). It is found that lowering the coefficient of restitution increases the efficiency of the SSVI NES to transfer and dissipate energy from a large-scale nine-story structure.

Keywords Nonlinear energy sink · Vibro-impact · Shock mitigation

1 Introduction

Structures are subject to destructive vibration amplitudes from impacts, collisions, earthquakes or wind. Hence, it is desirable to protect the structure by transferring energy to a dynamic absorber in a process known as passive targeted energy transfer (TET). It has recently gained increasing interest in many applications of

A. S. Saeed (✉) · M. A. Al-Shudeifat
Aerospace Engineering, Khalifa University of Science and Technology, Abu Dhabi, UAE

J. Awrejcewicz (ed.), *Perspectives in Dynamical Systems II: Mathematical and Numerical Approaches*, Springer Proceedings in Mathematics & Statistics 363,
https://doi.org/10.1007/978-3-030-77310-6_25

structural dynamics where a small essentially nonlinear lightweight attachment called nonlinear energy sink (NES) is usually added to the primary structure to enable passive energy transfer for wide range of frequencies through single or cascades of resonance captures. Recently, various NES designs have been numerically, analytically and experimentally investigated to achieve high percentages of energy transfer and dissipation. Depending on the way the nonlinearity is added, these NESs can be categorized into several types such as stiffness-based NESs, impact-based NESs, rotary NESs, and magnet-based NESs. The translational stiffness-based NESs employ an essentially nonlinear (usually cubic) stiffness coupling element to attach the NES mass to a floor in the primary structure [1]. Several enhancements have been proposed to this type where the addition of a linear or nonlinear damping elements [2], additional nonlinearly coupled mass [2], additional lateral stiffness elements [3] and negative, unsymmetrical or variable nonlinear stiffness components [4, 5] are analyzed for increasing the efficiency of the energy transfer and dissipation. Impact-based NESs employ a linearly coupled NES mass in addition to rigid barriers in the motion domain of the NES to engage in non-smooth impacts that integrate the essential non-linear property required for cascades of resonance captures. There are two main types of impact-based NESs: (1) double-sided vibro-impact (DSVI) NESs in which two rigid barriers are placed symmetrically from the initial position of the NES mass [6–9] and (2) single-sided vibro-impact (SSVI) NESs, which are the topic of this paper, realized by removing one of the rigid barriers to allow the NES mass gain considerable momentum during the non-impact phase [10–14]. Rotary NESs incorporate an inertially coupled NES mass through a rigid arm rotating about a vertical axis perpendicular to the direction of motion of the primary structure [15, 16]. Similarly, this type of NES has been further enhanced by employing an elastic arm instead of the rigid arm [17] or by adding a rigid barrier to incorporate non-smooth impacts with the associated floor of the primary structure [18]. Finally, the magnet-based NES mass is coupled to the associated floor of the primary structure through a nonlinear symmetric or asymmetric coupling magnetic force [19, 20]. Out of all the NES types, the SSVI NESs have been proven numerically and experimentally to be the most efficient for energy dissipation and shock mitigation. Hence, the focus of this paper is to investigate further enhancements to the SSVI NESs to improve its capability of engaging in rapid passive and nearly irreversible TET.

During an impact in a structure with SSVI NESs, the coefficient of restitution, a material property depending solely on the materials which get in contact, is defined as the ratio of the magnitude of restitutive impulse to deformative impulse. Most current works related to impact-based NESs consider steel-to-steel impacts which correspond to a coefficient of restitution of 0.7. The aim of this article is to investigate the effect of changing the coefficient of restitution on the performance of SSVI NES when attached to the top floor of a physical nine-story linear primary structure excited by an impulsive loading. The paper presents the system description and governing equations, a discussion of the numerical optimization process followed by results and concluded remarks.

2 System Description and Governing Equations

The physical large-scale nine-story structure analyzed in [11, 12, 14, 17] is considered here where a SSVI NES is attached to the top floor as shown in Fig. 1. The mass, damping and stiffness matrices of the primary structure identified by the modal analysis of the physical fixture for the system can be found in [11, 12, 14, 17]. Upon updating the mass **M**, damping **C** and stiffness **K** matrices of the whole structure to include the addition of the NES, the governing equations of motion are derived using Newtonian dynamics as

$$\mathbf{M}\ddot{\mathbf{x}} + \mathbf{C}\dot{\mathbf{x}} + \mathbf{k}\mathbf{x} = \mathbf{0} \tag{1}$$

The equations are numerically integrated using Runge-Kutta formulations. However, the numerical integration is continued until the impact condition for the SSVI NES given by

$$x_{nes} - x_1 \geq z_c \tag{2}$$

is satisfied where x_{nes} and x_1 are the displacement of the NES and the top floor respectively and z_c is the clearance as shown in Fig. 1. After calculating the precise time of the impact, the velocities of the top floor and the SSVI NES after the impact are calculated based on the conservation of momentum and the coefficient of restitution r_c principles as

$$\dot{x}_1^+ = \frac{m\dot{x}_{nes}^- + M_1\dot{x}_1^- - mr_c\left(\dot{x}_1^- - \dot{x}_{nes}^-\right)}{m + M_1} \tag{3}$$

$$\dot{x}_{nes}^+ = \dot{x}_1^+ + r_c\left(\dot{x}_1^- - \dot{x}_{nes}^-\right) \tag{4}$$

where M_1 and m are the masses of the top floor and NES respectively and the superscripts $+$ and $-$ indicate the velocities after and before the impacts respectively. Accordingly, there are three sources of energy transfer and dissipation in the coupled system. The first is due to the linear damping element, represented by λ_{nes} in Fig. 1, coupling the NES mass to the top floor of the structure. The second element is the kinetic energy lost during the inelastic impacts that occur when the NES collides with the rigid barrier attached to the top floor. The third is due to the non-smooth interference from the vibro-impacts which alters the global dynamic response of the structure causing energy to be transferred within its structural modes. Transferring energy from low-frequency high-energy destructive fundamental mode to high-frequency low-energy mode is another element of TET.

One way to quantify the performance of the coupled system for achieving efficient TET is by measuring the enhancement in the damping of the respective effective modal oscillators. This is quantified by the time-independent averaged

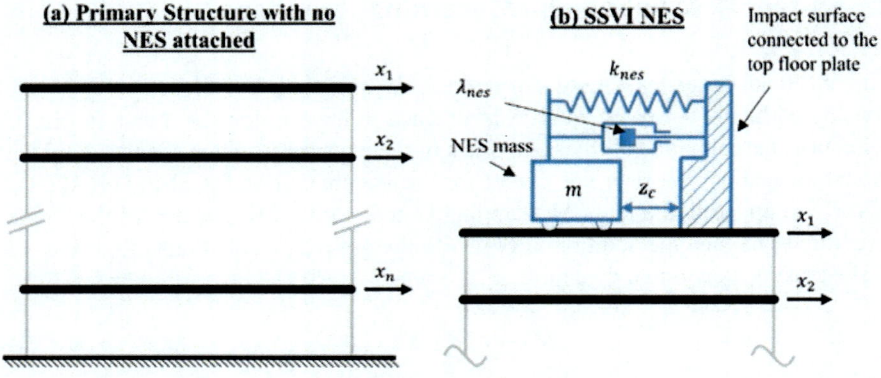

Fig. 1 Conceptual illustration of a SSVI NES coupled to the top floor of an n-story structure. (**a**) Primary structure with no NES attached. (**b**) SSVI NES

effective modal damping measures $\lambda_{eff,\,i}$ expressed as

$$\lambda_{eff,i} = \frac{\dot{q}_i(T_0)^2 - \dot{q}_i\left(T_f\right)^2 + \omega_i^2\left(\dot{q}_i(T_0)^2 - \dot{q}_i\left(T_f\right)^2\right)}{2\int_{T_0}^{T_f}\dot{q}_i^2 dt} = \frac{\mathrm{E}\left(T_0\right) - \mathrm{E}\left(T_\mathrm{f}\right)}{\int_{T_0}^{T_f}\dot{q}_i^2 dt} \tag{5}$$

where $\dot{q}_i$ is the modal velocity of the ith mode, T_0 and T_f are the initial and final simulation times, E is the instantaneous energy and ω_i^2 is the effective modal stiffness. Because the modal response of a linear primary structure without the NES leads to no energy transfer within its modes, the nominal modal damping λ_i of the ith structural mode is obtained from Eq. (5). The nonlinear interaction of the NES causes the energy dissipated by any mode plus its instantaneous energy to not sum up to its initial induced energy due to energy exchange between the structural modes and therefore the energy in each mode is not reserved. The ratio of the time-independent averaged effective damping measures $\lambda_{eff,\,i}$ to the nominal modal damping λ_i of the NES-free system is defined as the normalized weighted-averaged effective damping measures $\hat{\lambda}_{eff,i}$. If $\hat{\lambda}_{eff,i} < 1$, then energy is transferred into the ith mode and it eventually dissipates more energy than its initial energy and vice versa.

The SSVI NES parameters: k_{nes} and λ_{nes} are tuned in order to investigate the effect of changing the coefficient of restitution on achieving efficient and rapid transfer and dissipation of energy. The simulation time T_f is set to 5 s to ensure efficient and rapid TET, the clearance z_c is fixed to 0.015 m to ensure non-smooth vibro-impacts will occur and the NES mass m is taken as 500 kg and is assumed to be nonparasitic (i.e. do not add mass to the primary structure). We are interested in finding the optimum parameters of the SSVI NES that maximize normalized weighted-averaged effective damping measures of mode 1, $\hat{\lambda}_{eff,1}$, which indicates the maximum transfer of energy from the lowest fundamental (highly-energetic)

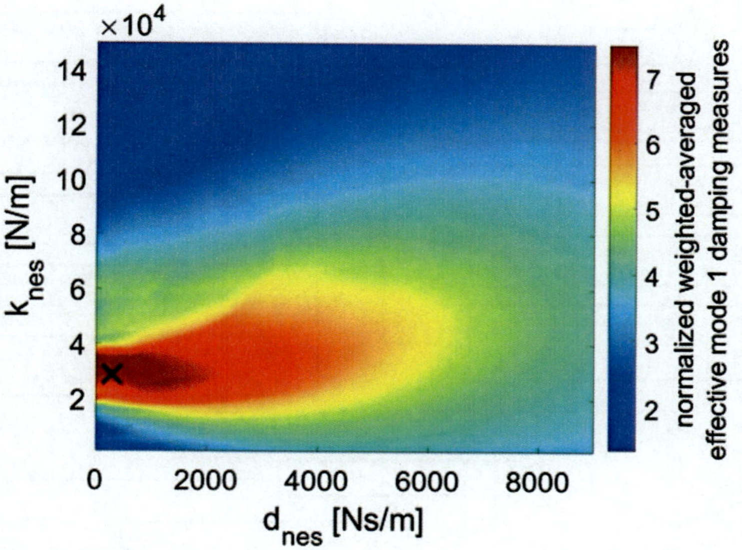

Fig. 2 Contour of $\hat{\lambda}_{eff,1}$ for varying k_{nes} and λ_{nes} using $r_c = 0.7$, black cross indicates optimal parameters used as initial guesses for the optimization algorithm

mode to higher frequency modes. Figure 2 shows $\hat{\lambda}_{eff,1}$for varying NES parameters for an impulsive excitation induced by identical initial velocity of 0.25 m/s equally to all floors of the primary structure using a coefficient of restitution of 0.7. It is noticed that a relatively weak stiffness-coupling element is required for the SSVI NES to transfer significant amount of energy from the first fundamental mode to be dissipated by the NES damping, inelastic impacts or higher structural modes. Further, the damping and stiffness coefficients, indicated by the black crosses in Fig. 2, giving the maximum achievable effective damping measure for $r_c = 0.7$ are used as initial guesses for optimizing $\hat{\lambda}_{eff,1}$ with different values of the coefficient of restitution using the optimization algorithm based on Nelder-Mead simplex method [21] as shown in Fig. 3.

3 Results and Discussion

The results of the numerical optimization are summarized in Fig. 4 which shows the maximum achievable normalized weighted-averaged effective mode 1 damping measures $\hat{\lambda}_{eff,1}$ by an optimized system at each specific value of the coefficient of restitution r_c for a clearance of 0.015 m and an impulsive loading through an identical initial velocity of 0.25 m/s induced equally to all floors of the primary structure. First, it is noticed that the optimized coupled systems for all values of r_c result in normalized weighted-averaged effective mode 1 damping measures $\hat{\lambda}_{eff,1}$

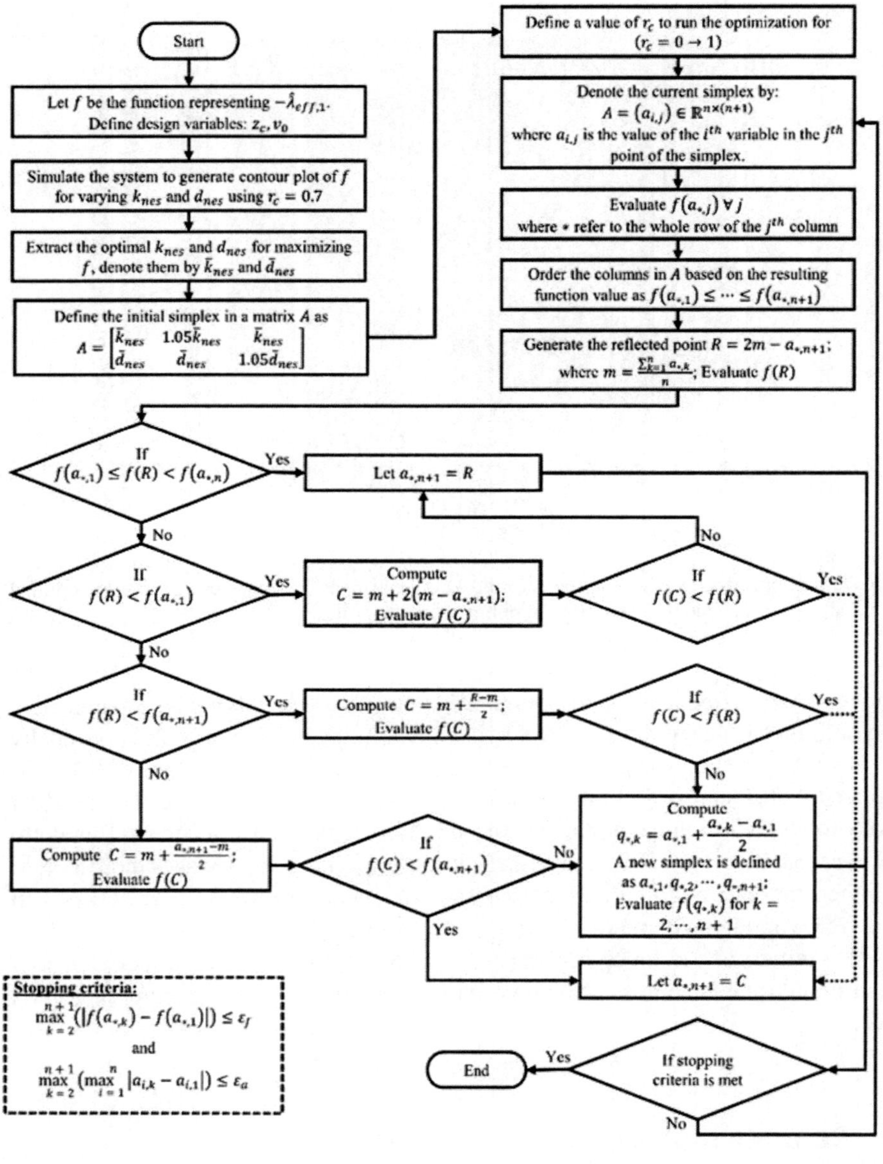

Fig. 3 Optimization algorithm implemented for investigating the effect of changing the coefficient of restitution

that are greater than unity. This indicates that in all cases, energy is being transferred from the high energy low frequency mode 1 to the NES itself or the other structural modes having lower energy and higher frequency. Consequently, this indicates that highly efficient TET has occurred by forcing energy to be transferred through single or cascades of resonance captures in a rapid and nearly irreversible way. In addition,

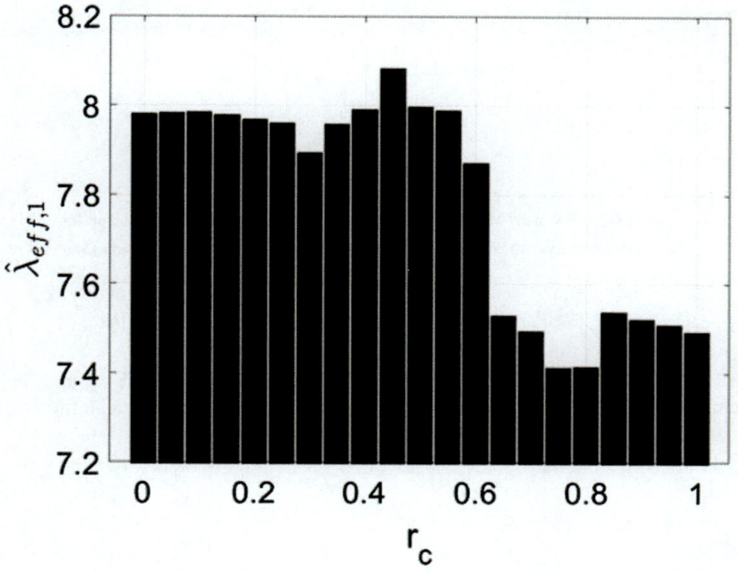

Fig. 4 Comparing $\hat{\lambda}_{eff,1}$ using optimized k_{nes} and λ_{nes} at each r_c

the goal of this paper is to investigate if changing the coefficient of restitution from its typical value of 0.7 which corresponds to steel-to-steel impacts will enhance the energy transfer and dissipation by resulting in higher values of $\hat{\lambda}_{eff,1}$. Figure 4 shows that the maximum normalized weighted-averaged effective mode 1 damping measure is 8.1 achievable with a coefficient of restitution of 0.45 where the optimized SSVI NES parameters are $k_{nes} = 29{,}482$ N/m and $d_{nes} = 281.5$ Ns/m compared to 7.5 achievable with a coefficient of restitution of 0.7 where the optimized SSVI NES parameters are $k_{nes} = 29{,}701$ N/m and $d_{nes} = 319.8$ Ns/m. Additionally, the performance of the optimized NESs at $r_c = 0.45$ and $r_c = 0.70$ are compared in Fig. 5a for varying initial impulsive energies induced equally to all floors. The modified system with coefficient of restitution of 0.45 shows higher normalized weighted-averaged effective mode 1 damping measures for a wide range of initial impulsive energies indicating that better energy transfer and dissipation is achievable at lower values of the coefficient of restitution as shown in Fig. 5b.

The response of the integrated structure is depicted in Fig. 6 for three cases: SSVI NES with optimal parameters at a coefficient of restitution of 0.45 and 0.7 as well as with the NES locked, where it interacts only through its mass. Figure 6a shows that the modal response of mode 1 of the structure with either of the attached NESs significantly suppresses its amplitude after few oscillation cycles indicating the efficiency of the nonlinear energy redistribution. Figure 6 also shows the time histories of the total energy dissipated E_{diss} and its contributors: energy dissipated by impacts E_{imp} (Fig. 6c), energy dissipated by damping E_{damp} (Fig. 6e) and energy dissipated through the inherent structural damping of the higher modes E_{high}

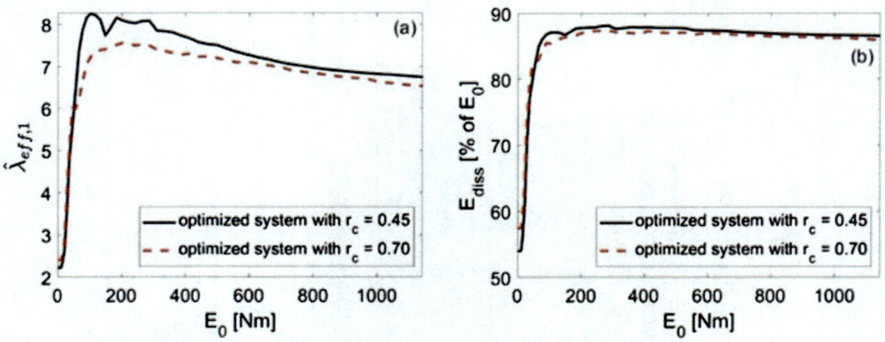

Fig. 5 **(a)** $\hat{\lambda}_{eff,1}$ and **(b)** Total energy dissipated E_{diss} of the structure-NES system for varying initial impulsive energy E_0 employed using identical initial velocity applied to all floors

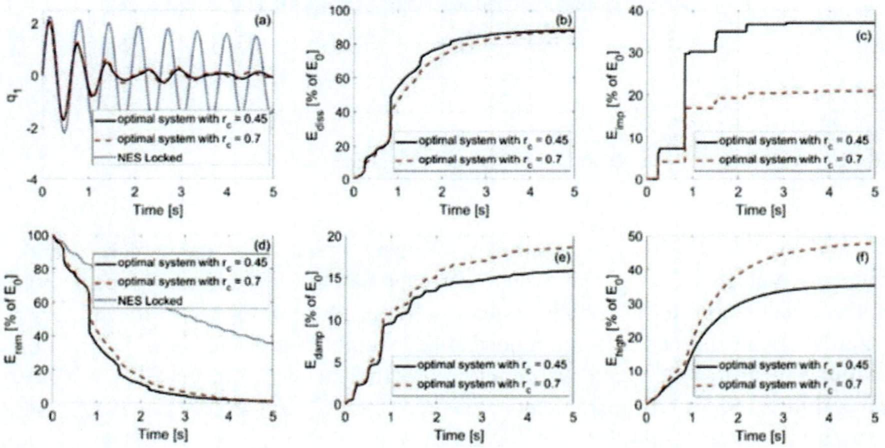

Fig. 6 Response of the structure-NES system with the optimal parameters for a coefficient of restitution of 0.45 and 0.7; for comparison the case of locked NES is also shown where appropriate show. The modal response of the first mode is shown in **(a)** and the time history of the total energy dissipated is shown in **(b)** with its breakdown showing energy dissipated by impacts in **(c)**, damping in **(e)** and higher modes in **(f)**. The time history of the energy remaining in the primary structure is shown in **(d)**

(Fig. 6f). Although a system with the modified NES gives less energy dissipation by damping and through higher structural modes, it gives enhanced total energy dissipation and quickly reduces the energy remaining in the primary structure (Fig. 6d) through the added dissipation from the impact signifying the importance of the proposed enhancement.

4 Conclusions

Many dynamical structures undergo continuous vibrations induced by shock or seismic excitations resulting from impacts, collisions, wind, earthquake, or fluid-structure interaction. Consequently, integrating dynamic vibration absorbers is becoming a high priority and essential requirement in many engineering applications. Nonlinear energy sinks (NESs) are the most efficient and robust passive attachments to act as rapid and passive device to transfer and dissipate energy from the primary structure. Hence, this paper investigated further improvements to the currently most efficient nonlinear energy sink (NES), the single-sided vibro-impact (SSVI) NES, by studying the effect of the coefficient of restitution, which is a ratio of restitutive to deformative impulses during impact, on its capability to irreversibly transfer induced impulsive energy out of the fundamental highly energetic mode. Currently, the coefficient of restitution used in most analytical, numerical and experimental analysis of impact-based NESs is 0.7 which corresponds to the typical steel-to-steel impacts. However, it is found in this paper that reducing the coefficient of restitution to 0.45 enhances the normalized weighted-averaged effective damping measure of the fundamental mode of a nine-story linear physical primary structure. The results are obtained for a simulation time of 5 s and using an NES mass of 5% of the whole structure indicating that rapid targeted energy transfer (TET) can occur with using a small mass of NES. The enhanced performance is obtained for a wide range of initial impulsive energies indicating the robustness of the proposed modification.

References

1. Vakakis, A.F., Gendelman, O.V., Bergman, L.A., McFarland, D.M., Kerschen, G., Lee, Y.S.: Nonlinear Targeted Energy Transfer in Mechanical and Structural Systems. Springer Science & Business Media, Dordrecht (2008)
2. Sapsis, T.P., Quinn, D.D., Vakakis, A.F., Bergman, L.A.: Effective stiffening and damping enhancement of structures with strongly nonlinear local attachments. J. Vib. Acoust. **134**(1), 11–16 (2012)
3. Saeed, A.S., Al-Shudeifat, M.A., Cantwell, W.J., Vakakis, A.F.: Two-dimensional nonlinear energy sink for effective passive seismic mitigation. Commun. Nonlinear Sci. Numer. Simulat. **99**, 105787 (2021)
4. Al-Shudeifat, M.A., Saeed, A.S.: Frequency-energy dependence of the bistable nonlinear energy sink. In: Proceedings of the ASME 2017 International Design Engineering Technical Conferences and Computers and Information in Engineering Conference (IDETC-CIE), Cleveland, OH, 6–9 Aug 2017
5. Al-Shudeifat, M.A., Saeed, A.S.: Analytical treatment for bistable nonlinearly coupled oscillator. In: Proceedings of the ASME 2017 International Design Engineering Technical Conferences and Computers and Information in Engineering Conference (IDETC-CIE), Cleveland, OH, 6–9 Aug 2017
6. Gendelman, O.V.: Analytic treatment of a system with a vibro-impact nonlinear energy sink. J. Sound Vib. **331**(21), 4599–4608 (2012)

7. Li, T., Seguy, S., Berlioz, A.: Dynamics of cubic and vibro-impact nonlinear energy sink: analytical, numerical, and experimental analysis. J. Vib. Acoust. **138**(3), 031010 (2016)
8. Gendelman, O.V., Alloni, A.: Dynamics of forced system with vibro-impact energy sink. J. Sound Vib. **358**(1), 301–314 (2015)
9. Li, T., Seguy, S., Berlioz, A.: On the dynamics around targeted energy transfer for vibro-impact nonlinear energy sink. Nonlinear Dyn. **87**(3), 1453–1466 (2017)
10. Al-Shudeifat, M.A., Wierschem, N., Quinn, D.D., Vakakis, A.F., Bergman, L.A., Spencer, B.F.: Numerical and experimental investigation of a highly effective single-sided vibro-impact nonlinear energy sink for shock mitigation. Int. J. Nonlinear Mech. **52**(1), 96–109 (2013)
11. Al-Shudeifat, M.A., Vakakis, A.F., Bergman, L.A.: Shock mitigation by means of low-to high frequency nonlinear targeted energy transfers in a largescale structure. J. Comput. Nonlinear Dyn. **11**(2), 021006 (2016)
12. Wierschem, N.E., Hubbard, S.A., Luo, J., Fahnestock, L.A., Spencer, B.F., McFarland, D.M., Quinn, D.D., Vakakis, A.F., Bergman, L.A.: Response attenuation in a large-scale structure subjected to blast excitation utilizing a system of essentially nonlinear vibration absorbers. J. Sound Vib. **389**(1), 52–72 (2017)
13. Li, W., Wierschem, N.E., Li, X., Yang, T.: On the energy transfer mechanism of the single-sided vibro-impact nonlinear energy sink. J. Sound Vib. **437**(1), 166–179 (2018)
14. Luo, J., Wierschem, N.E., Hubbard, S.A., Fahnestock, L.A., Quinn, D.D., McFarland, D.M., Spencer, D.F., Vakakis, A.F., Bergman, L.A.: Large-scale experimental evaluation and numerical simulation of a system of nonlinear energy sinks for seismic mitigation. Eng. Struct. **77**(1), 34–48 (2014)
15. Sigalov, G., Gendelman, O., Al-Shudeifat, M.A., Manevitch, L., Vakakis, A.F., Bergman, L.A.: Resonance captures and targeted energy transfers in an inertially-coupled rotational nonlinear energy sink. Nonlinear Dyn. **69**(4), 1693–1704 (2012)
16. Gendelman, O., Sigalov, G., Manevitch, L., Mane, M., Vakakis, A.F., Bergman, L.A.: Dynamics of an eccentric rotational nonlinear energy sink. J. Appl. Mech. **79**(1), 011012 (2012)
17. Saeed, A.S., Al-Shudeifat, M.A.: Rotary-oscillatory nonlinear energy sink of robust performance. Int. J. Non-Linear Mech. **177**, 103249 (2019)
18. Saeed, A.S., Al-Shudeifat, M.A., Vakakis, A.F., Cantwell, W.J.: Rotary-impact nonlinear energy sink for shock mitigation: analytical and numerical investigations. Archive Appl. Mech. **90**, 495–521 (2020)
19. Kremer, D., Liu, K.: A nonlinear energy sink with an energy harvester: transient responses. J. Sound Vib. **333**(20), 4859–4880 (2014)
20. Al-Shudeifat M.A.: Asymmetric magnet-based nonlinear energy sink. J. Comput. Nonlinear Dyn. **10**(1), 014502 (2015)
21. Nelder, J.A., Mead, R.: A simplex method for function minimization. Comput. J. **7**(4), 308–313 (1965)